Smart Materials and Devices for Energy Harvesting

Smart Materials and Devices for Energy Harvesting

Editor

Daniele Davino

MDPI • Basel • Beijing • Wuhan • Barcelona • Belgrade • Manchester • Tokyo • Cluj • Tianjin

Editor
Daniele Davino
University of Sannio
Italy

Editorial Office
MDPI
St. Alban-Anlage 66
4052 Basel, Switzerland

This is a reprint of articles from the Special Issue published online in the open access journal *Materials* (ISSN 1996-1944) (available at: https://www.mdpi.com/journal/materials/special_issues/smart_energy).

For citation purposes, cite each article independently as indicated on the article page online and as indicated below:

LastName, A.A.; LastName, B.B.; LastName, C.C. Article Title. *Journal Name* **Year**, *Volume Number*, Page Range.

ISBN 978-3-0365-3122-9 (Hbk)
ISBN 978-3-0365-3123-6 (PDF)

Cover image courtesy of Daniele Davino

© 2022 by the authors. Articles in this book are Open Access and distributed under the Creative Commons Attribution (CC BY) license, which allows users to download, copy and build upon published articles, as long as the author and publisher are properly credited, which ensures maximum dissemination and a wider impact of our publications.
The book as a whole is distributed by MDPI under the terms and conditions of the Creative Commons license CC BY-NC-ND.

Contents

About the Editor vii

Preface to "Smart Materials and Devices for Energy Harvesting" ix

Daniele Davino
Smart Materials and Devices for Energy Harvesting
Reprinted from: *Materials* **2021**, *14*, 4738, doi:10.3390/ma14164738 1

Hiroki Kurita, Kenichi Katabira, Yu Yoshida and Fumio Narita
Footstep Energy Harvesting with the Magnetostrictive Fiber Integrated Shoes
Reprinted from: *Materials* **2019**, *12*, 2055, doi:10.3390/ma12132055 5

Carmine Stefano Clemente and Daniele Davino
Modeling and Characterization of a Kinetic Energy Harvesting Device Based on Galfenol
Reprinted from: *Materials* **2019**, *12*, 3199, doi:10.3390/ma12193199 13

Stefano Palumbo, Mario Chiampi, Oriano Bottauscio and Mauro Zucca
Dynamic Simulation of a Fe-Ga Energy Harvester Prototype Through a Preisach-Type Hysteresis Model
Reprinted from: *Materials* **2019**, *12*, 3384, doi:10.3390/ma12203384 35

Lei Jin, Shiqiao Gao, Xiyang Zhang and Qinghe Wu
Output of MEMS Piezoelectric Energy Harvester of Double-Clamped Beams with Different Width Shapes
Reprinted from: *Materials* **2020**, *13*, 2330, doi:10.3390/ma13102330 47

Cong Du, Pengfei Liu, Hailu Yang, Gengfu Jiang, Linbing Wang and Markus Oeser
Finite Element Modeling and Performance Evaluation of Piezoelectric Energy Harvesters with Various Piezoelectric Unit Distributions
Reprinted from: *Materials* **2021**, *14*, 1405, doi:10.3390/ma14061405 61

Manuel Vázquez-Rodríguez, Francisco J. Jiménez, Lorena Pardo, Pilar Ochoa, Amador M. González and José de Frutos
A New Prospect in Road Traffic Energy Harvesting Using Lead-Free Piezoceramics
Reprinted from: *Materials* **2019**, *12*, 3725, doi:10.3390/ma12223725 79

Hailu Yang, Qian Zhao, Xueli Guo, Weidong Zhang, Pengfei Liu and Linbing Wang
Numerical Analysis of Signal Response Characteristic of Piezoelectric Energy Harvesters Embedded in Pavement
Reprinted from: *Materials* **2020**, *13*, 2770, doi:10.3390/ma13122770 95

Yuansheng Wang, Zhiyong Zhou, Qi Liu, Weiyang Qin and Pei Zhu
Harvesting Variable-Speed Wind Energy with a Dynamic Multi-Stable Configuration
Reprinted from: *Materials* **2020**, *13*, 1389, doi:10.3390/ma13061389 107

Guoxi Luo, Yunyun Luo, Qiankun Zhang, Shubei Wang, Lu Wang, Zhikang Li, Libo Zhao, Kwok Siong Teh and Zhuangde Jiang
The Radial Piezoelectric Response from Three-Dimensional Electrospun PVDF Micro Wall Structure
Reprinted from: *Materials* **2020**, *13*, 1368, doi:10.3390/ma13061368 127

Tiago Rodrigues-Marinho, Nelson Castro, Vitor Correia, Pedro Costa and Senentxu Lanceros-Méndez
Triboelectric Energy Harvesting Response of Different Polymer-Based Materials
Reprinted from: *Materials* **2020**, *13*, 4980, doi:10.3390/ma13214980 . . . . **137**

Mohammad Uddin, Shane Alford and Syed Mahfuzul Aziz
Evaluating Energy Generation Capacity of PVDF Sensors: Effects of Sensor Geometry and Loading
Reprinted from: *Materials* **2021**, *14*, 1895, doi:10.3390/ma14081895 . . . . **149**

George Karalis, Christos K. Mytafides, Lazaros Tzounis, Alkiviadis S. Paipetis and Nektaria-Marianthi Barkoula
An Approach toward the Realization of a Through-Thickness Glass Fiber/Epoxy Thermoelectric Generator
Reprinted from: *Materials* **2021**, *14*, 2173, doi:10.3390/ma14092173 . . . . **167**

Joel Joseph, Makoto Ohtsuka, Hiroyuki Miki and Manfred Kohl
Lumped Element Model for Thermomagnetic Generators Based on Magnetic SMA Films
Reprinted from: *Materials* **2021**, *14*, 1234, doi:10.3390/ma14051234 . . . . **181**

Sajid Naseem, Bianca R. Gevers, Frederick J. W. J. Labuschagné and Andreas Leuteritz
Preparation of Photoactive Transition-Metal Layered Double Hydroxides (LDH) to Replace Dye-Sensitized Materials in Solar Cells
Reprinted from: *Materials* **2020**, *13*, 4384, doi:10.3390/ma13194384 . . . . **195**

About the Editor

Daniele Davino Ph.D., is a Full Professor of electrical engineering at the Department of Engineering of the University of Sannio in Benevento, Italy. He teaches circuit theory, low-frequency electromagnetism and materials electromagnetic properties. Electrical engineering, engineering physics and materials engineering are his general research topics. Specifically, his research focusses on studying smart materials and their applications as smart actuation/sensors and energy harvesting devices. Additionally, he is expert on modeling hysteresis in magnetic, piezo and smart materials. He is very involved in the development of magnetostrictive and magneto-electric energy harvesters and in the development of magnetic polymeric foams.

Preface to "Smart Materials and Devices for Energy Harvesting"

This book is devoted to energy harvesting from smart materials and devices. It focusses on the latest available techniques recently published by researchers all over the world.

Energy harvesting is one of the most important enabling technologies for the "Internet of Things". It allows for the feeding of wireless sensors and low-power electronics in general, exploiting environmentally available energy that would otherwise be wasted. As a matter of fact, the limiting factor for wearable electronics or for wireless sensors in harsh environments is the finite energy stored in onboard batteries. Batteries give a finite duration to the stand-alone performances, but changing or recharging them, as often as necessary, is neither a practical, nor an economical or green-oriented strategy. Indeed, for example, wireless sensors located in strategic places in the environment require qualified technicians to reach and change the sensors, and this increases maintenance costs. On the other hand, energy harvesting can convert energy in the precise location where it is needed. This technique may also be exploited for other applications, such as for the powering of implantable medical/sensing devices for humans and animals.

Therefore, energy harvesting from smart materials will become increasingly important in the future. This book provides a broad perspective on this topic for researchers and readers with both physics and engineering backgrounds. Of the seventeen articles submitted, fifteen were accepted for publication after the peer-review process.

The published articles cover a range of topics and applications in energy harvesting. I hope that readers will enjoy the content of this book and benefit from it. This research is the result of the talented and hard work of many scholars.

Indeed, I would like to take this opportunity to express my appreciation for the MDPI Book staff, the editorial team of the Materials journal, and especially Ms. Fannie Xu, the Assistant Editor of the Special Issue. Additionally, I would like to thank all of the talented authors who contributed from many countries (Australia, China, Germany, Greece, Italy, Japan, Portugal, Spain, South Africa, and the USA) as well as all of the hardworking and professional reviewers. Finally, I thank my colleagues, Dr. Carmine Stefano Clemente and Vincenzo Paolo Loschiavo, for their lively and fruitful support of this work. Last but not least, I would like to thank my wife, Loredana, for her love and support.

Daniele Davino
Editor

Editorial

Smart Materials and Devices for Energy Harvesting

Daniele Davino

Department of Engineering, University of Sannio, 82100 Benevento, Italy; davino@unisannio.it

Citation: Davino, D. Smart Materials and Devices for Energy Harvesting. *Materials* 2021, 14, 4738. https://doi.org/10.3390/ma14164738

Received: 11 August 2021
Accepted: 20 August 2021
Published: 22 August 2021

Publisher's Note: MDPI stays neutral with regard to jurisdictional claims in published maps and institutional affiliations.

Copyright: © 2021 by the author. Licensee MDPI, Basel, Switzerland. This article is an open access article distributed under the terms and conditions of the Creative Commons Attribution (CC BY) license (https://creativecommons.org/licenses/by/4.0/).

Energy harvesting will be one of the key enabling technologies for the Internet of Things (IoT) world. This technique allows for the powering of wireless sensors and low-power electronics in general, exploiting environmentally available energy. As a matter of fact, the limiting factor for wearable electronics or for wireless sensors in harsh environments is the finite energy stored in the batteries on-board. Indeed, batteries gives a finite duration to the stand-alone performances. The most adopted solution is still to change or recharge the batteries as often as necessary, but this strategy is neither practical, nor economical, nor green-oriented. Indeed, in the case of wireless sensors, located in strategic places in the environment, the replacement, or the recharging of the batteries needs qualified technicians reaching the sensors and doing the operation and this increases the maintenance costs. On the other hand, energy harvesting can convert the energy, right in the place where it is needed. This technique may be exploited also for other applications, such as for the powering of implantable medical/sensing devices for humans and animals.

Energy harvesters based on magnetostrictive alloys are intrinsically robust and long-life. Indeed, these materials inherit most of the mechanical properties of iron, which is the main component of the alloy. The devices make use of kinetic energy and by having no moving parts, are robust and simple because the energy conversion takes place within the material. Indeed, by applying a time variable mechanical stress to the material, a time variable magnetization is obtained (Villari effect) and then a coil can link a variable magnetic flux and produce a voltage. Because of these reasons, they have been proposed for several tasks and even exploited for wearable energy harvesting, with focus on shoes, where high pressures are available because of walking. Piezoelectrics have also been considered for this purpose, but their energy harvesting performance can be reduced after being used several times, due to their brittleness. Fe-Co magnetostrictive alloys can be considered and produced in the form of fibers and integrated into the shoe heels. A relevant energy in the order of µJ can be recovered from a few thousands of steps of usual walking. It seems that the output energy is dependent on user habits of ambulation, not on their weight [1]. On the other hand, exploitation of magnetostrictive harvesters needs careful modeling and simulation because the magneto-mechanical coupling is strongly nonlinear and with hysteresis. A nonlinear model of magnetostrictive materials, such as Galfenol can be developed and its parameters can be determined by using measured magnetostrictives curves. Because of the mechanical, magnetic and electric quantities involved in the device, by using suitable analogies with voltages and currents, a three-port equivalent circuit can be developed. The characterization is performed by applying different compressive force profiles, resistor loads and permanent magnets for the magnetic bias. The modeling and the experiments confirm that the input force frequency and the magnet configuration strongly affect the output voltage and power, while an optimal resistive load, corresponding to the total equivalent coil resistance is needed to extract the maximum power [2]. If more accurate modeling is needed, then a hysteretic Preisach-type model can be considered with a finite element formulation and this approach allows good discrepancies between experimental and computational values of the output power [3].

Piezoelectric devices are the most common energy harvesters converting kinetic energy. Often, they are proposed as a cantilever beam, acting at a specific resonant frequency of the vibrations, or as cylinders directly exposed to the varying force. Since they do not need any coil or bias, they are among the simplest devices and many studies are devoted to the design and the optimization of different geometries to increase conversion [4]. Numerical studies are also useful to determine the best arrangement of the harvester composed of several basic piezo elements, exposed to a varying external force [5] or to model the effect of geometry parameters over the output power in order to get handy formulas [6,7]. Piezoelectrics can also be exploited to directly harvest the energy of variable-speed wind by means of a dynamic multi-stable flutter harvester [8]. Additionally, the search for new materials showing piezoelectricity is active. Electrospun polyvinylidene fluoride (PVDF) fibers can show piezoelectricity but the random arrangement of the fibers, fabricated by traditional electrospinning, reduces the performance and limits the applications. A newly developed 3D electrospinning technique can be exploited to fabricate a PVDF micro wall made up of densely stacked fibers in a fiber-by-fiber manner [9]. This material shows good piezoelectric performance over time and is promising for both energy harvesting and sensing applications.

Large deformations can be harvested by exploiting triboelectricity. Polymers and polymers composites can be arranged as triboelectric nanogenerators where a contact-separation system (10 N of force followed by 5 cm of separation per cycle) can harvest output power ranges from 0.2 to 5.9 mW, depending on the pair of materials, for an active area of 46.4 cm^2 by using Mica, polyamide (PA66) and styrene/ethylene-butadiene/styrene (SEBS) as positive electrode and polyvinylidene fluoride (PVDF) [10,11], polyurethane (PU), polypropylene (PP) and Kapton as a negative electrode. The highest performance is obtained with Mica with PVDF composites with 30 wt.% of barium titanate (BT) and PA66 with PU pairs.

Temperature gradients are a spread potential source of energy, even if poorly exploitable at a large scale, it is perfectly fitting for energy harvesting purposes. While conventional Thermoelectric Generators (TEGs) are already commercially available, the research frontier on this topic has two main goals: new materials, with improved figure of merits that could be exploited on large surfaces at reasonable costs. For example, a 10-ply glass fiber-reinforced polymer composite laminate can operate as a structural through-thickness TEG. For this purpose, inorganic tellurium nanowires have to be mixed with single-wall carbon nanotubes in a wet chemical approach [12]. This results in a flexible p-type thermoelectric material with a power factor value of 58.88 µW/m·K^2. Another new device is the thermomagnetic generator (TMG) based on magnetic shape memory alloy (MSMA) films. The TMG generators make use of the concept of resonant self-actuation of a cantilever, caused by a large abrupt temperature-dependent change of magnetization in the material and the rapid heat transfer inherent to the MSMA films [13]. A prototype based on Ni-Mn-Ga film, with a Curie temperature TC of 375 K, has shown a relevant power density of 80 mW/cm^3 for a heat source temperature of 443 K. This device is modellable with a lumped element model that can be used to estimate the effect of decreasing TC on the lower limit of the heat source temperature in order to predict the possible routes towards waste heat recovery near room temperature [13].

Photovoltaic is the only energy harvesting technique that is scalable from a few mW devices to MW plants. Nevertheless, the challenge is still to reduce the costs by keeping reasonable efficiencies and improving versatility. Thin-film solar cells are one of the solutions. In this framework, several techniques have been proposed and dye-sensitized solar cells (DSSCs) are one of those. The DSSC is relatively simple, semi-flexible and semi-transparent which offers a variety of applications not suitable for glass-based systems and most of the materials used are low-cost. However, few attempts have been tried to further eliminate expensive materials from the process. For example, the use of Fe-modified MgAl-layered double hydroxides (LDHs) to replace dye and semiconductor complexes has shown good results [14], with the MgFeAl-LDH that can act as a simultaneous photoabsorber and charge separator, effectively replacing the dye and semiconductor complex in DSSCs and still yielding an efficiency of 1.56%.

We have discussed several new materials and devices that allow energy harvesting from the environment: magnetostrictives and piezoelectrics, coupling mechanical and/or thermal variables, to electro- or magnetic- variables; materials and devices exploiting temperature gradients for direct conversion into electricity; new materials for more exploitable solar energy conversion and electro-active polymers (EAP) for energy harvesting. These are a few examples but surely there will be many other energy harvesting techniques in the future. Indeed, the field will advance as long as new multi-functional materials will be discovered.

Funding: This research received no external funding.

Institutional Review Board Statement: Not applicable.

Informed Consent Statement: Not applicable.

Data Availability Statement: Not applicable.

Acknowledgments: The Guest Editor thanks all the Authors that contributed with valuable works to this Special Issue from many countries (Australia, China, Germany, Greece, Italy, Japan, Portugal, Spain, South Africa, USA). Additionally, special thanks are due to all the reviewers involved: their constructive comments surely improved the papers. Finally, I am grateful to the Editorial Office of *Materials*, particularly to Fannie Xu. Their kind and competent assistance, together with a continuous effort, have been fundamental for the success of this Special Issue.

Conflicts of Interest: The author declares no conflict of interest.

References

1. Kurita, H.; Katabira, K.; Yoshida, Y.; Narita, F. Footstep Energy Harvesting with the Magnetostrictive Fiber Integrated Shoes. *Materials* **2019**, *12*, 2055. [CrossRef] [PubMed]
2. Clemente, C.; Davino, D. Modeling and Characterization of a Kinetic Energy Harvesting Device Based on Galfenol. *Materials* **2019**, *12*, 3199. [CrossRef] [PubMed]
3. Palumbo, S.; Chiampi, M.; Bottauscio, O.; Zucca, M. Dynamic Simulation of a Fe-Ga Energy Harvester Prototype Through a Preisach-Type Hysteresis Model. *Materials* **2019**, *12*, 3384. [CrossRef] [PubMed]
4. Jin, L.; Gao, S.; Zhang, X.; Wu, Q. Output of MEMS Piezoelectric Energy Harvester of Double-Clamped Beams with Different Width Shapes. *Materials* **2020**, *13*, 2330. [CrossRef] [PubMed]
5. Du, C.; Liu, P.; Yang, H.; Jiang, G.; Wang, L.; Oeser, M. Finite Element Modeling and Performance Evaluation of Piezoelectric Energy Harvesters with Various Piezoelectric Unit Distributions. *Materials* **2021**, *14*, 1405. [CrossRef] [PubMed]
6. Vázquez-Rodríguez, M.; Jiménez, F.; Pardo, L.; Ochoa, P.; González, A.; de Frutos, J. A New Prospect in Road Traffic Energy Harvesting Using Lead-Free Piezoceramics. *Materials* **2019**, *12*, 3725. [CrossRef] [PubMed]
7. Yang, H.; Zhao, Q.; Guo, X.; Zhang, W.; Liu, P.; Wang, L. Numerical Analysis of Signal Response Characteristic of Piezoelectric Energy Harvesters Embedded in Pavement. *Materials* **2020**, *13*, 2770. [CrossRef] [PubMed]
8. Wang, Y.; Zhou, Z.; Liu, Q.; Qin, W.; Zhu, P. Harvesting Variable-Speed Wind Energy with a Dynamic Multi-Stable Configuration. *Materials* **2020**, *13*, 1389. [CrossRef] [PubMed]
9. Luo, G.; Luo, Y.; Zhang, Q.; Wang, S.; Wang, L.; Li, Z.; Zhao, L.; Teh, K.; Jiang, Z. The Radial Piezoelectric Response from Three-Dimensional Electrospun PVDF Micro Wall Structure. *Materials* **2020**, *13*, 1368. [CrossRef] [PubMed]
10. Rodrigues-Marinho, T.; Castro, N.; Correia, V.; Costa, P.; Lanceros-Méndez, S. Triboelectric Energy Harvesting Response of Different Polymer-Based Materials. *Materials* **2020**, *13*, 4980. [CrossRef] [PubMed]
11. Uddin, M.; Alford, S.; Aziz, S. Evaluating Energy Generation Capacity of PVDF Sensors: Effects of Sensor Geometry and Loading. *Materials* **2021**, *14*, 1895. [CrossRef] [PubMed]
12. Karalis, G.; Mytafides, C.; Tzounis, L.; Paipetis, A.; Barkoula, N. An Approach toward the Realization of a Through-Thickness Glass Fiber/Epoxy Thermoelectric Generator. *Materials* **2021**, *14*, 2173. [CrossRef] [PubMed]
13. Joseph, J.; Ohtsuka, M.; Miki, H.; Kohl, M. Lumped Element Model for Thermomagnetic Generators Based on Magnetic SMA Films. *Materials* **2021**, *14*, 1234. [CrossRef] [PubMed]
14. Naseem, S.; Gevers, B.; Labuschagné, F.; Leuteritz, A. Preparation of Photoactive Transition-Metal Layered Double Hydroxides (LDH) to Replace Dye-Sensitized Materials in Solar Cells. *Materials* **2020**, *13*, 4384. [CrossRef] [PubMed]

Communication

Footstep Energy Harvesting with the Magnetostrictive Fiber Integrated Shoes

Hiroki Kurita, Kenichi Katabira, Yu Yoshida and Fumio Narita *

Department of Materials Processing, Graduate School of Engineering, Tohoku University, Aoba-yama 6-6-02, Sendai 980-8579, Japan
* Correspondence: narita@material.tohoku.ac.jp

Received: 1 May 2019; Accepted: 24 June 2019; Published: 26 June 2019

Abstract: Wearable energy harvesting devices attract attention as the devices provide electrical power without inhibiting user mobility and independence. While the piezoelectric materials integrated shoes have been considered as wearable energy harvesting devices for a long time, they can lose their energy harvesting performance after being used several times due to their brittleness. In this study, we focused on Fe–Co magnetostrictive materials and fabricated Fe–Co magnetostrictive fiber integrated shoes. We revealed that Fe–Co magnetostrictive fiber integrated shoes are capable of generating 1.2 µJ from 1000 steps of usual walking by the Villari (inverse magnetostrictive) effect. It seems that the output energy is dependent on user habit on ambulation, not on their weight. From both a mechanical and functional point of view, Fe–Co magnetostrictive fiber integrated shoes demonstrated stable energy harvesting performance after being used many times. It is likely that Fe–Co magnetostrictive fiber integrated shoes are available as sustainable and wearable energy harvesting devices.

Keywords: magnetostrictive; energy harvesting; wearable

1. Introduction

Miniaturization technology of electrical devices has allowed the development of various portable devices, such as watches, smartphones, etc. However, further downsizing of electrical devices, which is required with advances in electronic technology, has not yet been achieved because of the difficulty to downsize batteries for portable devices. Therefore, energy-harvesting devices, which generate electrical power from mechanical phenomena, attract attention for developing battery-free systems. Wearable energy harvesting devices have especially attracted a lot of interest as they can provide electrical power and ensure user mobility and independence [1].

Piezoelectric materials have been considered as wearable energy harvesting devices [2,3], and mainly integrated into shoes to harvest a large amount of mechanical energy [4–9]. For example, Turkman et al. have reported 1.4 mW of generated power by the applied mass of 90 kg [9]. However, it is well known that piezoelectric materials generally have low impact resistance and are easy to depolarize. Hence, piezoelectric materials integrated shoes have a high probability of fracturing, having electric fatigue, and becoming dysfunctional immediately; however, they show good energy harvesting performance for a temporary period of time [10].

Magnetostrictive materials have been considered as other potential candidate materials for energy harvesting [11,12]. Vibration energy harvesting by magnetostrictive materials allows one to obtain a representative high output electric current compared to piezoelectric materials [13]. In recent years, a new structure of magnetostrictive TbDyFe alloy (Terfenol-D) generator for harvesting the rotation of the human knee joint was presented [14]. However, the typical magnetostrictive materials, Terfenol-D and Galfenol alloys, are extremely brittle despite having outstanding magnetostriction (respectively 800–1200 ppm and 120–240 ppm [12]). Therefore, we have recently focused on iron cobalt (Fe–Co) magnetostrictive materials, which have acceptable toughness and are inexpensive [15–17]. It seems that Fe–Co magnetostrictive materials endure cyclic loads, and are a prime candidate material for an energy harvesting among magnetostrictive materials. Furthermore, the energy harvesting performance of magnetostrictive fiber integrated shoes has not yet been reported to the best of our knowledge. In this study, we fabricated Fe–Co magnetostrictive fiber integrated shoes and evaluated the output energy of footstep energy harvesting during ambulation activities.

2. Experimental Procedure

Figure 1 shows the preparation process of the Fe–Co magnetostrictive device. The Fe–Co continuous fibers were prepared by drawing, and the composition is $Fe_{29}Co_{71}$. 125 Fe–Co fibers with a diameter of 0.2 mm and a solenoid coil with 1050 turns (CA05310180, Takaha Kiko Co., Ltd., Iizuka, Fukuoka, Japan) were prepared as starting materials. The coil was covered with a polymer sheet and Fe–Co fibers were inserted into the center hole of the coil. After that, a mixed solution of Bisphenol F and curing agent were casted into the polymer sheet and cured at 80 °C for 180 min to fabricate the Fe–Co magnetostrictive device. The mixing weight ratio of Bisphenol F: Curing agent was determined to be 100:55. The Fe–Co magnetostrictive device was polished to obtain a diameter of 12 mm and a height of 34 mm, and integrated into a hollowed heel of the left pump, as shown in Figure 2. The volume fractions of Fe–Co fiber and other parts (epoxy matrix and solenoid coil) in the device were approximately 33% and 67%, respectively. The magnetostrictive coefficient of the Fe–Co magnetostrictive device (d_{33}^c) is expected to become smaller than that of the original Fe–Co fiber (d_{33}^f) because the coefficient of the device is expressed simply by $d_{33}^c = v^f d_{33}^f$, where v^f (=0.33) is the volume fraction of the Fe–Co fiber. A neodymium magnet with the surface magnetic induction of 360 mT was fixed at the bottom side of the Fe–Co magnetostrictive device, to orient the magnetization direction of Fe–Co fibers.

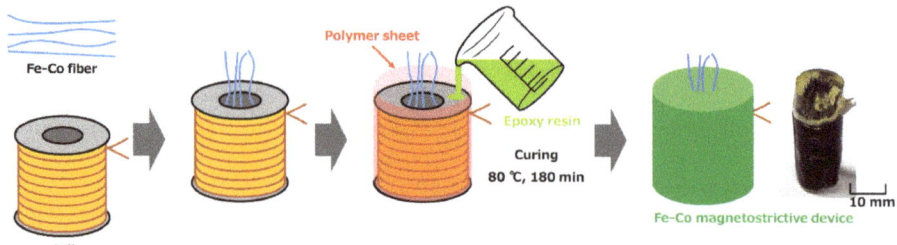

Figure 1. Schematic illustration of the preparation process of Fe–Co magnetostrictive device.

Figure 2. Integrated position of Fe–Co magnetostrictive device into a hollowed heel of the left pump. (**a**) side view, (**b**) bottom view, and (**c**) schematic illustration of Fe–Co magnetostrictive fiber integrated shoe.

The Fe–Co magnetostrictive device was connected to a resistance of 20 Ω, which agrees with the value of the device (19.7 Ω) to optimize output voltage and a data logger in parallel. Note that the input resistance value of the data logger was 14.5 MΩ. The output power was evaluated from a voltage and a resistance during two different ambulation activities of two subjects. Figure 3 shows the output power evaluation appearance during the usual walking of subject 1.

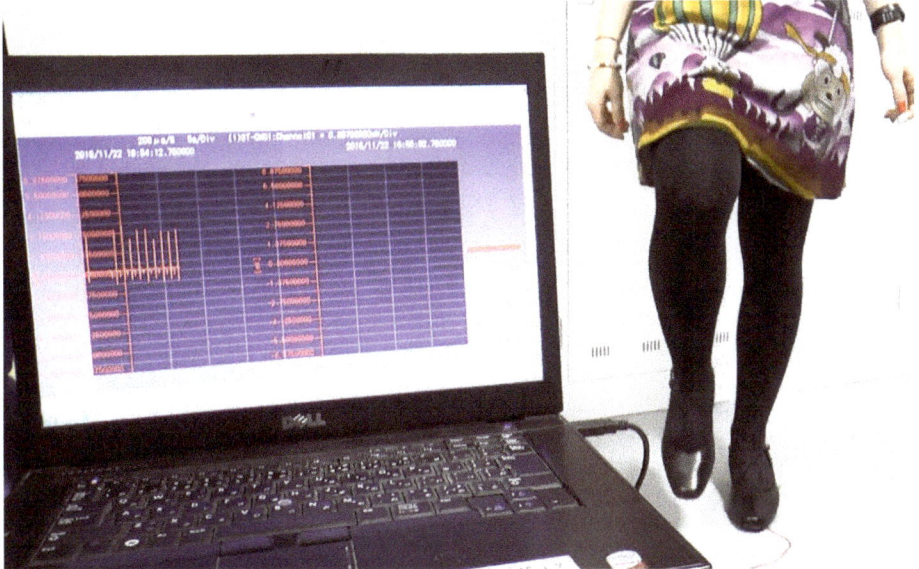

Figure 3. Output power evaluation appearance of Fe–Co magnetostrictive fiber integrated shoes during the usual walking of subject 1.

3. Theory

Let us now consider the system using rectangular Cartesian coordinates x_i (O-x_1, x_2, x_3). The easy axis for the magnetization of a magnetostrictive fiber is along the length direction (x_3-direction) [18]. For the magnetostrictive fiber, the changes of magnetic induction are decided by the stress amplitude applied in the direction of the fiber. The constitutive equations of the one-dimensional magnetostrictive fiber can be written as [16]

$$\varepsilon_{33} = s_{33}\sigma_{33} + d^{f}_{33}H_3 = s_{33}\sigma_{33} + d^{m}_{33}\left(\frac{B_0}{\mu_{33}}\right) + m_{33}\left(\frac{B_0}{\mu_{33}}\right)^2 \qquad (1)$$

$$B_3 = d^{m}_{33}\sigma_{33} + m_{33}\left(\frac{B_0}{\mu_{33}}\right)\sigma_{33} + B_0 \qquad (2)$$

where σ_{33} and ε_{33} are the stress and strain components, B_3 and H_3 are the magnetic induction and magnetic field intensity components, s_{33}, d^{m}_{33}, m_{33}, and μ_{33} are the elastic compliance, piezomagnetic constant, second-order magnetoelastic constant and magnetic permeability, respectively, and B_0 is the magnetic bias field. The output power is given using the resistance R as

$$P_{\text{out}} = \frac{V^2_{\text{out}}}{R} \qquad (3)$$

where the output voltage V_{out} for the magnetostrictive fiber is obtained as

$$V_{out} = -NA\frac{dB_3}{dt} \quad (4)$$

In Equation (4), N is the number of turns in the search coil, A is the cross-sectional area of all magnetostrictive fibers and t is the time. Substituting Equation (2) into Equation (4) gives

$$V_{out} = -NA\left[d_{33} + \left(\frac{m_{33}}{\mu_{33}}\right)B_0\right]\frac{d\sigma_{33}}{dt} \quad (5)$$

It is clear that the output voltage is proportional to the time derivative of the stress in the fiber. This mechanism differs from the piezoelectric material. That is, the output voltage of the piezoelectric material is proportional to the electric field amplitude associated with the stress amplitude [3].

4. Results and Discussion

Figure 4 shows the output power obtained by Fe–Co magnetostrictive fiber integrated shoes. The output power was calculated by Equation (3). The subjects walked at a velocity of 1 step/s (1 Hz). The output power was obtained when the left foot of subjects contacted to the floor. This result clearly shows that this power generation was attributed to the Villari (inverse magnetostrictive) effect. Figure 5 shows the output energy. Although there was no correlation between output energy and each step, the output energy during the usual walking of subject 2 (Figure 5b) was qualitatively larger than that of subject 1 (Figure 5a). Note that subject 1 and 2 have almost the same weight, therefore it is indicated that the output energy is not dependent on the weight. The output energy during the usual walking of subject 1 (Figure 5a) increased when she raised her leg higher (Figure 5c). It seems that larger impacts at the heel of the pump generate a larger output energy. In fact, although the magnetic induction is proportional to the stress (referred to as the Villari effect in Equation (2)), the voltage depends on the time derivative of the magnetic induction, i.e., stress-rate (see Equation (5)). Hence, the output power (proportional to the square of the voltage) is related to the speed of the heel rather than the weight. On the other hand, the output energy during the usual walking of subject 2 decreased when she raised her leg higher (Figure 5d). Moreover, a larger output energy was obtained when she walked at a velocity of 2 steps/s (2 Hz) without raising her legs higher (i.e., when the foot of subject 1 was contacted faster than her usual walk). Consequently, it is implied that the output energy is dependent on the subject's type of ambulation, not on the weight of subjects.

The maximum output energy achieved 2.6 nJ, and the instantaneous peak power was 2.4 µW. It is also capable of generating 1.2 µJ from 1000 steps of usual walking. In any case, it seems that Fe–Co magnetostrictive fiber integrated shoes generate low maximum output power, and it is necessary to generate larger output power for the practical realization of Fe–Co magnetostrictive fiber integrated shoes. The size of the coils dominate the output power from magnetostrictive materials, however, magnetostrictive devices should be small enough to be embedded inside shoes. It is likely that an important challenge in the coming years will be device size reduction as nanotechnology becomes increasingly prevalent. In addition, the magnetostrictive device was designed to fit into the heel of a woman's shoe. When the device is embedded in another shoe (e.g., a flat shoe), we have to consider a different design of the magnetostrictive device. For example, one conceivable design for a flat shoe is to adapt the configuration to the sole of a flat shoe, utilizing its flexural deformation.

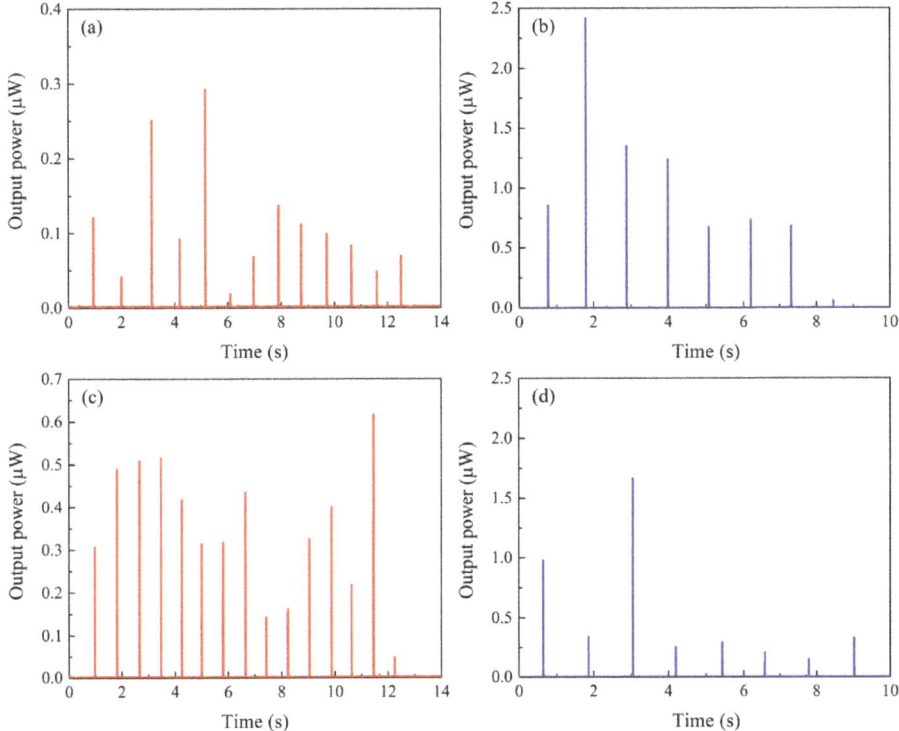

Figure 4. Output power obtained by Fe–Co magnetostrictive fiber integrated shoes; (**a**) during the usual walking of subject 1, (**b**) during the usual walking of subject 2, (**c**) during the walking with legs raised higher in subject 1, and (**d**) during the walking with legs raised higher in subject 2.

As mentioned above, general piezoelectric ceramics and typical magnetostrictive alloys are brittle. Therefore, these materials should lose their energy harvesting performance after being used several times or by an unheralded impact, even if they demonstrate outstanding energy harvesting for a temporary period of time. However, the Fe–Co magnetostrictive device did not break after 1 million compression tests (not shown here), whereas the coil was slightly damaged. While a fatigue test is required to investigate the accurate durability of Fe–Co magnetostrictive fiber integrated shoes, it is likely that Fe–Co magnetostrictive fiber integrated shoes permanently exercise an energy harvesting performance. From not only the mechanical point of view but also the functional one, Fe–Co magnetostrictive fiber integrated shoes have an advantage compared with piezoelectric materials integrated shoes. For piezoelectric materials, a polarization treatment is required to obtain excellent piezoelectric properties [3]. On the other hand, magnetostrictive materials do not require a polarization treatment; they provide stable magnetostrictive properties (i.e., energy harvesting performance). The Fe–Co magnetostrictive fiber integrated shoes used in this study actually demonstrated stable magnetostrictive properties. Hence, it seems that Fe–Co magnetostrictive fiber integrated shoes can be sustainable and wearable energy harvesting devices.

To achieve more stable and functional energy harvesting performance, we should explore every possibility for the best design of the shoes hereafter. For instance, it is necessary to consider the effect of a bias magnet interfering with the surface where the wearer walks, or the impact damage to a permanent magnet or electromagnet coil for biasing the magnetostrictive material.

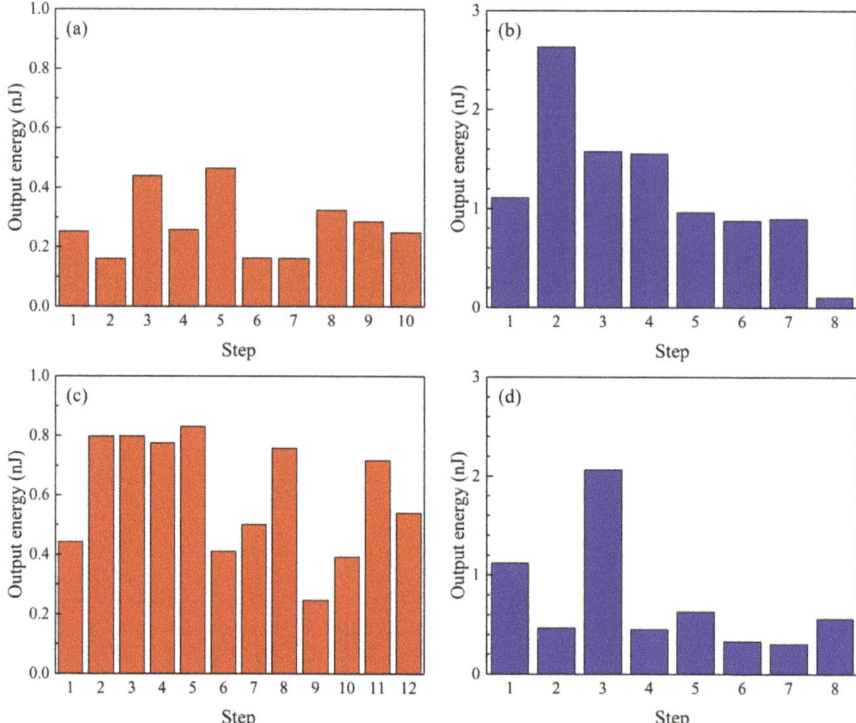

Figure 5. Output energy obtained by Fe–Co magnetostrictive fiber integrated shoes; (**a**) during the usual walking of subject 1, (**b**) during the usual walking of subject 2, (**c**) during walking with legs raised higher in subject 1, and (**d**) during the walking with legs raised higher in subject 2.

5. Conclusions

The footstep energy harvesting output power and energy of Fe–Co magnetostrictive fiber integrated shoes were evaluated during two different ambulation activities of two subjects. It was revealed that output power and energy were obtained during the ambulation activities generation by the Villari effect.

Despite subject 1 and 2 having almost the same weight, the output energies from their ambulation activities were extremely different. Furthermore, the output energy increased when the subjects applied larger impacts at the heel of the pump. Therefore, it was implied that the output energy is dependent on the habit of subjects on ambulation, not on their weight. Especially the impact when the pump contacts on the floor seems a dominant factor.

From both the mechanical and functional point of view, Fe–Co magnetostrictive fiber integrated shoes demonstrated stable magnetostrictive properties (i.e., energy harvesting performance). We believe that Fe–Co magnetostrictive fiber integrated shoes can be sustainable and wearable energy harvesting devices, with the improvement of maximum output power and energy by optimizing the magnetostrictive devices.

Author Contributions: Conceptualization, F.N.; methodology, F.N. and Y.Y.; validation, Y.Y.; formal analysis, H.K. and K.K.; investigation, K.K. and Y.Y.; resources, Y.Y.; data curation, Y.Y.; writing—original draft preparation, H.K.; writing—review and editing, K.K. and F.N.; visualization, H.K., K.K. and Y.Y.; supervision, F.N.; project administration, F.N.; funding acquisition, F.N.

Funding: This work was supported by the Matching Planner Program from the Japan Science and Technology Agency (JST).

Acknowledgments: We really appreciate Stéphanie KURITA and Yingmei XIE, who greatly assisted in obtaining the research data for evaluating energy harvesting performance of magnetostrictive fiber integrated shoes.

Conflicts of Interest: The authors declare that there are no potential conflict of interest with respect to the research, authorship and/or publication of this article.

References

1. Ylli, K.; Hoffmann, D.; Willmann, A.; Becker, P.; Folkmer, B.; Manoli, Y. Energy harvesting from human motion: Exploiting swing and shock excitations. *Smart Mater. Struct.* **2015**, *24*, 025029. [CrossRef]
2. Narita, F.; Nagaoka, H.; Wang, Z. Fabrication and impact output voltage characteristics of carbon fiber reinforced polymer composites with lead-free piezoelectric nano-particles. *Mater. Lett.* **2019**, *236*, 487–490. [CrossRef]
3. Wang, Z.; Narita, F. Corona poling conditions for barium titanate/epoxy composites and their unsteady wind energy harvesting potential. *Adv. Eng. Mater.* **2019**, in press. [CrossRef]
4. Mateu, L.; Moll, F. Appropriate charge control of the storage capacitor in a piezoelectric energy harvesting device for discontinuous load operation. *Sens. Actuators A* **2006**, *123*, 302–310. [CrossRef]
5. Rocha, J.G.; Goncalves, L.M.; Rocha, P.F.; Silva, M.P.; Lanceros-Mendez, S. Energy harvesting from piezoelectric materials fully integrated in footwear. *Trans. Ind. Electr.* **2010**, *57*, 813–819. [CrossRef]
6. Alumusallam, A.; Torah, R.N.; Zhu, D.; Tudor, M.J.; Beeby, S.P. Screen-printed piezoelectric shoe-insole energy harvester using an improved flexible PZT-polymer composites. *J. Phys. Conf. Ser.* **2013**, *476*, 012108. [CrossRef]
7. Jung, W.S.; Lee, M.J.; Kang, M.G.; Moon, H.G.; Yoon, S.J.; Baek, S.H.; Kang, C.Y. Powerful curved piezoelectric generator for wearable applications. *Nano Energy* **2015**, *13*, 174–181. [CrossRef]
8. Kalantarian, H.; Sarrafzadeh, M. Pedometers without batteries: An energy harvesting shoe. *Sens. J.* **2016**, *16*, 8314–8321. [CrossRef]
9. Turkman, A.C.; Celik, C. Energy harvesting with the piezoelectric material integrated shoe. *Energy* **2018**, *150*, 556–564. [CrossRef]
10. Siddiqui, S.; Kim, D.-I.; Roh, E.; Duy, L.T.; Trung, T.Q.; Nguyen, M.T.; Lee, N.-E. A durable and stable piezoelectric nanogenerator with nanocomposite nanofibers embedded in an elastomer under high loading for a self-powered sensor system. *Nano Energy* **2016**, *30*, 434–442. [CrossRef]
11. Deng, Z.; Dapino, M.J. Review of magnetostrictive vibration energy harvesters. *Smart Mater. Struct.* **2017**, *26*, 103001. [CrossRef]
12. Narita, F.; Fox, M. A review on piezoelectric, magnetostrictive, and magnetoelectric materials and device technologies for energy harvesting applications. *Adv. Eng. Mater.* **2018**, *20*, 1700743. [CrossRef]
13. Yang, Z.J.; Nakajima, K.; Onodera, R.; Tayama, T.; Chiba, D.; Narita, F. Magnetostrictive clad steel plates for high-performance vibration energy harvesting. *Appl. Phys. Lett.* **2018**, *112*, 073902. [CrossRef]
14. Yan, B.; Zhang, C.; Li, L. Magnetostrictive energy generator for harvesting the rotation of human knee joint. *AIP Adv.* **2018**, *8*, 056730. [CrossRef]
15. Narita, F. Inverse magnetostrictive effect in $Fe_{29}Co_{71}$ wire/polymer composites. *Adv. Eng. Mater.* **2017**, *19*, 1600586. [CrossRef]
16. Narita, F.; Katabira, K. Stress-rate dependent output voltage for $Fe_{29}Co_{71}$ magnetostrictive fiber/polymer composites: Fabrication, experimental observation and theoretical prediction. *Mater. Trans.* **2017**, *58*, 302–304. [CrossRef]
17. Katabira, K.; Yoshida, Y.; Masuda, A.; Watanabe, A.; Narita, F. Fabrication of Fe-Co magnetostrictive fiber reinforced plastic composites and their sensor performance evaluation. *Materials* **2018**, *11*, 406. [CrossRef] [PubMed]
18. Yang, Z.J.; Nakajima, K.; Jiang, L.; Kurita, H.; Murasawa, G.; Narita, F. Design, fabrication and evaluation of metal-matrix lightweight magnetostrictive fiber composites. *Mater. Des.* **2019**, *175*, 107803. [CrossRef]

© 2019 by the authors. Licensee MDPI, Basel, Switzerland. This article is an open access article distributed under the terms and conditions of the Creative Commons Attribution (CC BY) license (http://creativecommons.org/licenses/by/4.0/).

Article

Modeling and Characterization of a Kinetic Energy Harvesting Device Based on Galfenol

Carmine Stefano Clemente [1,*,†] and Daniele Davino [2,†]

1. Department of Energy, Systems, Territory and Construction Engineering, University of Pisa, 56122 Pisa, Italy
2. Department of Engineering, University of Sannio, 82100 Benevento, Italy; davino@unisannio.it
* Correspondence: carmine.clemente@ing.unipi.it
† These authors contributed equally to this work.

Received: 17 August 2019; Accepted: 25 September 2019; Published: 29 September 2019

Abstract: The proposal of Energy Harvesting (EH) techniques and devices has experienced a significant growth over the last years, because of the spread of low power electronic devices. Small ambient energy quantities can be recovered through EH and exploited to power Wireless Sensor Networks (WSN) used, for example, for the Structural Health Monitoring (SHM) of bridges or viaducts. For this purpose, research on EH devices based on magnetostrictive materials has significantly grown in the last years. However, these devices comprise different parts, such as a mechanical system, magnetic circuit and electrical connections, which are coupled together. Then, a method able to reproduce the performance may be a handy tool. This paper presents a nonlinear equivalent circuit of a harvester, based on multiple rods of Galfenol, which can be solved with standard circuit simulator. The circuital parameters are identified with measurements both on one rod and on the whole device. The validation of the circuit and the analysis of the power conversion performance of the device have been conducted with different working conditions (force profile, typology of permanent magnets, resistive electrical load).

Keywords: energy harvesting; magnetostrictive materials; Galfenol

1. Introduction

The goal of Energy Harvesting (EH) techniques is to scavenge small environmental energy quantities and convert them into electrical one in order to supply low-power consumption electronics [1–3]. Environmental energy is often present in form of mechanical vibrations, thermal gradients, pressure gradients and solar radiations. It is generally accepted that this energy go lost because of its relative expensiveness to be exploited on a large scale. On the other hand, ambient energy could represent a good candidate as a source for remote and harsh applications of Wireless Sensors Network (WSN).

Indeed, in recent decades, the technological development of Information and Communications Technology (ICT) has led to the massive diffusion of portable devices, such as mobile phones, music players, PDAs, tablets, wrist watches, smart watches and, lately, sensors devices used to monitor and control sensitive quantities (temperature, pressure, displacement, etc.), leading to the Internet of Things (IoT) devices. Among several applications, Wireless Sensors Network (WSN) has an important one in the Structural Health Monitoring (SHM) to monitor existing aged structures, such as bridges, viaducts or railways [4–11]. These may be placed in rural areas, where often the electrical network is not present. Then, these devices are supplied from batteries that should be regularly substituted or recharged from the power grid. This is a limitation in the application of WSN to continuous SHM because the maintenance cost of the sensors is considered high and this results in a detriment of users safety.

The advantages of EH is the possibility to minimize the frequency maintenance of batteries (charging or replacing) or even to eliminate this cost for the whole device's lifetime. This would be very beneficial to power several nodes in a network, used to monitor physical or environmental conditions of buildings [12–14]. In particular, in structure such as bridges, viaducts or railways, the energy is present in form of mechanical vibrations due to the ongoing traffic and, therefore, it can be extract and converted with suitable techniques [15–17] (Kinetic Energy Harvesting, KEH).

In order to power supply WSNs, a promising EH techniques could be the use of smart materials [18]. In particular, for KEH purposes, the most used are piezoelectric materials, which couple mechanical properties with electrical ones and magnetostrictive materials, which relate mechanical properties with magnetic ones. The latter are of a particular interest, because device based on magnetostrictives show better mechanical characteristics and higher energy density than piezoelectrics and other EH techniques. Moreover, they show a larger magneto-mechanical coupling, no depolarization and creep phenomena and longer life-time with respect to piezoelectrics [19]. The kinetic energy conversion [20,21] takes advantage of the inverse magnetostrictive, or Villari, effect. It consists of the variation of the magnetic induction inside the magnetostrictive material as a result of a variation in the applied mechanical stress. Then, according to Faraday's law, a coil placed around the "active" material shows a voltage at the terminals, because of the variation of the linkage magnetic flux. This effect can be used to supply in-situ a single node of a WSN for SHM, by coupling it to a rechargable battery or to a super-capacitor. For example, by considering a viaduct, the harvester could be installed, with some technical efforts, below the thermal expansions joints or into the structural bearings. In the first case, the ongoing traffic vehicles would be the source of vibration, with a behavior similar to a non-periodic train of impulses. The second typology would results in a low frequency behavior (related to the structure resonance, typically much lower than 100 Hz) but almost continuous in time.

Among the magnetostrictive materials, Galfenol is favored to be used with respect to Terfenol-D for EH purposes. In fact, the former has better magnetic and mechanical characteristics than Terfenol-D, which is fragile and shows lower saturation induction [22,23]. Moreover, Galfenol exhibits narrow hysteresis cycle and low coercive field [24–26], that is, low energy conversion losses.

The aim of this paper is to present, characterize and model a KEH system based on multiple stress-annealed (SA) Galfenol rods. In particular, the device belongs to the force-driven category and it is conceived to be located under the road paving of a bridge or viaduct [16].

In this paper, in Section 2 an experimental setup devoted to characterize the magnetostrictive materials is described and the general behaviors of SA Galfenol are highlighted. In Section 3 an analytical nonlinear fully-coupled model is obtained and used to develop an equivalent electrical circuit of a single rod KEH concept device where the whole mechanical, magnetic and electric quantities are suitably related. Finally, this three port equivalent circuit has been used on a more engineered KEH system based on three rods of SA Galfenol. Simulations and comparison with experimental tests in different working conditions have been performed and discussed in Section 4.

2. Experimental Setup and Static Magneto-Mechanical Characteristics of Galfenol

The active core of the KEH device studied in this work is Galfenol SA [24]. It is an Iron-Gallium alloy ($Fe_{81.6}Ga_{18.4}$) and it is textured polycrystalline, grown with free-standing zone melt (FSZM) technique by TdVib LLC. Table 1 lists some of Galfenol's magnetic and mechanical properties. It shows a good combination of magnetic, mechanical and magnetostrictive properties that common smart materials do not. In particular, it exhibits magnetostriction values of about 200–250 ppm but has better magnetic characteristics than Terfenol-D, which makes it particularly suitable for sensing and harvesting use. Gallium concentration strongly affects both the magnetic and magnetostrictive behavior, as reported in Reference [24]. Galfenol's ability to be used both in tension and compression, robustness, mechanical workability and high Curie temperature (~600 °C) is attracting interest for the alloy's use in harsh environments. Applications actively being investigated include transducers for down-hole use, fuel injectors, sensing and EH devices. Since the beginning of 2000, the use of stress annealing on Terfenol-D has been studied with the aim of impart a stress "frozen" into material. Stress annealing is a post-manufacturing process and it constitutes a company's sensitive information. However, the process mainly consists in the applications of high magnetic field and temperature to the material, for a certain time interval. After that, a built-in stress remains in the treated sampled which generates an uniaxial anisotropy that is desirable because it can greatly simplify the design of devices by obviating the need of a pre-stress mechanism [27–29]. It is worth to highlight that the Galfenol samples used in this work and the annealing process have been made and conducted, respectively, by TdVib LLC.

Table 1. Some properties of Galfenol [30].

Standard Composition	$Fe_{81.6}Ga_{18.4}$
Mechanical Properties	
Density	7800 kg/m^3
Young's Modulus at constant I	40–60 GPa
Young's Modulus at constant V	60–80 GPa
Bulk Modulus	125 GPa
Speed of Sound	2265–2775 m/s
Tensile Strength	350 MPa
Fatigue Strength	75 MPa fully reversed
Minimum Laminate Thickness	0.25 mm
Thermal Properties	
Thermal expansion coefficient	11 ppm/°C at 25 °C
Thermal Conductivity	15–20 W/(mK) at 25 °C
Melting Point	1450 °C
Electrical Properties	
Resistivity	85×10^{-8} Ωm
Curie Temperature	670 °C
Magnetostrictive Properties	
Strain (estimated linear)	200–250 ppm
Energy Density	0.3–0.6 kJ/m^3
Piezomagnetic Constant, d_{33}	20–30 nm/A
Magnetomechanical Properties	
Coupling Factor	0.6–0.7
Magnetic Properties	
Relative Permeability	75–100
Saturating Magnetic Field	8–20 kA/m
Coercivity, H_c	$\sim$800 A/m
Hysteresis (major loop)	1000 J/m^3
Saturation Flux Density	1.5–1.6 T

In order to model Galfenol SA, its experimental characterization is strictly necessary. Then, in the following it is described the experimental setup developed and realized for this purpose and the main characteristics and properties of Galfenol are measured and discussed. It is worth to note that, in an overall vision, magnetostrictive materials could be seen then as a 2 input–2 output system. In particular, by applying to the material one mechanical (*stress*, σ) and magnetic (*field*, H) input it responses with the corresponding mechanical (*strain*, ε) and magnetic (*magnetization*, M) output. A schematic blocks diagram of the experimental setup is shown in Figure 1a. In the central part of Figure 1a is the magnetostrictive material where the blue arrows represent the two input, while the red ones the output. The red dots and text box represent the measurements points and the measurement systems respectively. A 2-D sketch of the setup used for magnetostrictive materials characterization is represented in Figure 1b.

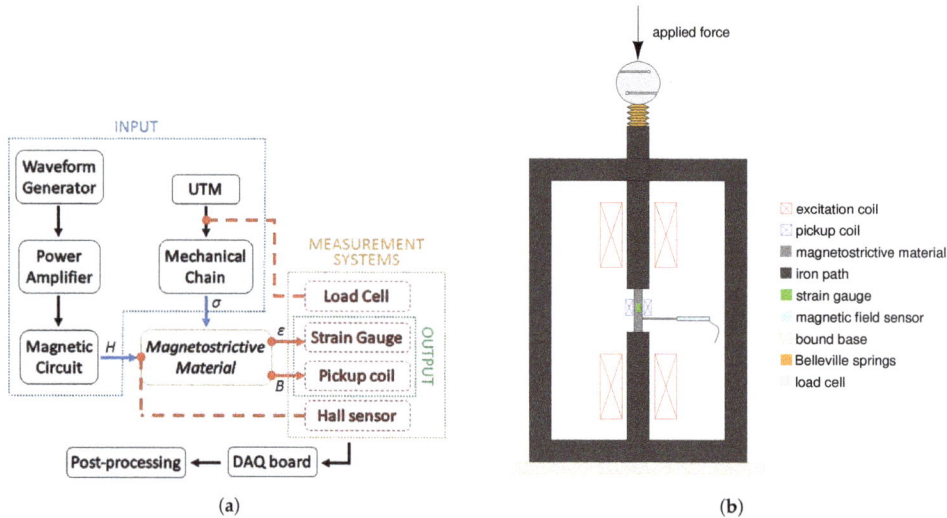

Figure 1. Illustration of the setup used for the characterization of magnetostrictive material. (**a**) Blocks diagram. (**b**) 2-D sketch.

More in details, the magnetic field is provided by two excitation coils placed in the central column of a 3-phase transformer-like iron path (cross section: 30 mm × 30 mm), which acts as a magnetic circuit. The excitation coils are supplied by a power amplifier, controlled by an arbitrary signals generator, as reported in Figure 1a. The central column is cut in two halves—the bottom part is bound with the remaining iron path, while the top half part is connected, through Belleville springs and a load cell, to the Universal Testing Machine (UTM), constituting thus a mechanical chain able to provide the stress to the magnetostrictive sample. The springs are installed because they allow to mitigate the magnetostrictive effect when a constant stress is applied in the mechanical chain. Moreover, the top half part can slide in a square hole made in the upper horizontal arm of the iron part; so that it allows both the stress transfer and the magnetic circuit closing. A Galfenol cylindrical sample (30 mm length and 5 mm diameter) is placed between the two parts of the central column and it is equipped with two strain gauges located on the same external surface but on opposite faces. The measured strain is the average of the two measurement points and any non-uniform stress transfer to the material, due to imperfections, is compensated. A pickup coil is wounded around the sample, while a Hall effect probe (connected to a gaussmeter) is placed perpendicularly in contact with the external surface of the sample, in order to measure the applied magnetic field [25,31]. Both the inputs and outputs are measured by the above mentioned sensors and acquired with a DAQ. The strain, stress and magnetic field are directly measured while the magnetic flux density is measured indirectly. Since in the used experimental setup the cross section of the sample, namely S and the pickup coil turns number, namely N, do not change, is possible to express the Faraday's law of induction as:

$$v = -\frac{d\Phi}{dt} = -NS\frac{dB}{dt} \quad (1)$$

where v is the measured pickup coil voltage (so the electromotive force, emf), Φ is the linked magnetic flux and B the magnetic induction. As a consequence it is possible to write B as follows:

$$B = -\frac{1}{NS}\int_t v\, dt \quad (2)$$

The integration in Equation (2) is numerically executed in post-processing, in order to obtain the magnetic induction. Finally, being the relation among flux density, magnetization and magnetic field as follows:

$$B = \mu_0(H + M) \qquad (3)$$

the magnetization can be obtained.

The characterization of a magnetostrictive materials consists in the measurement of the output when one input changes and the other is kept constant. Therefore, four different types of plots are measurable: $\mu_0 M - H$ and ε-H curves at different constant σ and ΔB-σ and ε-σ curves at different constant H. On the other hand, the first and third ones are interesting for EH purposes. At this point it is important to note that the UTM is a device controlling the strain velocity. As a consequence, the set values of the force are applied to the material by a feedback on the force sensor measurement and a null cross-head speed. The magnetostrictive effect counteract the UTM applied stress, by increasing the latter because the crosshead speed control is not sufficient to compensate this variation of force in a short time. To reduce this effect the Belleville springs are introduced in the mechanical chain in order to have a more stable constant applied stress. On the other hand, it should be pointed out that the curves with constant magnetic field are obtained in constant drive excitation coils current, unlike what has been proposed in Reference [32], where the UTM device is directly connected with the sample but a closed loop feedback on the magnetic field is applied.

Magnetic characteristics, at different constant compressive stress, are plotted in Figure 2a. The curves shown very narrow hysteresis loops, low coercive and saturation field and high saturation magnetization (around 1.7–1.8 T). Furthermore, it is possible to note three distinct regions—the first where the curves are quite overlapped and with a straight slope, the second is characterized by a sudden increasing of slope (so-called "knee bending point") where the relative permeability is higher than before and, finally, the last region where the material approaches and reaches magnetic saturation. The particular shape of the magnetic cycles, at zero stress, is quite different from common soft ferromagnetic materials, such as non stress annealed (NSA) Galfenol, especially about the presence of the first region. From a phenomenological point of view, this behavior can be ascribed to the annealing process. Indeed, the built-in stress, obtained after the stress annealing process, acts as a pre-stress applied to the annealing-less material (NSA), as reported also in References [24,25]. At this point, it is worth noting that the magnetic curves are quite similar each other with respect to the applied stress. This property, named "*self-similarity*" [26,33,34], has been used to help the material modeling, as shown in the following section. Figure 2b shows the magnetic flux density variation in SA Galfenol, when constants magnetic field and cyclic compressive stress are applied. It should be pointed out that the maximum variation is available at 8 kA/m, that is the optimal bias. Indeed, by comparing this characteristic with the magnetic one, the wider magnetization range between the curves with lower and higher applied stress is obtained after the knee point. Conversely, by increasing the applied constant magnetic bias, the flux variation decreases being negligible at magnetic field larger than 20 kA/m because of the saturation conditions. It is worth to note that the maximum variation is about 1.2 T and the hysteresis is almost negligible.

In conclusion, in this section the usual behaviors of SA Galfenol characteristics are discussed. In particular, some general properties have been recognized and are listed below [34]:

- $\mu_0 M$-H are nonlinear and show narrow hysteresis loop;
- $\mu_0 M$-H show saturation when $|H| \to \infty$;
- $\mu_0 M$ is odd function of H;
- the $\mu_0 M$-H cycles are self-similar with respect to the stress;
- regarding $\mu_0 M$-H, if the compressive stress σ increases then the $\mu_0 M$-H cycles drop down.
- regarding ΔB-σ, there is an optimum H_{opt} that makes ΔB largest.

In the next section, these results are exploited to build up an analytical model and an equivalent circuit of a KEH device.

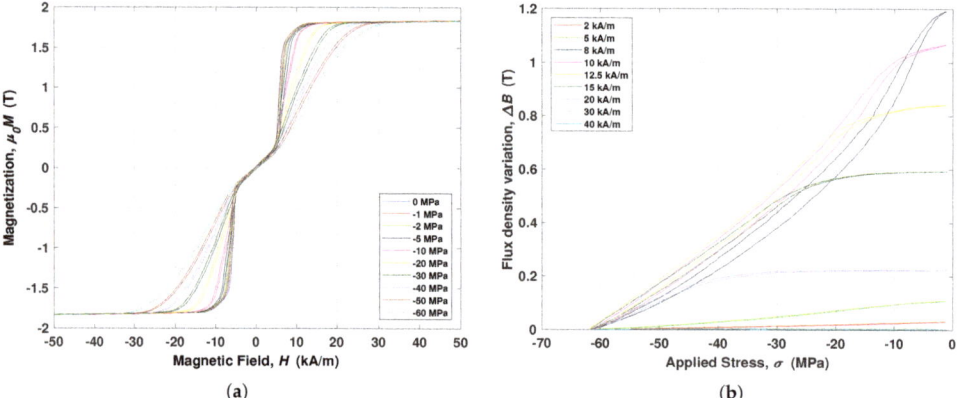

Figure 2. Magneto-mechanical characteristics of SA Galfenol. (**a**) Magnetic characteristics at different compressive applied stress. (**b**) Flux density variation by applying cyclic compressive stress for different magnetic bias conditions (harvesting loops).

3. Modelling

In order to analyze the EH conversion mechanism and to design a reliable energy recovering system, the material modeling or rather the constitutive relations that relate the mechanical and magnetic inputs and outputs, is of great importance [35]. As shown in the previous section, such links are related with strong non-linearity and hysteresis. However, the linear modeling is widely present in literature to describe harvesters, because it allows to obtain analytic expressions concerning the EH phenomenon. On the other hand, it offers acceptable results only when there are small variations of the inputs. These conditions are quite unreal for standard applications. Indeed, when such approximation is adopted, notable errors in outputs, with respect to the magnetic bias, are obtained [36–38].

For EH purposes, the constitutive relations should consider the losses, due to hysteresis or eddy currents. The latter process, for suitable small dimensions of the active sample with respect to the frequency ranges, could be neglected. Conversely, hysteresis loop shape does not depend on the input rate. Moreover, its area represents a loss energy density and, under periodic input excitation, the dissipated power is proportional to the frequency. On the other hand, when Galfenol is considered, the coercive field is very small and the hysteresis loop is narrow. Then, the hysteresis losses can be not taken into account while maintaining the model accuracy. This implicates the construction of several memory-less nonlinear models, to describe the magnetostrictive relationships [21,39]. However, none of them have been adopted in the complete modeling of a KEH device, as described hereafter. In particular, in the following a nonlinear fully-coupled model for SA Galfenol and an equivalent circuit of a single rod KEH device are shown.

3.1. Analytical Model of SA Galfenol

The modeling of a cylinder presented here is based on the following hypothesis:

- all mechanical input and fields are parallel to the cylinder axis;
- any transformations is isothermal;
- hysteresis phenomena is negligible;

then, the *Gibbs* free energy density can be considered as [25,26,34,40,41]:

$$G(\sigma, H) = \frac{\sigma^2}{2E} + \frac{\mu_0 H^2}{2} + \Psi(\sigma, H) \tag{4}$$

where E and μ_0 are, respectively, the Young's Modulus and the vacuum magnetic permeability.

The first two terms on the right hand side of Equation (4) constitute pure linear elastic and magnetic energy contributions, respectively. While the third term is the Gibbs free energy contribution due to the magneto-mechanical coupling of magnetostrictive materials and the function $\Psi(\sigma, H)$ is properly determined in accordance with the material characteristics. H and σ are the state variables, while $\varepsilon = \varepsilon(\sigma, H)$ and $B = B(\sigma, H)$ are the state functions. It is worth to note that if hysteresis is considered, then the memory state is an additional state variable [40]. Conversely, by neglecting the hysteresis, any transformation inside the material can be considered as *lossless*. From a thermodynamic point of view, this brings to the following constrain [39,42]:

$$\left.\frac{\partial \varepsilon}{\partial H}\right|_{\sigma} = \left.\frac{\partial B}{\partial \sigma}\right|_{H} \tag{5}$$

where the variables σ and H as suffix indicate that the derivatives are taken by assuming those variables as constants.

The Gibbs free energy, previously identified in Equation (4), is a state function, that is, it is a quantity that depends uniquely on the state of a system, since it admits an exact differential in a close range of an equilibrium state and, therefore, does not depend on previous thermodynamic transformations. At this stage, the conditions expressed in Equation (5) allow to relate the σ and H state variables within the corresponding ε and B state functions as follows:

$$\begin{cases} \varepsilon = \left.\dfrac{\partial G}{\partial \sigma}\right|_{H} \\ B = \left.\dfrac{\partial G}{\partial H}\right|_{\sigma} \end{cases} \tag{6}$$

The relations expressed in Equations (4) and (6) indicate that the determination of the function $\Psi(\sigma, H)$ allows to obtain the nonlinear model. It is assumed here that the function $\Psi(\sigma, H)$ is expressed as [26,34,40,41]:

$$\Psi(\sigma, H) = f(\sigma) \cdot u(z) \tag{7}$$

where $z = z(\sigma, H)$. The functions $f(\sigma)$, $u(z)$ and $z(\sigma, H)$ must be suitably determined, according to the material characteristics, in order to mimic the main physical behaviors of the material, as saturation effect and magnetization dependence by the stress [34,41]. Consequently, the Gibbs free energy is:

$$G(\sigma, H) = \frac{\sigma^2}{2E} + \frac{\mu_0 H^2}{2} + f(\sigma) \cdot u(z) \tag{8}$$

By deriving $G(\sigma, H)$ once respect to σ at constant H and once respect to H at constant σ, as in the relations expressed in Equation (6), it can be obtained ε and B, respectively, in the nonlinear case [26]:

$$\begin{cases} \varepsilon = \left.\dfrac{\partial G}{\partial \sigma}\right|_{H} = \dfrac{\sigma}{E} + f'(\sigma) \cdot u(z) + f(\sigma) \cdot u'(z) \cdot \dfrac{\partial z}{\partial \sigma} \\ B = \left.\dfrac{\partial G}{\partial H}\right|_{\sigma} = \mu_0 H + f(\sigma) \cdot u'(z) \cdot \dfrac{\partial z}{\partial H} \end{cases} \tag{9}$$

where $f'(\sigma)$ and $u'(z)$ represent the derivatives of the functions $f(\sigma)$ and $u(z)$ respectively, that is,

$$f'(\sigma) = \frac{df(\sigma)}{d\sigma}, \qquad u'(z) = \frac{du(z)}{dz}$$

By starting from the general expressions of Equation (9) and by using the assumption of $z = H/f(\sigma)$, in order to take into account the dependence of the magnetic response from the applied stress [40,41], it is possible to re-arrange the general nonlinear model as follows [34]:

$$\begin{cases} \varepsilon = \left.\dfrac{\partial G}{\partial \sigma}\right|_H = \dfrac{\sigma}{E} - f'(\sigma) \cdot [z \cdot u'(z) - u(z)] \\ B = \left.\dfrac{\partial G}{\partial H}\right|_\sigma = \mu_0 H + u'(z) \end{cases} \quad (10)$$

It should be noted that the second of Equation (10) implies that $u'(z)$ constitutes the magnetic polarization $M(H, \sigma)$ of the material. Then, the terms $u'(z)$, its integral $u(z)$, $f(\sigma)$ and z have to be chosen to fit Equation (10) for the adopted material. To this scope, for SA Galfenol, it has been assumed [26]:

$$M(H, \sigma) = u'(z) = \dfrac{\alpha \cdot z}{\beta + z^4} + M_s \cdot \tanh(z) \quad (11)$$

$$f(\sigma) = \gamma \cdot (\sigma + \sigma_b) \quad (12)$$

$$z = \dfrac{H}{f(\sigma)} = \dfrac{H}{\gamma \cdot (\sigma + \sigma_b)} \quad (13)$$

where M_s is the magnetic saturation, σ_b is the built-in stress due to the stress annealing process, while α, β and γ are some model's parameters.

The Equations (11)–(13) are found on a phenomenological base, with the aim to reproduce the magnetic characteristic of SA Galfenol discussed in Section 2. In particular, the three distinct regions of the magnetic induction at constant stress and its slope varying by increasing the mechanical stress are taken into account. The function expressed by the Equation (11) is odd and consists of two terms:

- a fractional fourth order function in the z variable, which describes SA Galfenol behavior at low magnetic field applied and the above mentioned kinking phenomena;
- an hyperbolic function able to mimic the response for increasing magnetic field and saturation.

Finally, as described by the first equations of the sets (9) and (10), the strain $\varepsilon(H, \sigma)$ is derived starting from Equations (11)–(13). Then, it follows [26]:

$$\varepsilon(H, \sigma) = \dfrac{\sigma}{E} - \gamma \cdot \left\{ z \cdot \left[\dfrac{\alpha \cdot z}{\beta + z^4} + M_s \cdot \tanh(z)\right] - \left[\dfrac{\alpha}{2\sqrt{\beta}} \cdot \arctan\left(\dfrac{z^2}{\sqrt{\beta}}\right) + M_s \cdot \ln(\cosh(z))\right] \right\} \quad (14)$$

The parameters M_s and σ_b can be directly obtained by a comparison with experimental data, while the remaining parameters are determined by using a standard minimization technique. All the steps about the identification procedure are described in Reference [26]. More in detail, the parameters M_s and σ_b are correlated to the M-H curves at different constant stresses, that are shown in Figure 2a. From these measurements the magnetic saturation can be assumed to be $M_s = 1.72$ T. By examining the above mentioned curves, by increasing the stress at the same magnetic field, the magnetic polarization M decreases. Moreover, if the magnetic field is enhanced, the magnetic polarization also increases. Consequently, the material gets to be magnetically harder when larger compressive stress are applied and, furthermore, all curves shown a similar profile, as discussed in Section 2. Consequently, it is said that the curves are self-similar [26,33,34,43] and, by looking for a suitable scaling methodology between input variables (H, σ), all curves almost drop into a single one. In this study, the condition of self-similarity is attained by assuming the magnetic field H divided by the complete stress applied to the Galfenol rod, that is, $\sigma + \sigma_b$, where σ_b is the built-in stress and σ the external applied stress. Since the built-in stress is a proper characteristic of the material, it is constant for all curves and can be achieved as consequence. Then, the value of σ_b is matched in such a way that the upshot curves of the M vs. $H/(\sigma + \sigma_b)$ are overlapped. For the considered SA Galfenol sample, σ_b is found to be -50 MPa such that the curves of M vs. $H/(\sigma + \sigma_b)$ present a matching profile as shown in Reference [26].

This value agrees with what is found in the literature [27], where the built-in stress values are obtained with an energy-based model minimization method and from manufacturer [30].

The other three parameters in Equations (11)–(13) have been achieved with a nonlinear least mean squares method applied between measurements and the nonlinear model. The procedure, shown in Reference [26], is to find α, β and γ such that the error between model and measurements is minimum. The method gave as results $\alpha = -1.1233$ T, $\beta = 0.8415$ and $\gamma = -9.7927 \times 10^{-5}$ T^{-1}, with a normalized residual of 0.0938. The comparison among the simulated magnetic characteristics and the measured ones is reported in Figure 3. The agreement is quite remarkable. It is worth to highlight that the curves of Figure 3 have been achieved by starting from the cycles of Figure 2a. In particular, the descending and ascending branches of the cycles are averaged point by point into the hysteresis-less curves.

In conclusion, the Equations (11)–(14) constitute a fully coupled nonlinear model, capable to mimic the typical behavior of SA Galfenol.

3.2. Three-Port Equivalent Circuit of a Single Rod KEH Device

In a KEH devices based on magnetostrictive materials, mechanical, magnetic and electric quantities are involved and coupled together [44]. Therefore, it may be difficult to study and design the different parts without a model that takes into account all the interactions. Then, an equivalent circuit, where the electrical components are related with the above-mentioned quantities, could be a handy tool able to simulate the working conditions of the KEH with a circuit simulator, like LTspice [45].

The nonlinear model previously developed has been used in this section to build up an equivalent lumped parameters circuit of a KEH device with a single rod of SA Galfenol. In such a way, it is possible to simulate a force-driven harvester in different conditions. In particular, the KEH device belongs to the force driven category, that is, the force source is directly put in contact with the active material, along the longitudinal axis, as reported in Figure 4.

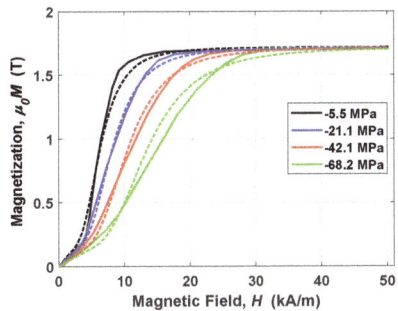

Figure 3. Magnetic characteristic (first quarter) of SA Galfenol, at different compressive stress—comparison between measurements (solid line) and simulation of Equation (11) (dashed line).

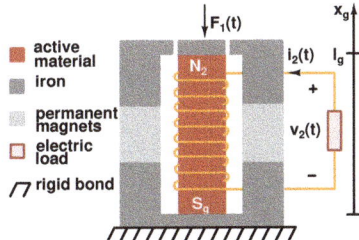

Figure 4. 2-D sketch of a force driven magnetostrictive KEH concept device and its main elements.

In spite of the arrangements, the main elements of a force driven KEH device are cataloged in:

- the active material, in this case SA Galfenol;
- a mechanical frame devoted to transfer the force to the active material;
- the magnetic circuit providing the magnetic bias to improve the harvested power;
- the coil, generally wounded around the active material, in order to exploit the Villari effect and Faraday's law.

Then, by considering the sketch of the KEH concept device shown in Figure 4, the Galfenol rod, with length l_g and a cross section S_g, is located in the central part. The iron frame has the twofold role as mechanical frame and magnetic circuit, where l_{fe} is the center line length and S_{fe} is the cross section. From the topside of the iron frame, through an air gap, a force $F_1(t)$ is applied to the sample. The iron structure hosts the PMs with length l_m and cross section S_m. Finally, the conversion of vibrational-mechanical energy into electrical one is possible by exploiting both the inverse magnetostriction effect and Faraday's law, by means of a N_2-turn coil placed around the active rod. The analysis presented in the following has been performed under these assumptions [26,34,41]:

1. all the fields are coaxial and directed along the magnetostrictive material axis, in such a way to have a scalar representation of the constitutive relationships;
2. the frequency of the applied force is small enough in such a way that it can be neglected the mechanical propagation phenomena in the rod. As a consequence, any mechanical element can be modeled as lumped mass, spring, damper. Moreover, vibrations are much lower than the mechanical resonance of the structure.
3. the previous point implies that the stress is assumed uniformly distributed along the rod axis;
4. the electromagnetic field propagation is neglected, then the electric load is considered as a lumped element;
5. the magnetic circuit is assumed to be a flux tube for the magnetic induction;
6. eddy currents are negligible and magnetic circuit theory can be used to model the magnetic part.

The complete system can be now modeled through circuit tools giving an explicit analysis of the harvester. The equivalent three-port circuit, through particular analogies, relates to each other all the mechanical, magnetic and electric quantities of the global system, such as the force, the output voltage and the magnetization of PM. Moreover, all the possible external elements involved in the harvester can be suitably added to the model, including the mechanical system model, the biasing model and the electric load. At this point it is worth to note that, from the assumptions previously taken into account, the equivalent circuit has lumped elements [21,26,34,41,46,47]. Then, regarding the magnetostrictive sample, it is possible to express the applied stress as $\sigma = F_1/S_g$, the strain as $\varepsilon = x_g/l_g$ and the magnetic flux as $\phi_g = B_g S_g$, where B_g is the magnetic flux in the rod and x_g is the magnetostrictive rod tip displacement.

Let's start with the representation of the active rod. Let's consider the set of Equation (10), which is rewritten in order to make them more appropriate for equivalent circuit modeling. By following the same line of reasoning adopted in References [26,34,41], the first of the Equation (10) is multiplied by l_g and the second equation by S_g, as follows:

$$\begin{cases} x_g = \dfrac{F_1}{\zeta} + l_g \cdot g(F_1, H_g) \\ \phi_g = \mu_0 H_g S_g + S_g \cdot m(F_1, H_g) \end{cases} \quad (15)$$

where:

- $\zeta = E S_g / l_g$ is the mechanical stiffness of the magnetostrictive sample;
- $g(F_1, H_g) = -f'(\sigma) \cdot [z \cdot u'(z) - u(z)]$;
- ϕ_g and H_g are, respectively, the magnetic flux and the magnetic field in the magnetostrictive rod;
- $m(F_1, H_g) = u'(z)$.

In the second equation of the set of Equation (15), the term $\mu_0 H_g S_g$ can be rewritten as $l_g H_g / \mathcal{R}_g$, where $\mathcal{R}_g = l_g/(\mu_0 S_g)$ represents the linear contribution of the total *reluctance* of the magnetoelastic rod. Indeed, generally, the magnetic permeability of a magnetostrictive material, such as Galfenol, can be split into two parts—obviously a nonlinear part, represented in this case by the model and a linear contribution represented by $\mathcal{R}_g$. The system of Equation (15) can be arranged as follows:

$$\begin{cases} F_1 = \xi\, x_g - l_g\, \xi \cdot g(F_1, H_g) \\ H_g\, l_g = \mathcal{R}_g\, \phi_g - \mathcal{R}_g\, S_g \cdot m(F_1, H_g) \end{cases} \tag{16}$$

The following analogies between the mechanical and electric quantities are adopted to convert the mechanical equation in the first of the set of Equation (16), into the electrical equivalent. In particular:

- the force corresponds to a voltage, $F_1(t) \iff v_1$;
- the rod tip velocity corresponds to a current, $\dot{x}_g(t) \iff i_1$.

The following set of equations are consequently achieved:

$$\begin{cases} v_1(t) = \xi \int_t i_1\, \mathrm{d}t - l_g\, \xi \cdot g(v_1, H_g) \\ H_g\, l_g = \mathcal{R}_g\, \phi_g - \mathcal{R}_g\, S_g \cdot m(v_1, H_g) \end{cases} \tag{17}$$

For the second of the set of Equation (17), that is, the magnetic part and by applying magnetic circuit theory, the quantities $H_g\, l_g$ corresponds to a magnetic voltage. As a consequence, the analogies among magnetic voltage and electrical one can be made, that are:

- $H_g\, l_g \Leftrightarrow v_g$;
- the reluctance $\mathcal{R}_g$ is modeled by a resistance R_g;
- i_g is the flux ϕ_g, which flows in the reluctance rod $\mathcal{R}_g$.

$$\begin{cases} v_1(t) = \xi \int_t i_1\, \mathrm{d}t - l_g\, \xi \cdot g\!\left(v_1, \dfrac{v_g}{l_g}\right) \\ v_g = R_g\, i_g - R_g\, S_g \cdot m\!\left(v_1, \dfrac{v_g}{l_g}\right) \end{cases} \tag{18}$$

The previous set of Equation (18) constitutes the loop Kirchhoff's voltage law for a well-posed two-port circuit, which is similar to the one reported in Reference [41]. The first port is composed by a voltage source, that is, the external applied force, with a series capacitance, $C = 1/\xi = l_g/(E\, S_g)$ and a nonlinear dependent voltage source, which represents the nonlinear magnetostrictive response of the rod. The second equation can be modeled as a voltage source, that is, the source of magnetic bias, with a series resistance, which represents the linear reluctance and a nonlinear dependent voltage source that takes into account the nonlinear magnetic response of rod, as shown in Figure 5.

The last missing part is constituted by the electric energy conversion, which is not yet expressed in Equation (18). In particular, the output voltage across the pickup coil, wounded around the active material, is due from the exploitation of the Villari effect and the Faraday's law. The first is considered in the $m(v_1, \frac{v_g}{l_g})$ function of the nonlinear model, while the latter reads:

$$v_{out}(t) = -\frac{\mathrm{d}}{\mathrm{d}t} \Phi_g(t) \tag{19}$$

where $\Phi_g(t) = N_2\, \phi_g$ is flux linkage in N_2-turn pickup coil.

In the equivalent circuit, the coil flux linkage is modeled by $N_2\, i_g$; then when a current flows in the pickup coil, this current will influence the total flux in the magnetostrictive element by the transformer effect. As a consequence, a voltage source, corresponding to the $N_2\, i_2$ of the pickup coil circuit, is added in series in the second port of Figure 5. Finally, a third port, representing the electric side, is added in the circuit model where R_{coil} is the resistance of the pickup coil, i_2 is the current which flows in the pickup coil and v_2 is the output voltage. In conclusion, Figure 5 reports the *general* three port circuit which represents the bare magnetostrictive material rod with a N_2 winding.

The set of Equation (18) and the above mentioned different analogies point out that, with the aim to model a whole device, further lumped circuit elements can be inserted now to the three ports. For example, by considering the concept device of Figure 4, if the harvester undergoes a vibrating force generator and the Galfenol mass m_g is considered then these can be simply solved by connecting the series of a voltage generator v_{force} and a induct or m_g to the first port. While a viscous friction is represented by a resistor:

$$F = m\frac{d^2x}{dt^2} \iff v = m\frac{di}{dt}$$
$$F = r\frac{dx}{dt} \iff v = r\,i \quad (20)$$

Moreover, by considering the magnetic circuit theory, the iron frame and the PM are represented by a reluctance R_{fe} and the series of a reluctance R_g and a magnetomotive force $M_m l_m$, as reported in Figure 6 where also the generic electric load is connected to the electric port.

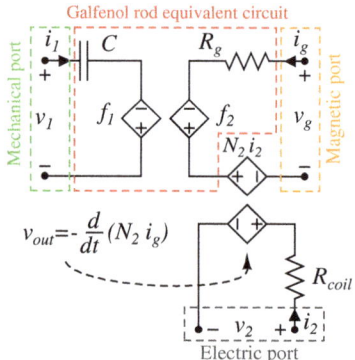

Figure 5. Equivalent three-port nonlinear model circuit of a Galfenol rod. In the first port, voltages and currents correspond to forces and velocities, respectively, while in the second port are magnetic voltages and flux. Furthermore, voltages and currents of the third port correspond to real quantities delivered to the load. The nonlinear functions are: $f_1 = l_g\,\zeta \cdot g(v_1, \frac{v_g}{l_g})$ and $f_2 = R_g\,S_g \cdot m(v_1, \frac{v_g}{l_g})$.

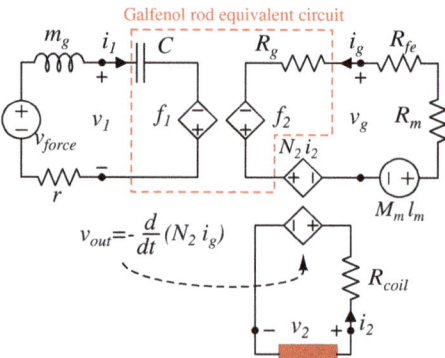

Figure 6. Equivalent three port nonlinear model circuit of the concept device in Figure 4.

4. Characterizations and Simulations of the Three Rod Device

The three-port equivalent circuit has been validated on a real KEH device based on a single rod of SA Galfenol, as shown in Reference [26]. More in details, the model and the equivalent circuit are able to receive a force profile as input and to provide an output voltage with good accuracy, with the exception of very low and high biasing fields, that are not effective for EH aims. Then, in this section, the effectiveness of the equivalent circuit is tested on a KEH system, conceived for a realistic application, which has been proposed in References [16,17].

The core of the system, consisting in a three-rods SA Galfenol KEH device, is presented, characterized and modeled. The device is enclosed within two steel plates, with a height of 6 mm and diameter of 52 mm, that connect three Galfenol rods, with a length of 21 mm and diameter of 6 mm, 120° spaced and 18 mm far from the center. The rods have pilot pins (6 mm height) entering in the disks. Furthermore, the top plate hosts a steel sphere (10 mm diameter) to provide a single contact point with the external force source, in order to equally transfer the stress to each rod. A column of neodymium disk PMs is placed in the center to provide a magnetic bias to the Galfenol rods. Indeed, the two steel plates have a twofold purpose—they are a low reluctance path for the PMs flux and they grant the mechanical stress transfer to the active elements.

A PCB (70 mm × 70 mm and 3 mm thick), shown in Figure 7c,d has been designed and carried out to have reliable electrical connections for three 2000-turn pickup coils. Figure 7 shows the three-rods Galfenol KEH device developed. In particular, in the Figure 7a,b two drafts regarding the mechanical connections are depicted, while in Figure 7c, two photos of the device are shown.

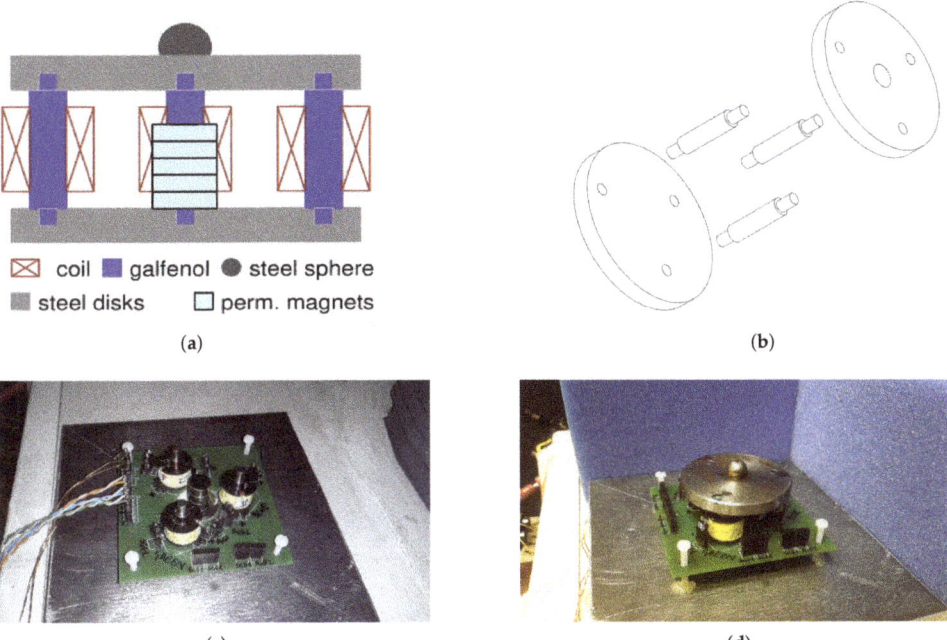

Figure 7. KEH device based on three Galfenol rods. PMs are inserted to provide the magnetic bias. (**a**) 2-D sketch. (**b**) 3-D sketch. (**c**) Photo of the PCB, PMs and Galfenol rods during installation. (**d**) Photo of the whole 3-rod kinetic energy harvesting (KEH).

The proposed three-rods KEH system can be used with different coils connection, PMs configurations and force types, then permitting different electrical, magnetic and mechanical operative conditions. Consequently, design work is important to identify the system's parameters, such as the electric load, the pickup coils connections and the typology of magnets, in order to optimize the harvested energy. Periodic compressive forces, at three different low-frequencies, have been applied by the UTM. Then, discs PM (20 mm length) with three different diameters (10, 15 and 20 mm) have been tested. In addition to the three-rods KEH device's characterization, a whole equivalent circuit has been performed and simulated. In such a way, the experimental data and the simulations can be compared in order to estimate the circuital model goodness.

In the light of results shown in Section 3.2, the equivalent circuit of the three-rods KEH device is depicted in Figure 8. On the behalf of clarity, the three rods are distinguished with different colors—green, red and blue. The three-port equivalent circuit of each rod has been highlighted by the corresponding color. The mechanical ports are connected in series because the force applied to the steel sphere is divided to each rod. The applied force is represented by a voltage generator. On the magnetic side, each port is connected, through its steel reluctance ($\mathcal{R}_{fe}$), to the neodymium magnets. As for the single rod case, the equivalent circuit of the PMs stack is a real voltage generator. The PM height is 1 mm lower than the rod ones, then the corresponding air gap reluctance, $\mathcal{R}_{air}$, is added in series. By increasing the diameter of PMs, the value $M_m l_m$ remains constant, while the reluctance $\mathcal{R}_m$ decreases. Finally, $\mathcal{R}_{leak}$ represents any leakage of the magnetic flux. It has been placed in parallel with $\mathcal{R}_{air}$ and PM circuit series because, for this device, it is reasonable to consider the leakage flux closing from the top to the bottom disc. Details of the above circuit elements are reported in Appendix A.

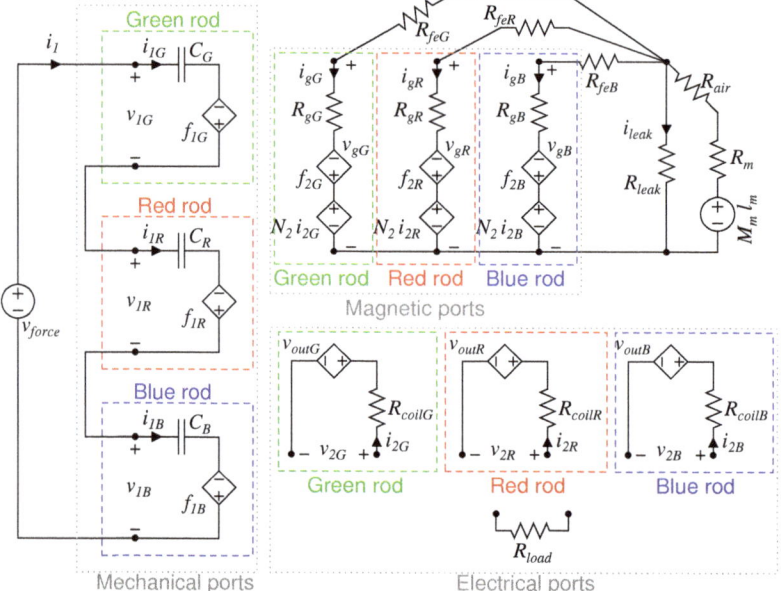

Figure 8. Equivalent three port nonlinear model circuit of the three-rods Galfenol KEH. The mechanical ports are connected in series because the universal testing machine (UTM) speed is the same for each rod, while the electrical ports can be connected together in series or parallel on a resistive load.

Low frequency mechanical stress variations can be generated by the UTM, producing a compressive force with a period larger than 1 s. A load cell is used to measure the instantaneous force applied to the device. The applied force is constituted by the same profiles but repeated with different time rates and each profile consists in ten complete cycles. Consequently, all profiles are a compressive force varying from 0 up to −2000 N, although with three distinct UTM crosshead speeds, that are of 150, 100 and 50 mm/min respectively, corresponding in frequencies of about 0.8, 0.55 and 0.3 Hz.

The measured stress profile applied to the three-rods KEH device is plotted in Figure 9. The force measurement is acquired and can be exploited as the input equivalent force generator (v_{force}), in order to make simulations in LTspice.

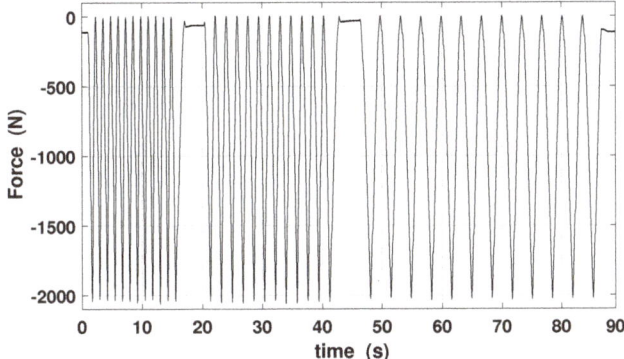

Figure 9. Force profile applied to the 3-rod KEH device. It consists in ten compressive cycles from 0 to −2000 N at different UTM speeds—150, 100, 50 mm/min.

The time-profiles of measured and simulated output voltages for different coils connection are shown in Figure 10. For the sake of shortness, only the 15 mm diameter PM and 150 mm/min strain velocity case has been reported for different resistive loads. The equivalent circuit is able to mimic quite well the shape of the output voltage. Furthermore, as expected, the measured voltage increases for larger electric load values, both for series and parallel pickup coil connections.

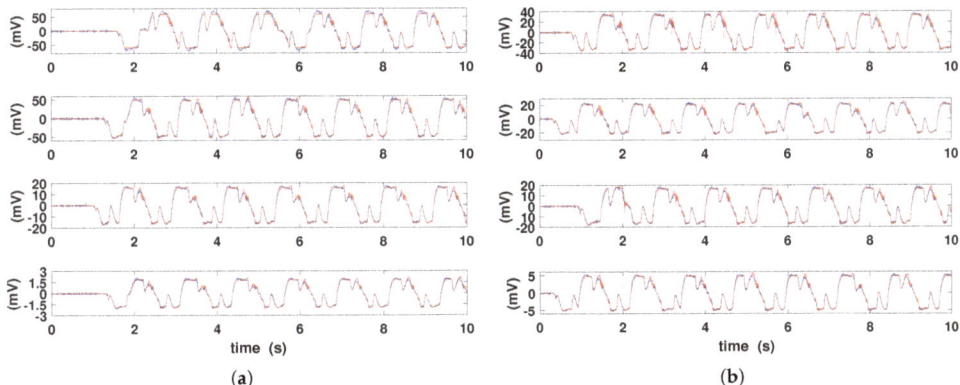

Figure 10. Measured (color red) and simulated (color blue) output voltages. 15 mm diameter PM and 150 mm/min strain velocity are applied. (**a**) Coils connected in series on 10 k, 525, 100 and 10 Ω (from top to bottom pane). (**b**) Coils connected in parallel on 10 k, 100, 58 and 10 Ω (from top to bottom pane).

The measured and simulated average power, RMS and peak-to-peak voltage with respect to different electric loads, magnets diameter and UTM speeds, are shown in Figures 11–13, respectively. Each measurement point is obtained by averaging over ten cycles and by calculating the standard deviation, that is represented with error bars, in order to define the confidence interval measurements.

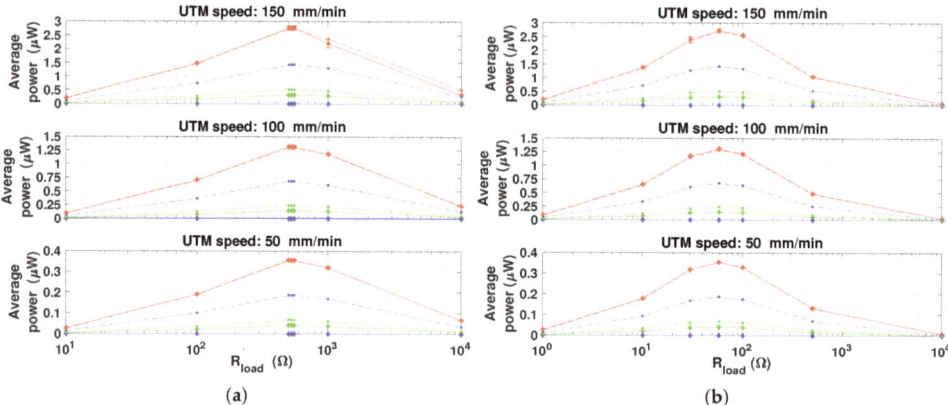

Figure 11. Comparison between the average measured (diamonds with solid line) and simulated (dots with dashed line) power vs. the applied electric load. PMs with different diameters are considered—10 (green), 15 (red) and 20 mm (blue). (**a**) Coils connected in series. (**b**) Coils connected in parallel.

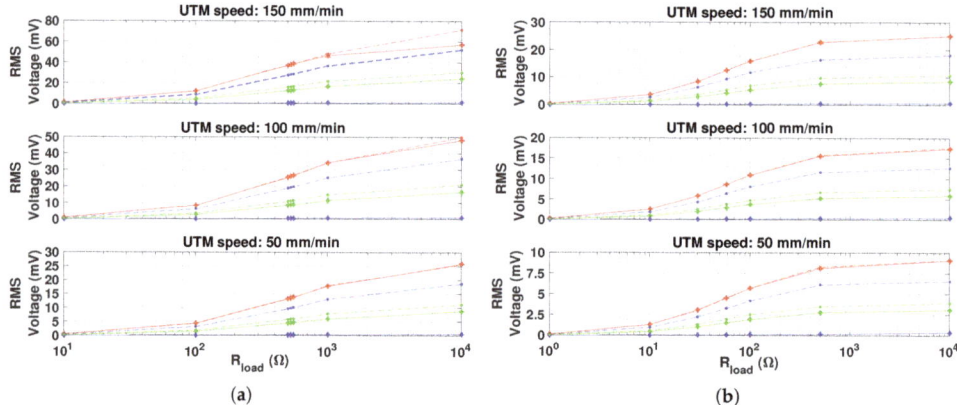

Figure 12. Comparison between the RMS measured (diamonds with solid line) and simulated (dots with dashed line) voltage vs. the applied electric load. PMs with different diameters are considered—10 (green), 15 (red) and 20 mm (blue). (**a**) Coils connected in series. (**b**) Coils connected in parallel.

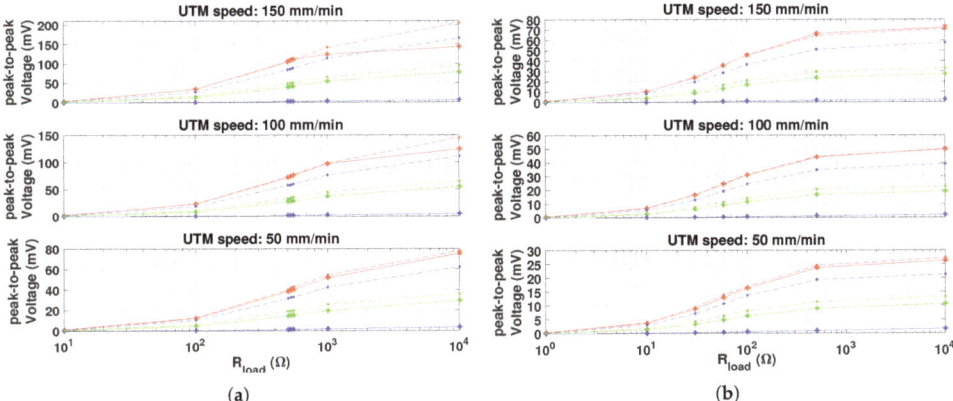

Figure 13. Comparison between the peak-to-peak measured (diamonds with solid line) and simulated (dots with dashed line) voltage vs. the applied electric load. PMs with different diameters are considered—10 (green), 15 (red) and 20 mm (blue). (**a**) Coils connected in series. (**b**) Coils connected in parallel.

About the performance of the device, the maximum power is obtained when the load is equal to the total coil resistance. Power peaks are present around R_{load} = 525 Ω, in case of coils connected in series and around R_{load} = 58 Ω, in case of parallel connection. Indeed, these two values represent, respectively, the total equivalent coil resistance (the single coil resistance is 175 Ω, as reported also in Appendix A). This is in agreement of the optimum energy transfer theorem.

The input frequency strongly affects the harvested power. Indeed, it is noticeable that by doubling the force input frequency, the average power quadruples. Moreover, the peaks average power both for series and parallel connections are equal, when the same force profile and optimal resistive load are applied. In other terms, from the energy conversion point of view, when optimal electric load condition is applied, the two connections are equivalent [17]. By increasing the force frequency and the output load, the RMS and peak-to-peak voltages also increase [16,17], as a explicit consequence of Ohm's law and Faraday's law. As expected, the harvested power is depending on the diameter of PM used as they provide the magnetic bias. In particular, the 15 mm diameter case gave the largest power values, about 8 times larger than the 10 mm diameter case. The curves referred to the 20 mm diameter case show that the average power, RMS and peak-to-peak voltage are very low. This may suggest that the magnetic saturation condition occurs in the Galfenol rods.

The comparison between measurements and simulations shows that the circuital model is able to predict the average power with an error lower than 5% when optimal PM is considered. On the other hand, errors around 60% and larger than 100% occur when 10 mm and 20 mm diameter PM, respectively, are applied. In the latter case, magnets and Galfenol rods are very close each other, then magnetic leakage increases. As a consequence, the comparison among measurements and simulations suggests that, due to lumped parameters, the model is not able to reproduce these conditions.

Finally, the power harvested by the KEH device under study is comparable with other EH device presented in literature. For example, the measured output power in Reference [48] is 450 mW with 60 Hz frequency of the applied force, while it is about 16 mW at 100 Hz in Reference [49] and 0.1 mW at 5 Hz for the device presented in Reference [50]. In order to make a comparison, by considering that in first approximation the output power scales with the square of the input frequency, it is possible to estimate a KEH device's output powers of about 0.75 mW at 5 Hz, 11 mW at 60 Hz and 30 mW at 100 Hz, that are comparable by considering the volume of active material.

5. Conclusions

In this work, the characterization and modeling of a KEH device based on Galfenol has been concerned. With this aim, a nonlinear model of SA Galfenol have been adopted and its parameters have been determined by using the measured magnetostrictives curves. The model takes into account the most relevant experimental behaviors, that is, nonlinearity, the magneto-mechanical coupling and saturation effects. Because of the mechanical, magnetic and electric quantities involved in a KEH device, a three-port equivalent circuit of a concept device based on a single Galfenol rod is reported and described. In sight on the results obtained, a more complex KEH system based on three Galfenol rods, has been experimentally characterized and modeled. The characterization has been carried out by applying different compressive force profiles, electric loads and permanent magnets. As expected, the input force frequency and the magnets configuration strongly affect the output voltage and power, while an optimal resistive load, corresponding to the total equivalent coil resistance, is needed to extract the maximum power (about 3 µW at 0.8 Hz of input force frequency and optimal bias and electric load). From the energy conversion point of view, no differences have been found among series and parallel coils connection. Furthermore, the comparison among experimental data and simulations have confirmed the ability of the circuital model to predict the output voltage and harvested power for different loads and pickup coils connection when the optimal bias condition is concerned.

Author Contributions: Conceptualization, C.S.C. and D.D.; Investigation, C.S.C.; Methodology, C.S.C.; Supervision, D.D.; Validation, C.S.C.; Writing—original draft, C.S.C.; Writing—review and editing, D.D.

Funding: This research received no external funding.

Conflicts of Interest: The authors declare that there are no potential conflicts of interest with respect to the research, authorship and/or publication of this article.

Abbreviations

The following abbreviations are used in this manuscript:

ICT	Information and Communications Technology
PDA	Personal Digital Assistant
WSN	Wireless Sensors Network
SHM	Structural Health Monitoring
EH	Energy Harvesting
IoT	Internet of Things
KEH	Kinetic Energy Harvesting
NSA	Non Stress Annealed
SA	Stress Annealed
PCB	Printed Circuit Board
PM	Permanent Magnet
RMS	Root Mean Square

Appendix A. Identification Procedure of the Three-Rods Galfenol Keh Device Equivalent Circuit Parameters

This appendix is devoted to present more in detail and give hints on the identification of the equivalent circuit of the three-rods device, represented in Figure 8.

By considering $x \in \{G, R, B\}$, the parameters are: M_s, σ_b, α, β, γ, E, C_x, f_{1x}, f_{2x}, R_{fex}, R_{leak}, R_{air}, R_m, M_m, l_m, R_{gx}, N_2, v_{outx}, R_{coilx}, R_{load}.

In the following a list of the parameter's values is presented:

1. M_s, σ_b, α, β, γ, are related to the proposed nonlinear model for SA Galfenol and are detailed in Section 3.1.
2. $C_G = C_R = C_B = 1/\xi = l_g/(ES_g)$, are equals because same rods have been used.
3. $l_g = 21$ mm and $S_g = \pi(3)^2$ mm^2 are geometric quantities of the employed rods, so directly measurable.

4. $E = 60$ GPa is the Galfenol Young Modulus, which can be determined from the material datasheet.
5. $f_{1G} = l_g \zeta \cdot g(v_{1G}, \frac{v_{gG}}{l_g})$, $f_{1R} = l_g \zeta \cdot g(v_{1R}, \frac{v_{gR}}{l_g})$ and $f_{1B} = l_g \zeta \cdot g(v_{1B}, \frac{v_{gB}}{l_g})$ are the nonlinear dependent voltage sources on the mechanical ports.
6. $f_{2G} = R_g S_g \cdot m(v_{1G}, \frac{v_{gG}}{l_g})$, $f_{2R} = R_g S_g \cdot m(v_{1R}, \frac{v_{gR}}{l_g})$ and $f_{2B} = R_g S_g \cdot m(v_{1B}, \frac{v_{gB}}{l_g})$ are the nonlinear dependent voltage sources on the magnetic ports.
7. $R_{feG} = R_{feR} = R_{feB} = l_{fe}/(\mu_{r,fe}\mu_0 S_{fe})$, are the reluctance of the steel magnetic circuit for each rod and for the symmetry of the object have been considered equals. It can be achieved from the magnetic circuit length, section and from the magnetic characteristics of the steel. In particular, $l_{fe} = 49$ mm, $\mu_{r,fe} = 500$ and $S_{fe} = 94.2$ mm^2.
8. $R_{leak} = 1.216 \times 10^6$ H^{-1} represents the leakage of magnetic flux between the top e bottom discs. Its value have been obtained with a minimization of the error between measured and simulated output voltages in case of optimal magnetic bias and one specific electric load inserted.
9. $R_{air} = l_{air}/(\mu_0 S_{air})$ is the reluctance of air-gap between the top steel disc and the permanent magnet. In particular, $l_{air} = 1$ mm while the section has been considered equal to PM disc section, $S_{air} = S_m$.
10. $R_m = l_m/(\mu_0 S_m)$ is the reluctance of the permanent magnet. It depends only by directly measurable quantities.
11. $l_m = 20$ mm is the discs permanent magnet length, then directly measurable.
12. S_m is the discs permanent magnet section, then directly measurable, in particular it have been used 10, 15 and 20 mm diameter of PM.
13. M_m is the magnetization of neodymium PM and it is obtained from material datasheed (in this work it is considered equal to 1000 kA/m).
14. $R_{gG} = R_{gR} = R_{gB} = l_g/(\mu_0 S_g)$, are the linear contribution of the total reluctance of the rods and have been considered equals.
15. $N_2 = 2000$ is the turn number of pickup coils, which is the same for each coil.
16. $v_{outG} = -\frac{d}{dt}N_2 i_{gG}$, $v_{outR} = -\frac{d}{dt}N_2 i_{gR}$ and $v_{outB} = -\frac{d}{dt}N_2 i_{gB}$, are the nonlinear dependent voltage sources on the electrical ports.
17. $R_{coilG} = R_{coilR} = R_{coilB} = 175$ Ω is the pickup coil resistance are directly measurable and equals because same coils have been used.
18. R_{load} is the applied electrical load.

References

1. Priya, S., Inman, D.J. (Eds.) *Energy Harvesting Technologies*; Springer: Boston, MA, USA, 2009. [CrossRef]
2. Kaźmierski, T.J., Beeby, S. (Eds.) *Energy Harvesting Systems*; Springer: New York, NY, USA, 2011. [CrossRef]
3. Elvin, N.; Erturk, A. *Advances in Energy Harvesting Methods*; Springer: Berlin/Heidelberg, Germany, 2013; p. 455.
4. Chang, P.; Flatau, A.; Liu, S. Review paper: Health monitoring of civil infrastructure. *Struct. Health Monit.* **2003**, *2*, 257–267. [CrossRef]
5. Lynch, J. An overview of wireless structural health monitoring for civil structures. *Philos. Trans. R. Soc. A Math. Phys. Eng. Sci.* **2007**, *365*, 345–372. [CrossRef] [PubMed]
6. Balageas, D.; Fritzen, C.P.; Güemes, A. *Structural Health Monitoring*; John Wiley & Sons: Hoboken, NJ, USA, 2010; pp. 1–495. [CrossRef]
7. Chang, F.; Markmiller, J.; Yang, J.; Kim, Y. Structural Health Monitoring. In *System Health Management: With Aerospace Applications*; John Wiley & Sons: Hoboken, NJ, USA, 2011; pp. 419–428. [CrossRef]
8. Torfs, T.; Sterken, T.; Brebels, S.; Santana, J.; Van Den Hoven, R.; Spiering, V.; Bertsch, N.; Trapani, D.; Zonta, D. Low power wireless sensor network for building monitoring. *IEEE Sens. J.* **2013**, *13*, 909–915. [CrossRef]
9. Davino, D.; Pecce, M.; Visone, C.; Clemente, C.; Ielardi, A. Dynamic monitoring of guardrails: Approach to a low-cost system. In Proceedings of the 2015 IEEE Workshop on Environmental, Energy, and Structural Monitoring Systems (EESMS), Trento, Italy, 9–10 July 2015; pp. 56–60. [CrossRef]
10. Bhuiyan, M.; Wang, G.; Cao, J.; Wu, J. Deploying Wireless Sensor Networks with Fault-Tolerance for Structural Health Monitoring. *IEEE Trans. Comput.* **2015**, *64*, 382–395. [CrossRef]

11. Bhuiyan, M.; Wang, G.; Wu, J.; Cao, J.; Liu, X.; Wang, T. Dependable Structural Health Monitoring Using Wireless Sensor Networks. *IEEE Trans. Dependable Secur. Comput.* **2017**, *14*, 363–376. [CrossRef]
12. Park, G.; Rosing, T.; Todd, M.D.; Farrar, C.R.; Hodgkiss, W. Energy Harvesting for Structural Health Monitoring Sensor Networks. *J. Infrastruct. Syst.* **2008**, *14*, 64–79. [CrossRef]
13. Basagni, S.; Naderi, M.Y.; Petrioli, C.; Spenza, D. Wireless Sensor Networks with Energy Harvesting. In *Mobile Ad Hoc Networking*; John Wiley & Sons: Hoboken, NJ, USA, 2013; pp. 701–736. [CrossRef]
14. Shaikh, F.K.; Zeadally, S. Energy harvesting in wireless sensor networks: A comprehensive review. *Renew. Sustain. Energy Rev.* **2016**, *55*, 1041–1054. [CrossRef]
15. Davino, D.; Visone, C.; Giustiniani, A. Vibration energy harvesting devices based on magnetostrictive materials. In Proceedings of the Sixth International Conference on Bridge Maintenance, Safety and Management, Stresa, Italy, 2–12 July 2012; pp. 1511–1518.
16. Clemente, C.S.; Davino, D.; Maddaloni, G.; Pecce, M.R.; Visone, C. A Magnetostrictive Energy Harvesting System for Bridge Structural Health Monitoring. *Adv. Sci. Technol.* **2016**, *101*, 20–25. [CrossRef]
17. Clemente, C.S.; Davino, D.; Visone, C. Experimental Characterization of a Three-Rod Magnetostrictive Device for Energy Harvesting. *IEEE Trans. Magn.* **2017**, *53*, 1–4. [CrossRef]
18. Varadan, V.; Vinoy, K.; Gopalakrishnan, S. *Smart Material Systems and MEMS: Design and Development Methodologies*; John Wiley & Sons: Hoboken, NJ, USA, 2006; pp. 1–404. [CrossRef]
19. Engdahl, G. (Ed.) *Handbook of Giant Magnetostrictive Materials*; Academic Press: San Diego, CA, USA, 2000; pp. 1–386. [CrossRef]
20. Zhao, X.; Lord, D. Application of the Villari effect to electric power harvesting. *J. Appl. Phys.* **2006**, *99*. [CrossRef]
21. Davino, D.; Giustiniani, A.; Visone, C. Capacitive load effects on a magnetostrictive fully coupled energy harvesting device. *IEEE Trans. Magn.* **2009**, *45*, 4108–4111. [CrossRef]
22. Bottauscio, O.; Roccato, P.; Zucca, M. Modeling the dynamic behavior of magnetostrictive actuators. *IEEE Trans. Magn.* **2010**, *46*, 3022–3028. [CrossRef]
23. Bottauscio, O.; Roccato, P.; Zucca, M. Micropositioning through magnetostrictive actuators. *Sens. Lett.* **2013**, *11*, 87–90. [CrossRef]
24. Atulasimha, J.; Flatau, A. A review of magnetostrictive iron-gallium alloys. *Smart Mater. Struct.* **2011**, *20*. [CrossRef]
25. Davino, D.; Giustiniani, A.; Visone, C.; Adly, A. Energy harvesting tests with galfenol at variable magneto-mechanical conditions. *IEEE Trans. Magn.* **2012**, *48*, 3096–3099. [CrossRef]
26. Clemente, C.; Mahgoub, A.; Davino, D.; Visone, C. Multiphysics circuit of a magnetostrictive energy harvesting device. *J. Intell. Mater. Syst. Struct.* **2017**, *28*, 2317–2330. [CrossRef]
27. Restorff, J.; Wun-Fogle, M.; Clark, A.; Hathaway, K. Induced magnetic anisotropy in stress-annealed galfenol alloys. *IEEE Trans. Magn.* **2006**, *42*, 3087–3089. [CrossRef]
28. Wun-Fogle, M.; Restorff, J.; Clark, A. Magnetostriction of stress-annealed Fe-Ga and Fe-Ga-Al alloys under compressive and tensile stress. *J. Intell. Mater. Syst. Struct.* **2006**, *17*, 117–122. [CrossRef]
29. Apicella, V.; Clemente, C.; Davino, D.; Visone, C. Experimental evaluation of external and built-in stress in Galfenol rods. *Phys. B: Condens. Matter* **2018**, *549*, 53–57. [CrossRef]
30. TdVib LLC. TdVib Website. Available online: http://tdvib.com/ (accessed on 28 August 2019).
31. Moffett, M.; Linberg, J.; McLaughlin, E. Characterization of Terfenol-D for magnetostrictive transducers. *J. Acoust. Soc. Am.* **1991**, *89*, 1448–1455. [CrossRef]
32. Atulasimha, J.; Flatau, A. Experimental actuation and sensing behavior of single-crystal iron-gallium alloys. *J. Intell. Mater. Syst. Struct.* **2008**, *19*, 1371–1381. [CrossRef]
33. Jin, Y.; Gu, S.; Bennett, L.; Della Torre, E.; Provenzano, V.; Zhao, Q. Self-similarity in $(\partial M/\partial T)_H$ curves for magnetocaloric materials with ferro-to-paramagnetic phase transitions. *J. Appl. Phys.* **2012**, *111*. [CrossRef]
34. Davino, D.; Giustiniani, A.; Visone, C. Magnetoelastic Energy Harvesting: Modeling and Experiments. In *Smart Actuation and Sensing Systems—Recent Advances and Future Challenges*; InTech: London, UK, 2012. [CrossRef]
35. Erturk, A.; Inman, D. Issues in mathematical modeling of piezoelectric energy harvesters. *Smart Mater. Struct.* **2008**, *17*. [CrossRef]
36. Berbyuk, V. Towards dynamics of controlled multibody systems with magnetostrictive transducers. *Multibody Syst. Dyn.* **2007**, *18*, 203–216. [CrossRef]

37. Davino, D.; Giustiniani, A.; Visone, C. Analysis of a magnetostrictive power harvesting device with hysteretic characteristics. *J. Appl. Phys.* **2009**, *105*. [CrossRef]
38. Cao, S.; Zheng, J.; Guo, Y.; Li, Q.; Sang, J.; Wang, B.; Yan, R. Dynamic characteristics of galfenol cantilever energy harvester. *IEEE Trans. Magn.* **2015**, *51*. [CrossRef]
39. Wan, Y.; Fang, D.; Hwang, K.C. Non-linear constitutive relations for magnetostrictive materials. *Int. J. Non-Linear Mech.* **2003**, *38*, 1053–1065. [CrossRef]
40. Davino, D.; Krejčí, P.; Visone, C. Fully coupled modeling of magneto-mechanical hysteresis through 'thermodynamic' compatibility. *Smart Mater. Struct.* **2013**, *22*. [CrossRef]
41. Davino, D.; Giustiniani, A.; Visone, C. A two-port nonlinear model for magnetoelastic energy-harvesting devices. *IEEE Trans. Ind. Electron.* **2011**, *58*, 2556–2564. [CrossRef]
42. Sun, L.; Zheng, X. Numerical simulation on coupling behavior of Terfenol-D rods. *Int. J. Solids Struct.* **2006**, *43*, 1613–1623. [CrossRef]
43. Ovichi, M.; Elbidweihy, H.; Torre, E.; Bennett, L.; Ghahremani, M.; Johnson, F.; Zou, M. Magnetocaloric effect in NiMnInSi Heusler alloys. *J. Appl. Phys.* **2015**, *117*. [CrossRef]
44. Palumbo, S.; Rasilo, P.; Zucca, M. Experimental investigation on a Fe-Ga close yoke vibrational harvester by matching magnetic and mechanical biases. *J. Magn. Magn. Mater.* **2019**, *469*, 354–363. [CrossRef]
45. Technology, L. LTspice Circuit Simulator. 2017. Available online: http://www.linear.com/designtools/software/ (accessed on 28 August 2019).
46. Williams, C.; Yates, R. Analysis of a micro-electric generator for microsystems. *Sens. Actuators A Phys.* **1996**, *52*, 8–11. [CrossRef]
47. Yang, Y.; Tang, L. Equivalent circuit modeling of piezoelectric energy harvesters. *J. Intell. Mater. Syst. Struct.* **2009**, *20*, 2223–2235. [CrossRef]
48. Berbyuk, V. Vibration energy harvesting using Galfenol based transducer. In Proceedings of the Active and Passive Smart Structures and Integrated Systems, San Diego, CA, USA, 10–14 March 2013; Volume 8688. [CrossRef]
49. Ahmed, U.; Jeronen, J.; Zucca, M.; Palumbo, S.; Rasilo, P. Finite element analysis of magnetostrictive energy harvesting concept device utilizing thermodynamic magneto-mechanical model. *J. Magn. Magn. Mater.* **2019**, *486*. [CrossRef]
50. Viola, A.; Franzitta, V.; Cipriani, G.; Di Dio, V.; Raimondi, F.; Trapanese, M. A Magnetostrictive Electric Power Generator for Energy Harvesting From Traffic: Design and Experimental Verification. *IEEE Trans. Magn.* **2015**, *51*. [CrossRef]

© 2019 by the authors. Licensee MDPI, Basel, Switzerland. This article is an open access article distributed under the terms and conditions of the Creative Commons Attribution (CC BY) license (http://creativecommons.org/licenses/by/4.0/).

Article

Dynamic Simulation of a Fe-Ga Energy Harvester Prototype Through a Preisach-Type Hysteresis Model

Stefano Palumbo [1,2,3], Mario Chiampi [1], Oriano Bottauscio [1] and Mauro Zucca [1,*]

1. Istituto Nazionale di Ricerca Metrologica, INRIM, Strada delle Cacce 91, 10135 Torino, Italy; s.palumbo@inrim.it (S.P.); m.chiampi@inrim.it (M.C.); o.bottauscio@inrim.it (O.B.)
2. Politecnico di Torino, Department DET, Corso Duca degli Abruzzi 24, 10129 Torino, Italy
3. Istituto Italiano di Tecnologia, IIT, Graphene Labs, Via Morego 30, 16163 Genova, Italy
* Correspondence: m.zucca@inrim.it; Tel.: +39-011-3919-827

Received: 11 September 2019; Accepted: 15 October 2019; Published: 17 October 2019

Abstract: This paper presents the modeling of an Fe–Ga energy harvester prototype, within a large range of values of operating parameters (mechanical preload, amplitude and frequency of dynamic load, electric load resistance). The simulations, based on a hysteretic Preisach-type model, employ a voltage-driven finite element formulation using the fixed-point technique, to handle the material nonlinearities. Due to the magneto–mechanical characteristics of Fe–Ga, a preliminary tuning must be performed for each preload to individualize the fixed point constant, to ensure a good convergence of the method. This paper demonstrates how this approach leads to good results for the Fe–Ga prototype. The relative discrepancies between experimental and computational values of the output power remain lower than 5% in the entire range of operating parameters considered.

Keywords: energy harvesting; finite element model; iron–gallium; measurements; preisach model

1. Introduction

There has long been interest in giant magnetostrictive materials (GMMs). The first real application of a magnetostrictive material for actuators with marine sonar dates to the late 1970s. In the mid-1980s, Terfenol-D became commercially available and new high-energy-density devices were developed, along with design tools suitable for predicting the dynamic performance of GMMs. In the 1990s, the first attempts to model GMMs, in which the material is simply considered as nonlinear, were proposed [1–5]. In more detail, the model was considered 1D [2], or the magneto–elastic coupling was neglected [4,5].

In the 2000s, along with an improvement in the devices, came the development of more accurate general mathematical models of magnetostrictive materials [6,7]. The piezomagnetic equations were presented in early papers, such as [7]. Even though these early papers were only focused on models of the magnetostrictive material, new approaches showed the coupling of electromagnetic and magnetoelastic phenomena, together with the development of advanced hysteresis models [8–12]. Finally, the Armstrong model-based approach was applied, to simulate an entire device [13], in which the coupling variables were magnetostriction and magnetic permeability. A more general approach was based on hysteretic energy-weighted models [14], where magnetic field and stress were assumed as the state variables. Other approaches, based on phenomenological models of magnetostriction, adopted genetic algorithms [15,16] or used classical Preisach operators and a memoryless bivariate function, including a pure elastic contribution [17].

In the last decade, interest has shifted from actuators to energy-harvesting devices. An approach based on a Preisach-type magnetoelastic model was proposed by the authors of [18], for the analysis of giant magnetostrictive actuators based on Terfenol-D. From this, the authors proposed a magneto–mechanical (MM) hysteresis model of the magnetostrictive material [19], in which the

Preisach model is modified, introducing an effective field that is the sum of the applied field, and a corrector term, which is a function of the mechanical stress. Such a model was employed for the design of an energy harvester based on a Terfenol-D rod.

Despite its larger magnetostriction (λ_s ~1200–1600 ppm), Terfenol-D has been gradually surpassed by Fe–Ga alloys (λ_s ~250–350 ppm). The reasons for this overtaking, which are analysed in [20], can be briefly summarized as follows: high tensile strength, ductile nature, and low hysteresis losses, combined with strong magneto–elastic coupling. In addition, Fe–Ga alloys show steel-like structural properties, and can be machined easily, either welded or rolled.

Several approaches to Fe–Ga alloy modelling can be found in the literature, differing in the respective constitutive equations they are based on. A static finite element method (FEM) approach, based on the Armstrong model and presented in [21], is characterized by a so-called weak coupling between the magnetic and mechanical problem. A phenomenological model, proposed in [22], considers the effects of the hysteresis phenomena on the mechanical and magnetic energy exchanges, and is consistent with classical non-equilibrium thermodynamics. An improved Preisach-based phenomenological model is the core of [23]. Here, emphasis was placed on the effect of hysteresis, such as energy dissipation and non-differentiability of the Preisach operator. The non-hysteretic model in [24] is based on the Gibbs free energy, and characterised by a realistic magneto–mechanical modelling. The paper is unusual in its representation of the mechanical, magnetic, and electrical parts of the device in a nonlinear three-port circuit. In [25], an interesting comparison between three mathematical models is reported, one based on the Preisach operator and the other two on a non-hysteretic model, with or without a feedback loop. Finally, a thermodynamic approach based on Helmholtz free energy density is presented in [26], where the magnetic flux density and mechanical strain are assumed as state variables. This choice leads to a simplified implementation of FEM models, avoiding the problem of constitutive law inversion, which is time-consuming.

In this paper, drawing from the model developed in [19], a specific identification has been carried out for a Fe–Ga rod, in order to reproduce its non-linear magnetic characteristics, starting from a set of experimental curves. Then, the behaviour of a harvester generator prototype, coupled to an electric circuit with a load resistance, has been simulated in dynamic conditions through a voltage-driven finite element model. Particular attention is devoted to the convergence of the fixed point (FP) method, adopted to handle the Fe–Ga non-linearity. Since the convergence is strongly affected by the preload, a preliminary tuning of the correct FP constant is adopted, to significantly improve the accuracy of the device output quantities (voltage and power). The proposed approach has been tested in a large range of values of operating parameters (mechanical preload, frequency and amplitude of the dynamic mechanical load, electric load resistance). The comparison of the measurements with a harvester prototype, specifically realized with the simulation results, show a more than satisfactory agreement.

2. Materials and Methods

The principle of the magnetostrictive generator is well known: in the presence of a mechanical preload and a magnetic bias, a dynamic force is applied to a Fe–Ga rod, usually cylindrical. The applied stress dynamically modifies the magnetic permeability of the magnetostrictive material, giving rise to time variations of the magnetic flux flowing in the rod, so that an electromotive force is induced in a pick-up coil wrapped around the rod. The coil is connected to an electric circuit, where the circulating current transfers power to a suitable load. A scheme of the device is presented in Figure 1.

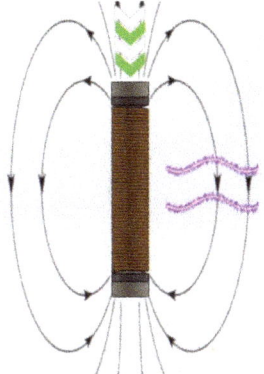

Figure 1. Schematic representation of the magnetostrictive principle. By applying a dynamic stress on an Fe–Ga rod within a mechanical preload and a magnetic bias, it is possible to generate a sinusoidal current inside the pick-up coil.

In the generator, the independent quantities are the mechanical preload σ_0, the amplitude and frequency of the dynamic applied stress $\sigma(t)$ and the magnetic bias, and these are usually produced by permanent magnets chosen during the design phase. The output are the electrical quantities (voltage and current), or simply the electric power transferred to the load.

2.1. Magneto-Mechanical Preisach Model of the Fe–Ga Rod

Magnetostrictive materials are characterized by a strong link between magnetic and mechanical properties, both characterized by hysteresis. A good model of this complex behaviour is essential for an accurate simulation of a magnetostrictive device.

The present work adopts a modified Preisach model, described in [27], which is based on the introduction of an effective field, H_e. The latter is the sum of the applied external field H and a corrector term $\xi(J,\sigma)$ [28,29], which, in turn, is a function of the magnetic polarization J and of the mechanical stress σ:

$$H_e = H + \xi(J,\sigma) \tag{1}$$

The magnetic flux density B is expressed as:

$$B = \Psi(H) = \mu_0 H + J = \mu_0 H + \varsigma(H_e) \tag{2}$$

where the magnetic polarization J is a function of H_e and includes both reversible and irreversible contributions. The hysteresis contribution is introduced by the function ς, described through the Classical Preisach Model (CPM). In Equations (1) and (2), and in the following relations, dependence on time is omitted from all field quantities for the sake of simplicity.

The effective field H_e introduces a correction in the MS-Preisach model that makes it implicit. The nonlinear relation Ψ between B and H is handled by the Fixed Point (FP) technique, which splits the nonlinearity into a linear term with constant coefficient η and a residual **R**, to be iteratively updated. The iterative numerical scheme is described in detail in [18].

2.2. Finite Element Model of the Generator

The equations, which govern the dynamic behaviour of the generator, involve both the electromagnetic field and the electric circuit connected to the pick-up coil, requiring the adoption of a voltage-driven formulation able to simultaneously solve the two sets of equations.

Drawing on the harvester geometry, the device is described in a cylindrical reference system of coordinates r, z, ϑ, leading to a 2D axis-symmetric problem. The electromagnetic field is developed in terms of a magnetic vector potential $\mathbf{A}$ ($B = curl\mathbf{A}$), with only the ϑ-component, whose distribution is the unknown in the electromagnetic problem.

Having applied the FP technique, the electromagnetic field equation in the weak form becomes:

$$\int_\Omega \nu curl\mathbf{A} \cdot curl\mathbf{w} ds = -\int_\Omega \mathbf{R} \cdot curl\mathbf{w} ds + \int_{\Omega_c} \frac{N_c i_c}{S_c} w ds - \frac{1}{\rho}\int_{\Omega_r} \dot{\mathbf{A}} w ds + \frac{1}{\rho}\int_{\Omega_r} \mathbf{M}_{\Omega_r}(\dot{\mathbf{A}}) w ds \quad (3)$$

where $\mathbf{w}$ is the test function, ν is the magnetic permittivity (η in the magnetostrictive rod and vacuum value elsewhere), $\mathbf{R}$ is the FP residual and Ω is the 2D full domain under study. Subdomain Ω_r is the trace on Ω of the rod with electrical conductivity ρ, and $\mathbf{M}_\Omega$ denotes the mean time derivative of $\mathbf{A}$ over Ω_r. Subdomain Ω_c is the trace (with area S_c) on Ω of the pick-up coil, which has N_c turns where the unknown current i_c is flowing.

The problem is completed by the electric loop equation of the pick-up coil:

$$R_l i_c + L_l \frac{di_c}{dt} + e_c + R_c i_c = 0 \quad (4)$$

where R_c is coil resistance ($R_c = 32.6\ \Omega$), R_l and L_l are resistance and stray inductance of the electric load and the electromotive force induced at the coil terminals is expressed in terms of $\mathbf{A}$, as:

$$e_c = \frac{2\pi N_c}{S_c} \int_{\Omega_c} r\dot{A} ds \quad (5)$$

The field equation is solved through the Finite Element Method (FEM), suited for the analysis of these kind of devices (see, for example, [30–32]). The domain Ω is suitably enlarged, to allow homogeneous Dirichlet conditions on its boundaries, and discretized using first-order elements. The time evolution is handled by adopting a step-by-step scheme.

From a mechanical point of view, the mechanical force at each time step is assumed to be applied normally to the rod cross-section with only the axial-component. Taking advantage of the cylindrical geometry, the applied stress can be considered uniformly distributed at each point of the rod and computed simply as the ratio of the instantaneous force to the rod cross-section. This is a simplification, which avoids the solution of a mechanical problem by the finite element method to compute the spatial distribution of stress within the rod. It has been verified in other studies [18], related to magnetostrictive actuators, that this hypothesis is acceptable. The elongation of the magnetostrictive rod, defined as $\Delta L = \Gamma(\sigma, J)L$, is computed by averaging the local strain λ over a rod cross-section Σ_r and length L:

$$\Gamma = \frac{2\pi}{V_r} \int_{\Sigma_r} \lambda r ds \quad (6)$$

where V_r is the volume of the rod.

However, the limited size of the strain inside the rod (a few hundred ppm) does not justify a remeshing of the FEM domain during the solution of the electromagnetic field problem. Therefore, it is assumed that the rod shape is remains unchanged throughout the entire computational process. All computations required for the device simulation are performed using a homemade code (*Sally2D*), which is able to manage hysteretic materials.

2.3. Magnetostrictive Properties

The Preisach distribution $\Phi(\alpha,\beta)$, together with the corrective function ξ (used in Equation (1)), and the strain function λ (as appears in Equation (6)), are necessary to fully identify the MS-model.

These functions must be experimentally determined, starting with Fe–Ga material characterization. The procedure for quasi-static material characterization, described in [10], includes the measurement of magnetic hysteresis loops (see Figure 2a) and rod elongation at different mechanical loads.

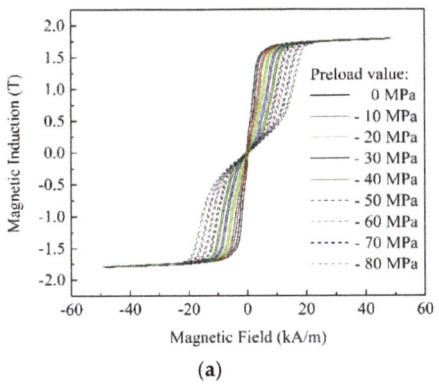
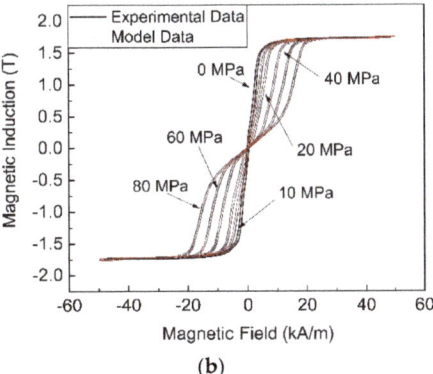

Figure 2. (a) Measured quasistatic B–H loops for different levels of the constant preload; (b) superposition of model data B–H loops on experimental B–H loops.

Starting from an experimental loop (H_{exp}, B_{exp}), the simulated effective field H_e is computed and then compared (see Figure 2b) with the experimental waveform:

$$\begin{aligned} J_{exp} &= B_{exp} - \mu_0 H_{exp} \\ H_e &= \zeta^{-1}(J_{exp}) \\ \xi(J_{exp}, \sigma_{exp}) &= H_e - H_{exp} \end{aligned} \quad (7)$$

where the inversion of the Preisach model (ζ^{-1}) is performed through a numerical procedure. The application of this procedure provides a set of discretized values of the experimental curves of ξ and λ in the plane J, σ. Since, unlike Terfenol-D, analytical functions able to interpolate the curves in the whole plane cannot be determined, the simulated data must be reconstructed through a four-point spatial interpolation.

This approach, based on interpolation, without the possibility of reasonably extrapolating data, limits the maximum stress applicable, since the sum of mechanical preload and dynamic load cannot overcome the highest static load applied during the experimental material characterisation (in this work, 80 MPa).

2.4. Harvester Prototype Setup and Modeling

The direct force harvester is based on the use of a polycrystalline $Fe_{81}Ga_{19}$ sample (cubic grains with <100> easy axes) in the shape of a cylindrical rod, with a length of 60 mm and a diameter of 12 mm. Thanks to the material workability, the rod was machined, reducing the external diameter to 6 mm from a length of 53 mm, in order to host a 2000-turns pick-up coil with 32.6 Ω electric resistance. The magnetic bias was generated by a couple of permanent magnets (PMs) placed at the extremities of the rod. A picture of the system, including the rod with the pick-up coil coated with white Teflon, is shown in Figure 3a, while a scheme is presented in Figure 3b.

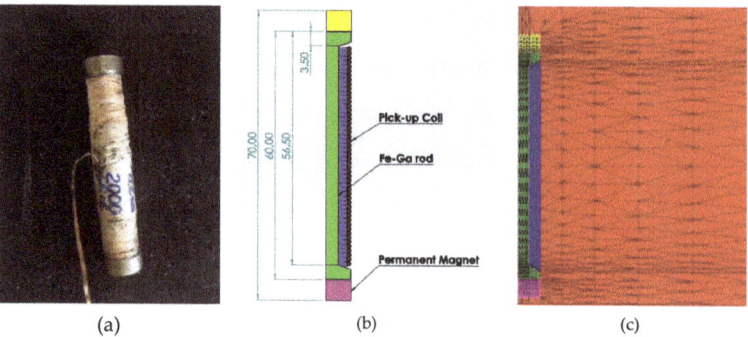

Figure 3. EH system: (**a**) Actual picture of the Fe–Ga rod with a pick-up coil of 2000 turns and a resistance of 32.6 Ω; (**b**) Sketch of the rod aligned section with permanent magnets and pick-up coil. Dimension in mm; (**c**) 2D mesh for a limited portion of the computational domain Ω used for the FEM analysis (Fe–Ga in green, pick-up coil in blue, PMs in yellow and purple, air in red).

The axis-symmetric structure of the prototype is discretized into triangular finite elements. A portion of the considered domain Ω is shown in Figure 3c, where each colour indicates a different material: air (red), Fe–Ga (green), pick-up coil (blue), PMs (yellow and purple).

2.5. Characterization and Experimental Setups

The setup for the experimental characterization of the magnetostrictive material is based on a three-legged magnetizer and a test machine, and provides the hysteresis loops of Figure 2a, as described in [33,34].

The experimental setup for the measurements on the Fe–Ga direct-force harvester (Figure 4) makes use of many of the experimental components described in [33]. In particular, a 10 kN-100 Hz fatigue-testing machine and a control software have been used to generate fully controlled harmonic vibrations. A programmable resistor (Pickering PXI 40-297-002 programmable precision Resistors, Pickering Interfaces Ltd., Clacton-on-Sea, Essex, UK) works as an electric load. Output power is measured, along with other electrical parameters, by a wattmeter Yokogawa WT 3000 (Yokogawa Electric Co., Musashino, Tokyo, Japan) using a 2-ampere channel, with a minimum range of 5 mA.

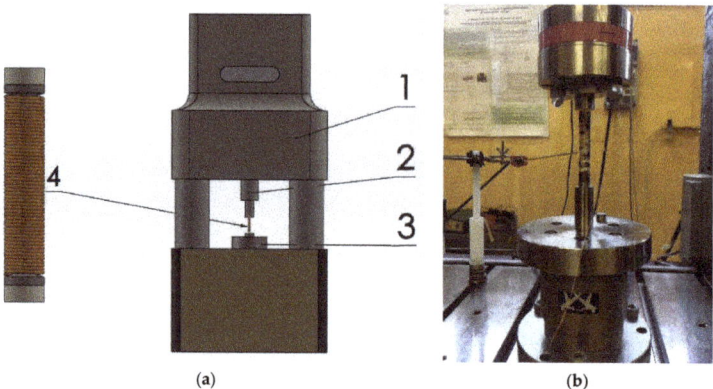

Figure 4. (**a**) 1. Fatigue testing machine applying the mechanical load through 2. plunger; using a 3. load cell for the feedback loop. The magnetic field bias is applied to the Fe–Ga rod 4. through a pair of PMs; (**b**) Picture of the Fe–Ga rod, with PMs inserted in the fatigue-testing machine for experimental analysis.

3. Results

3.1. Preliminary Results

In the experimental setup, the magnetic bias was provided by two Nd–Fe–B PMs with 955 kA/m coercive field and 1.2 T remanence and kept constant during all measurements. The load resistance was set to 160 Ω (near the peak output power), to be added to the 32.6 Ω coil resistance. The vibration force magnitude and the mechanical bias can be modified. However, in preliminary comparisons between experiments and numerical predictions, the dynamic vibration force peak was set to 8 MPa, with a fixed frequency of 100 Hz. Thus, in the preliminary tests, the mechanical preload σ_0 was the only parameter left free. This choice comes from the experimental analysis carried out in [34], where it was highlighted that, for a given magnetic field bias, only one value of the mechanical bias is able to maximize the output power. The mechanical preload σ_0 ranges from 30 MPa to 70 MPa, since the dynamic load is kept to 8 MPa.

As stated previously, the FEM model adopts the FP technique as an iterative method to handle the nonlinearities. This implies the use of a suitable FP constant η, and of a limit value of the convergence index together with the maximum number of iterations.

The value of η is an essential parameter, because it affects both the convergence of the iterative method and its speed. In the presence of invariable magnetic characteristics (or quite close to them, as for Terfenol-D), the optimal value of the FP constant is chosen, starting from the maximum and minimum slopes of the B-H curve. The average value between the two slopes guarantees a regular convergence of the iterative process. However, in the Fe–Ga alloy, the magnetic behavior significantly varies with the mechanical load (sum of the preload and dynamic load), as illustrated by the curves in Figure 2a, which show an evident bending at the preload increase. In addition, the presence of a magnetic bias makes the definition of the FP constant even more critical.

The main consequence of the strong dependence of the Fe–Ga magnetic properties on the mechanical preload is that a single FP constant cannot ensure convergence for the whole range of the considered values, as shown in Figure 5. Here, the experimental data (points in blue) are compared with the results provided by simulation using two FP constants—$\eta = 3400$ and $\eta = 50$—having imposed 800 iterations. Experimental measurement points have been repeated and verified, taking into account the effect of wrong positioning evident in [30]. Besides accurate positioning, in order to obtain good measurement repeatability, a long warming-up period (approximatively 40 min) is required for each measurement point. The estimated measurement uncertainty is about $\pm 5\%$.

The first value of η would be suitable for values of reduced preloads (about 40 MPa), while the second one fits the behavior of higher values (about 70 MPa). However, it is evident that the use of a single FP constant fails, leading to calculated results not always in agreement with experiments. Thus, for the simulation of this material in dynamic conditions, a new identification procedure is needed.

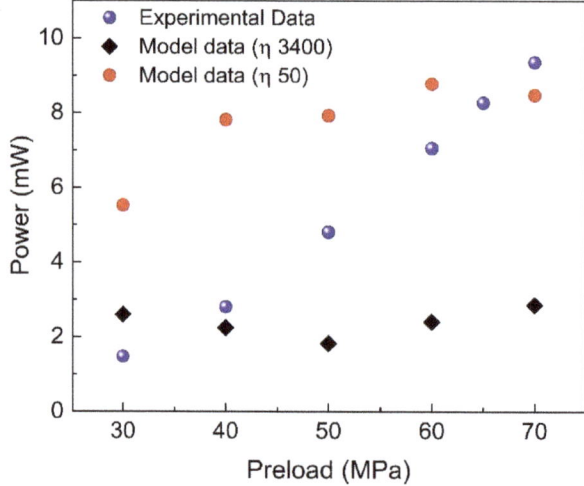

Figure 5. Comparison of the output power given by the model data (red circles and black diamonds) and the experimental data (blue spheres) for different values of mechanical preload.

3.2. Model Tuning

The model presented in this paper is phenomenological and the first tuning is performed through identification, carried out starting from the magnetic characteristics of the material. The second tuning, which concerns the convergence of the iterative method based on the FP technique, does not require additional measurements, but considers only the value reached by the convergence index (an L2 norm of the relative difference of the solutions at two successive steps) after a stated number of iterations. For each preload value, the evaluation of the correct FP constant is obtained through a process of trial and error. Figure 6 summarizes the FP constant selection process.

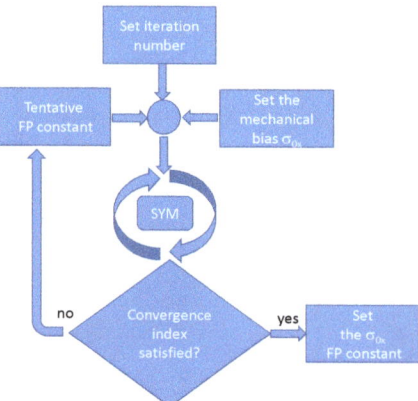

Figure 6. Scheme for the choice of the FP constant for a generic σ_{0x} mechanical preload.

Preliminary computations show that a convergence index equal to 4×10^{-3} is a good compromise between accuracy of the results and calculation times (around 10 h).

Figure 7 illustrates the convergence index and the simulation total time as a function of the number of iterations, for the mechanical preload value of 70 MPa.

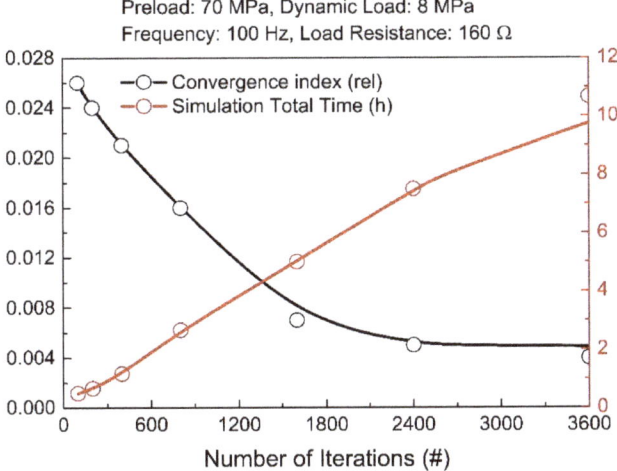

Figure 7. Evolution of convergence index and simulation time as a function of the number of iterations. Trend lines are just a guide for the eyes.

3.3. Results

The correct identification of the model, and the correct choice of the constant FP described in Section 3.2, lead to an excellent agreement between the experimental and computational values of the output power transferred to the load, as shown in Figure 8.

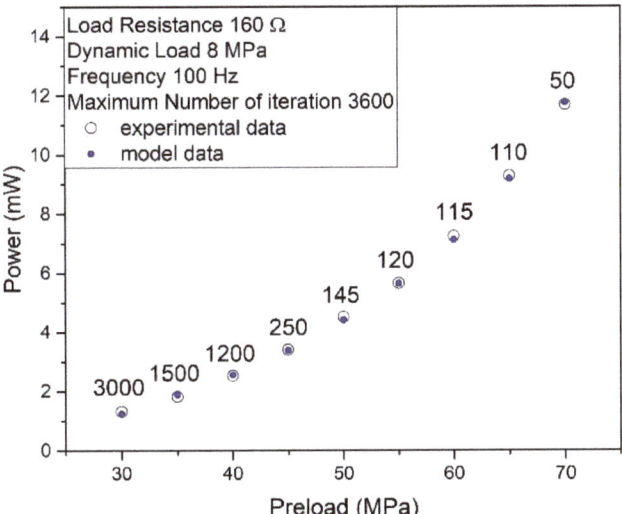

Figure 8. Comparison of experimental data (empty circles) and model data (solid circles) for different values of mechanical preload. The label indicates the value of the parameter η. Load resistance 160 Ω, dynamic load peak 8 MPa, frequency 100 Hz.

The values of the FP constant previously identified for each mechanical preload (see described Figure 8) provide accurate output powers, varying all other parameters. Notably, in Figure 9a, the vibration frequency ranges from 20 Hz to 100 Hz; in Figure 9b, the dynamic load amplitude varies from 4 MPa to 10 MPa; finally, the load resistance varies from 10 Ω to 10 kΩ in Figure 9c.

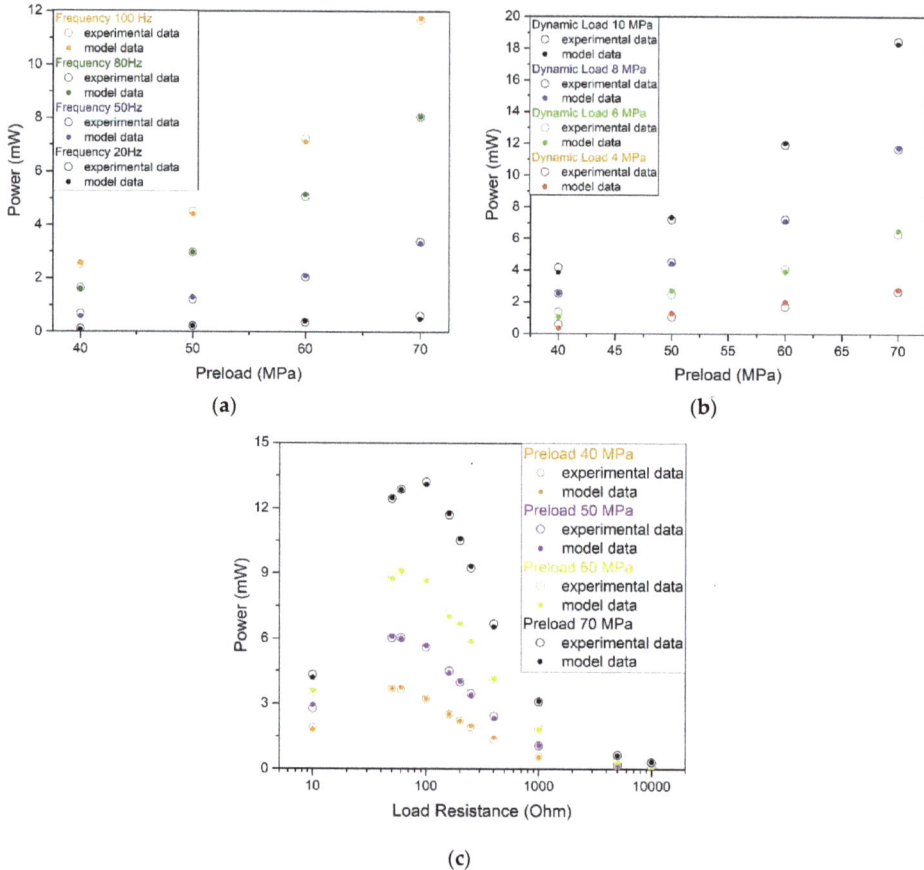

Figure 9. Comparison of the output power evaluated by the model, and results of experiment upon varying the following parameters: (**a**) Value of the dynamic load (from 4 MPa to 10 MPa), in which the output power is a function of mechanical preload; (**b**) frequency (from 20 Hz to 100 Hz), where the output power is a function of mechanical preload; (**c**) load resistance (from 10 Ω to 10 kΩ), for several values of mechanical preload.

The direct dependence of the output power vs frequency (Figure 9a), or amplitude of the mechanical stress applied (Figure 9b), is easy to understand. The diagram of the output power vs load resistance (Figure 9c) shows a maximum load resistance in the range 80–90 Ω. The presence of a peak complies with the maximum power transfer theorem, which states that the maximum power provided by a source is obtained when external resistance is equal to the internal resistance of the source. In this case, the internal resistance includes both the coil electric resistance R_c (32.6 Ω) and an equivalent resistance R_e (estimated around 50 Ω), which takes into account the magnetic losses inside the rod.

The satisfactory agreement between experiments and computations, which is reached in all analysed situations, provides evidence that the FP constant values are not affected by parameters other than the mechanical preload. The influence of the magnetic bias, imposed by the permanent magnets, was not analysed.

4. Conclusions

A Preisach-type model is applied here to simulate the behaviour of an Fe–Ga harvester prototype. This approach, making a preliminary material identification, has been implemented in a 2D axis-symmetric finite element software. Non-linearities in the computational code are handled with the FP technique, whose convergence is heavily influenced by the mechanical bias values. The model requires a specific tuning, in order to determine the suitable values of η_i for each preload σ_{0i}. The tuning ensures a satisfactory result compared to validation measurements. However, the model results are unsatisfactory when using a unique value of the FP constant.

The model is able to provide effective results in a wide range of mechanical preloads and, with the experimental apparatus available for this investigation, the results have been validated up to 70 MPa. The model is also able to provide good results with an associated variation in frequency, vibration amplitude, or electrical load.

The model has, therefore, proven to be a valid design tool for harvesters based on magnetostrictive materials, and will be used for the design of an optimized EH with magnetic yoke.

Author Contributions: S.P and M.Z. conceived the paper and designed the experiments; O.B. and M.C. developed the software; S.P. and M.C. performed the formal analysis; S.P. and M.Z. performed the investigation and the analysis; S.P. and M.Z. wrote the original draft while all the authors contributed to the final manuscript and approved the final paper.

Funding: This research received no external funding.

Conflicts of Interest: The authors declare no conflict of interest.

References

1. Mandayam, S.; Udpa, L.; Udpa, S.S.; Sun, Y.S. A fast iterative finite element model for electrodynamic and magnetostrictive vibration adsorbers. *IEEE Trans. Magn.* **1994**, *30*, 3300–3303. [CrossRef]
2. Adly, A.A.; Mayergoyz, I.D.; Bergqvist, A. Utilizing anisotropic Preisach-type models in the accurate simulation of magnetostriction. *IEEE Trans. Magn.* **1997**, *33*, 3931–3933. [CrossRef]
3. Armstrong, W.D. Magnetization and magnetostriction processes in Tb (0.27–0.30) Dy (0.73–0.70) Fe (1.9–2.0). *J. Appl. Phys.* **1997**, *81*, 2321. [CrossRef]
4. Gros, L.; Reyne, G.; Body, C.; Meunier, G. Strong coupling magneto mechanical methods applied to model heavy magnetostrictive actuators. *IEEE Trans. Magn.* **1998**, *34*, 3150–3153. [CrossRef]
5. Stillesjö, F.; Engdahl, G.; Bergqvist, A. A design technique for magnetostrictive actuators with laminated active materials. *Simul. Model. Pract. Theory* **1998**, *32*, 2141–2143.
6. Besbes, M.; Ren, Z.; Razek, A. A generalized finite element mode of magnetostriction phenomena. *IEEE Trans. Magn.* **2001**, *37*, 3324–3328. [CrossRef]
7. Pèrez-Aparicio, J.L.; Sosa, H. A continuum three-dimensional, fully coupled, dynamic, non-linear finite element formulation for magnetostrictive materials. *Smart Mater. Struct.* **2004**, *13*, 493–502. [CrossRef]
8. Davino, D.; Natale, C.; Pirozzi, S.; Visone, C. Rate-dependent losses modeling for magnetostrictive actuators. *J. Magn. Magn. Mat.* **2004**, *272*, e1781–e1782. [CrossRef]
9. Bottauscio, O.; Chiampi, M.; Lovisolo, A.; Roccato, P.E.; Zucca, M. Dynamic modelling and experimental analysis of Terfenol-D rods for magnetostrictive actuators. *J. Appl. Phys.* **2008**, *103*, 07F121. [CrossRef]
10. Bottauscio, O.; Lovisolo, A.; Roccato, P.E.; Zucca, M.; Sasso, C.; Bonin, R. Modelling and experimental analysis of magnetostrictive devices: from the material characterisation to their dynamic behaviour. *IEEE Trans. Magn.* **2008**, *44*, 3009–3012. [CrossRef]
11. Stuebner, M.; Atulasimha, J.; Smith, R.C. Quantification of hysteresis and nonlinear effects on the frequency response of ferroelectric and ferromagnetic materials. *Smart Mater. Struct.* **2009**, *18*, 104019. [CrossRef]
12. Kaltenbacher, M.; Meiler, M.; Ertl, M. Physical modeling and numerical computation of magnetostriction. *Int. J. Comp. Math. Electr. Electron. Eng.* **2009**, *48*, 819–832. [CrossRef]

13. Graham, F.C.; Mudivarthi, C.; Datta, S.; Flatau, A.B. Modeling of a Galfenol transducer using the bidirectionally coupled magnetoelastic model. *Smart Mater. Struct.* **2009**, *18*, 104013. [CrossRef]
14. Evans, P.G.; Dapino, M.J. Efficient magnetic hysteresis model for field and stress application in magnetostrictive Galfenol. *J. Appl. Phys.* **2010**, *107*, 3009–3012. [CrossRef]
15. Almeida, L.A.L.; Depp, G.S.; Lima, A.M.N.; Neff, H. Modeling a magnetostrictive transducer using genetic algorithm. *J. Magn. Magn. Mat.* **2001**, *266–230*, 1262–1264. [CrossRef]
16. Cao, S.C.; Wang, B.; Zheng, J.; Huang, W.; Sun, Y.; Yang, Q. Modeling dynamic hysteresis for giant magnetostrictive actuator using hybrid genetic algorithm. *IEEE Trans. Magn.* **2006**, *42*, 911–914.
17. Davino, D.; Giustiniani, A.; Visone, C. Experimental properties of an efficient stress-dependent magnetostriction model. *J. Appl. Phys.* **2009**, *105*, 07D512. [CrossRef]
18. Bottauscio, O.; Roccato, P.E.; Zucca, M. Modeling the dynamic behavior of magnetostrictive actuators. *IEEE Trans. Magn.* **2010**, *46*, 3022–3028. [CrossRef]
19. Zucca, M.; Bottauscio, O. Hysteretic modeling of electrical micro-power generators based on Villari effect. *IEEE Trans. Magn.* **2012**, *48*, 3092. [CrossRef]
20. Berbyuk, V. Vibration energy harvesting using Galfenol-based transducer. *Act. Passiv. Smart Struct. Integr. Syst.* **2013**, *8688*, 86881F.
21. Rezaeealam, B. Finite element analysis of magnetostrictive vibration energy harvester. *Compel Int. J. Comput. Math. Electr. Electron. Eng.* **2012**, *31*, 1757–1773. [CrossRef]
22. Davino, D.; Krejčí, P.; Visone, C. Fully coupled modeling of magneto-mechanical hysteresis through thermodynamic compatibility. *Smart Mater. Struct.* **2013**, *22*, 095009. [CrossRef]
23. Davino, D.; Krejčí, P.; Pimenov, A.; Rachinskii, D.; Visone, C. Analysis of an operator-differential model for magnetostrictive energy harvesting. *Commun. Nonlinear Sci. Numer. Simul.* **2016**, *39*, 504–519. [CrossRef]
24. Clemente, C.S.; Mahgoub, A.; Davino, D.; Visone, C. Multiphysics circuit of a magnetostrictive energy harvesting device. *J. Intell. Mater. Syst. Struct.* **2017**, *28*, 2317–2330. [CrossRef]
25. Kholmetskaa, I.; Chlebouna, J.; Krejcí˘, P. Numerical modeling of Galfenol magnetostrictive response. *Appl. Math. Comput.* **2018**, *319*, 527–537. [CrossRef]
26. Ahmed, U.; Jeronen, J.; Zucca, M.; Palumbo, S.; Rasilo, P. Finite element analysis of magnetostrictive energy harvesting concept device utilizing thermodynamic magneto-mechanical model. *J. Magn. Magn. Mater.* **2019**, *486*, 165275. [CrossRef]
27. Bertotti, G.; Mayergoy, I.D. *The Science of Hysteresis*; Academic Press: Oxford, UK, 2006; Volume 1.
28. Sablik, M.J.; Jiles, D.C. Coupled magnetoelastic theory of magnetic and magnetostrictive hysteresis. *IEEE Trans. Magn.* **1993**, *29*, 429–439. [CrossRef]
29. Calkins, F.T.; Smith, R.C.; Flatau, A.B. Energy-based hysteresis model for magnetostrictive transducers. *IEEE Trans. Magn.* **2000**, *36*, 3931–3933. [CrossRef]
30. Cho, C.; Hong, Y.; Jiang, C.; Chen, Y. Study of Dynamic Behavior of Magnetostrictively Patterned Flexible Micro-Wings. *IEEE Trans. Magn.* **2019**, *55*, 1–4. [CrossRef]
31. Backman, G.; Lawton, B.; Morley, N.A. Magnetostrictive Energy Harvesting: Materials and Design Study. *IEEE Trans. Magn.* **2019**, *55*. [CrossRef]
32. Song, Y. Finite-Element Implementation of Piezoelectric Energy Harvesting System from Vibrations of Railway Bridge. *J. Energy Eng.* **2018**, *145*, 04018076. [CrossRef]
33. Zucca, M.; Mei, P.; Ferrara, E.; Fiorillo, F. Sensing dynamic forces by Fe-Ga in compression. *IEEE Trans. Magn.* **2017**, *53*, 1–4. [CrossRef]
34. Palumbo, S.; Rasilo, P.; Zucca, M. Experimental investigation on a Fe-Ga close yoke vibrational harvester by matching magnetic and mechanical biases. *J. Magn. Magn. Mater.* **2019**, *469*, 354–363. [CrossRef]

© 2019 by the authors. Licensee MDPI, Basel, Switzerland. This article is an open access article distributed under the terms and conditions of the Creative Commons Attribution (CC BY) license (http://creativecommons.org/licenses/by/4.0/).

Article

Output of MEMS Piezoelectric Energy Harvester of Double-Clamped Beams with Different Width Shapes

Lei Jin, Shiqiao Gao *, Xiyang Zhang and Qinghe Wu

School of Mechatronic Engineering, Beijing Institute of Technology, Beijing 100081, China; jinlei@bit.edu.cn (L.J.); zhangxiyang2012@163.com (X.Z.); wuqh_123@163.com (Q.W.)
* Correspondence: gaoshq@bit.edu.cn

Received: 16 April 2020; Accepted: 15 May 2020; Published: 19 May 2020

Abstract: For a microelectromechanical system (MEMS) piezoelectric energy harvester consisting of double-clamped beams, the effects of both beam shape and electrode arrangement on the voltage outputs are analyzed. For two kinds of harvester structures including millimeter-scale and micro-scale, and different shapes including rectangular, segmentally trapezoidal and concave parabolic are taken into account. Corresponding electric outputs are calculated and tested. Their results are in good agreement with each other. The experimental results validate the theoretical analysis.

Keywords: double-clamped; width shapes; piezoelectric energy harvester; electrodes pair; MEMS structure

1. Introduction

In general, a piezoelectric energy harvester is composed of beam structures. Because the bending curvature of beam is not a constant along the length direction whether for a cantilever beam or a double-clamped beam, the strain distributions is not uniform. Therefor the efficiency of energy harvesting is not good enough. To improve the output efficiency, in recent years, much effort has been put into the aspect of the optimization of the shapes. Based on studies of the trapezoidal shapes [1–4], arrayed trapezoidal beams [5], reversed trapezoidal beams [6], triangular shapes [7], and optimized shapes [8,9], Jin, Gao et al. [10] explained the shape effects of cantilever beams on electric outputs by an analytical method. At the same time, non-linear trapezoidal shapes of cantilever beams [11] and composite cantilever beams [12] were also investigated.

In addition to the cantilever beam, the other typical forms are the double-clamped beam structures. These structures are usually used in hybrid piezoelectric-electromagnetic energy harvesters [13–17]. They have better stability and symmetry than the cantilever beam so as to be more conducive to the stable movement of the magnet mass. By contrast with the unidirectional curvature of cantilever beam, there are different directional curvatures for a double-clamped beam. Optimization concerns not only the beam shapes but also the arrangement of the electrodes. However, so far research on the shapes of the beam and the arrangement of the electrode pair, especially for a microelectromechanical system (MEMS) energy harvester, is still rare. Therefore, it is necessary to study the optimal shapes of the beam and the arrangements of the electrode pair for a harvester with a double-clamped beam.

In this paper, a dynamic analytical method is presented to analyze the piezoelectric energy harvester consisting of double-clamped beams. The effects of both beam shape and the electrode arrangement for the MEMS energy harvester are studied analytically. At the same time, corresponding experiments are conducted both for millimeter-scale and micro-scale structures.

2. Governing Equations of the Piezoelectric Energy Harvester

For a bending beam, the strain ε_1 along the length direction can be written by $\frac{z}{\rho(x)}$, where $\frac{1}{\rho(x)} = \frac{\partial^2 w}{\partial x^2}$ is the bending curvature of the beam. According to the direct effect and the converse effect described by Meeker [18] (IEEE Standard on Piezoelectricity), the constitutive equations of composite beam consisting of a base structure and a piezoelectric layer shown as in Figure 1 can be written by:

$$\begin{cases} T_1 = Y_p \frac{z}{\rho(x)} - e_{31} E_3 & \xi \leq z \leq \xi + \delta \\ T_1 = Y_s \frac{z}{\rho(x)} & -(h-\xi) \leq z < \xi \\ D_3 = e_{31} \frac{z}{\rho(x)} + \varepsilon_{33} E_3 & \xi \leq z \leq \xi + \delta \end{cases} \quad (1)$$

where x is the coordinate in length direction of composite beam, z is the coordinate in thickness direction, T_1 is the stress in the length direction x, E_3 and D_3 are the electric field and the electric displacement in the thickness direction z respectively, Y_p and Y_s are elastic modulus of the piezoelectric layer and substructure respectively, e_{31} is the piezoelectric constant, ε_{33} is the permittivity of the piezoelectric layer, h and δ are thicknesses of the substructure and piezoelectric layer respectively, ξ is the distance from the interface to the neutral surface of the composite beam which is derived by balancing the internal force of the cross section as $\xi = \frac{Y_s h^2 - Y_p \delta^2}{2(Y_p \delta + Y_s h)}$ [9].

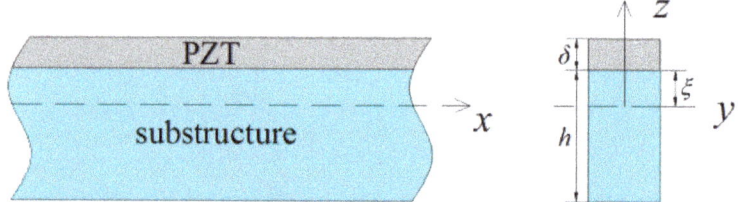

Figure 1. A composite beam.

Integrating the moment of stress around the neutral surface of the composite beam for all the cross section and integrating the electric field and the electric displacement for all the volume of piezoelectric layer, leads to:

$$\begin{cases} M(x) = \frac{K_D \varphi(x)}{\rho(x)} - e_{31}\left(\xi + \frac{\delta}{2}\right)\varphi(x) V \\ Q = e_{31}\left(\xi + \frac{\delta}{2}\right) \int_0^L \frac{1}{\rho(x)} \varphi(x) dx + \frac{\varepsilon_{33} S}{\delta} V \end{cases} \quad (2)$$

where $M(x) = \int_{-(h_s-\xi)}^{\xi+\delta} T_1 z \varphi(x) dz$ is the internal moment of a cross section of the beam, $\varphi(x)$ is the width of the composite beam which is a function along the x-direction, $K_D = \int_{-(h_s-\xi)}^{\xi} Y_s z^2 dz + \int_{\xi}^{\xi+\delta} Y_p z^2 dz$ is the stiffness per width of the composite beam against bending, $V = E_3 \delta$ is the voltage of the piezoelectric layer in the thickness direction z, $Q = \frac{1}{\delta} \int_0^L \int_{\xi}^{\xi+\delta} D_3 dz \varphi(x) dx$ is the average charges on the surface of the piezoelectric layer, S is the surface area of piezoelectric layer.

For analysis of the symmetric deformation of a double-clamped beam with a concentrated proof mass shown as in Figure 2, only a half part of both the beam and the mass needs to be considered.

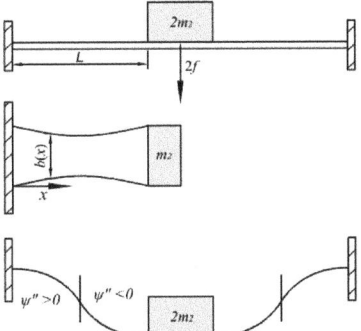

Figure 2. A double-clamped beam with a concentrated proof mass.

The deflection can be written by:

$$w(x,t) = \frac{\psi(x)}{\psi(L)} w_m(x,t) \qquad (3)$$

where $w_m(t) = w(L,t)$ is the displacement of the proof mass and $\psi(x)$ is the mode function of double-clamped beam with a concentrated proof mass. The bending curvature of the neutral surface can be expressed by $\frac{1}{\rho(x)} = \frac{\psi''(x)}{\psi(L)} w_m(t)$.

Substituting it into Equation (2), multiplying the former one by $\psi''(x)$ and then integrating it along the length direction x, leads to:

$$\begin{cases} \int_0^L M(x)\psi''(x)dx = \frac{K_D}{\psi(L)}\{\int_0^L \varphi(x)[\psi''(x)]^2 dx\}w_m - e_{31}\left(\xi + \frac{\delta}{2}\right)[\int_0^L \varphi(x)\psi''(x)dx]V \\ Q = e_{31}\left(\xi + \frac{\delta}{2}\right)\frac{1}{\psi(L)}[\int_0^L \psi''(x)\varphi(x)dx]w_m + \frac{\varepsilon_{33}^S}{\delta}V \end{cases} \qquad (4)$$

where L is the half length of the beam.

The virtual work principle can be written by:

$$f\delta w_m = \delta \int_0^L M(x) \frac{\partial^2 w}{\partial x^2} dx = \frac{\delta w_m}{\psi(L)} \int_0^L M(x)\psi''(x) dx \qquad (5)$$

where f is generalized force of the proof mass.

Substituting it into Equation (4), leads to:

$$\begin{cases} f = \frac{K_D}{\psi^2(L)}\{\int_0^L \varphi(x)[\psi''(x)]^2 dx\}w_m - e_{31}\left(\xi + \frac{\delta}{2}\right)\frac{1}{\psi(L)}[\int_0^L \varphi(x)\psi''(x)dx]V \\ Q = e_{31}\left(\xi + \frac{\delta}{2}\right)\frac{1}{\psi(L)}[\int_0^L \psi''(x)\varphi(x)dx]w_m + \frac{\varepsilon_{33}^S}{\delta}V \end{cases} \qquad (6)$$

In the dynamic case, there is:

$$f = ma(t) - m\ddot{w}_m - c\dot{w}_m \qquad (7)$$

where $m = m_2 + \int_0^L \psi_T \psi_T \mu \varphi(x) dx$, μ is the mass per unit width and per unit length of the composite beam, c is damping efficient of the beam, $a(t)$ is the exciting acceleration.

Substituting Equation (7) into Equation (6) and making a derivative of the second one with respect to time t, leads to:

$$\begin{cases} m\ddot{w}_m + c\dot{w}_m + kw_m - \Theta V = ma(t) \\ I = \Theta \dot{w}_m + C_p \dot{V} \end{cases} \qquad (8)$$

where:

$$k = \frac{K_D}{\psi^2(L)} \int_0^L \varphi(x)[\psi''(x)]^2 dx \tag{9}$$

$$\Theta = e_{31}\left(\xi + \frac{\delta}{2}\right)\frac{1}{\psi(L)} \int_0^L \varphi(x)\psi''(x) dx \tag{10}$$

$$C_p = \frac{\varepsilon_{33} S}{\delta} \tag{11}$$

where I is electric current, k is called the stiffness, Θ is called the converting factor (or coupling factor), C_p is the capacitance.

The energy efficiency depends on the converting factor. If the electrode is fully covered on the piezoelectric layer to form one electrode pair, integrating this in Equation (10), leads to:

$$\Theta = e_{31}\left(\xi + \frac{\delta}{2}\right)\frac{1}{\psi(L)}[-\varphi'(L)\psi(L) + \int_0^L \varphi''(x)\psi(x) dx] \tag{12}$$

For the rectangular beam, $\varphi'(x) = 0$, $\varphi''(x) = 0$, there is $\Theta = 0$. For a trapezoidal beam $\varphi''(x) = 0$, there is $\Theta = -e_{31}\left(\xi + \frac{\delta}{2}\right)\varphi'(L)\psi(L)$. It can be seen that most charges are cancelled out because the charge sign of the left part is opposite the right one that resulted from changing the sign of the curvature. This phenomenon does not occur on the cantilever beam reported in [10] because there is no problem caused by the sign changing of bending curvature.

To stack all the positive charges on one electrode and all the negative charges on the other electrode, the electrode on the upper surface should be divided into two parts and the lower surface can share one electrode. Through the positive and negative reverse connection $V_{12} = V_{10} - V_{20}$ shown as in Figure 3b, where the enlargement of the area A is shown as in Figure 3c, the positive and negative charges cannot cancel out each other so that more charges can be effectively collected.

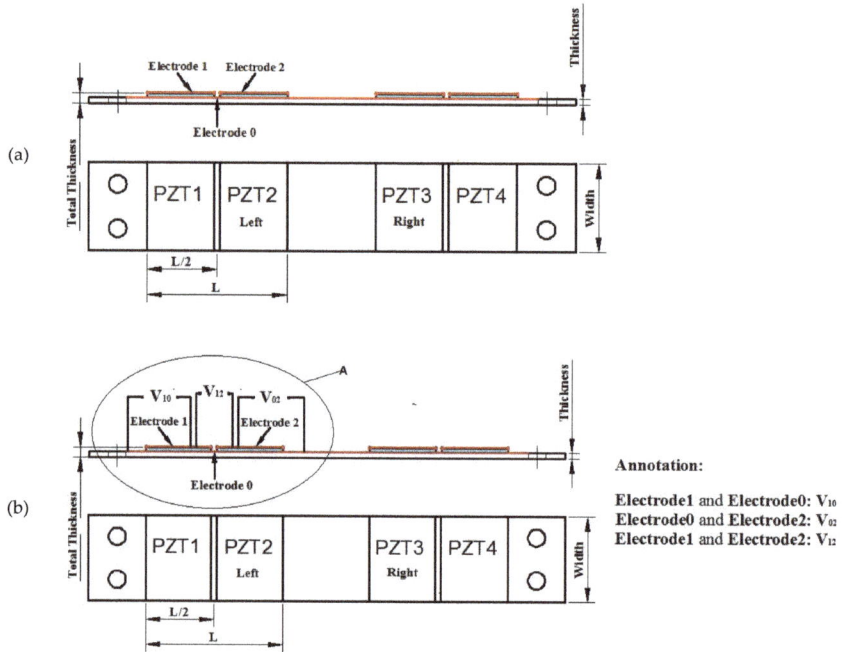

Figure 3. *Cont.*

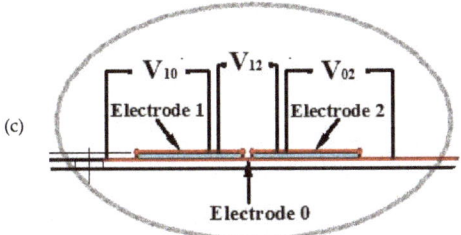

Figure 3. Double discrete electrodes and voltage outputs, (**a**) places of PZT and electrodes; (**b**) connections of electrodes for output voltages; (**c**) enlargement of the area A in (**b**).

In this case, the converting factor should be expressed as:

$$\Theta = e_{31}\left(\xi + \frac{\delta}{2}\right)\frac{1}{\psi(L)}\left[\int_0^{\frac{L}{2}} \varphi(x)\psi''(x)dx - \int_{\frac{L}{2}}^{L} \varphi(x)\psi''(x)dx\right] \quad (13)$$

The energy efficiency depends on the converting factor. From its expression, it can be seen that, converting factor Θ depends on the width function $\varphi(x)$. For different width shapes, there are different converting factors. To obtain a maximum converting factor, the width shape has to be optimized. For the symmetrical shape to the center point $x = \frac{L}{2}$ of the half beam, the above equation of the converting factor can be further derived as:

$$\Theta = e_{31}\left(\xi + \frac{\delta}{2}\right)\frac{1}{\psi(L)}\left[2\varphi\left(\frac{L}{2}\right)\psi'\left(\frac{L}{2}\right) + \varphi'(L)\psi(L) + \int_0^{\frac{L}{2}} \varphi''(x)\psi(x)dx - \int_{\frac{L}{2}}^{L} \varphi''(x)\psi(x)dx\right] \quad (14)$$

For a beam with double rectangular parts shown in Figure 4a, $\varphi'(x) = 0$ and $\varphi''(x) = 0$, there is:

$$\Theta = e_{31}\left(\xi + \frac{\delta}{2}\right)\frac{2}{\psi(L)}\varphi\left(\frac{L}{2}\right)\psi_1'\left(\frac{L}{2}\right) \quad (15)$$

For a beam with double trapezoidal parts shown in Figure 4b, $\varphi''(x) = 0$, there is:

$$\Theta = e_{31}\left(\xi + \frac{\delta}{2}\right)\frac{1}{\psi(L)}\left[2\varphi\left(\frac{L}{2}\right)\psi'\left(\frac{L}{2}\right) + \varphi'(L)\psi(L)\right] \quad (16)$$

For a beam with concave parabolic shape shown in Figure 4c, $\varphi''(x)$ is a constant, and there is:

$$\Theta = e_{31}\left(\xi + \frac{\delta}{2}\right)\frac{1}{\psi(L)}\left\{2\varphi\left(\frac{L}{2}\right)\psi'\left(\frac{L}{2}\right) + \varphi'(L)\psi(L) + \varphi''\left[\int_0^{\frac{L}{2}} \psi(x)dx - \int_{\frac{L}{2}}^{L} \psi(x)dx\right]\right\} \quad (17)$$

From calculation of the converting factors corresponding to the above three shapes with the same width at the clamped end, it can be seen that, the double trapezoidal shape is better than the rectangular, whereas the concave parabolic shape is better than the double trapezoidal.

For an open circuit where there is no electric current, there is:

$$V_{12} = \frac{\Theta}{C_p} \quad (18)$$

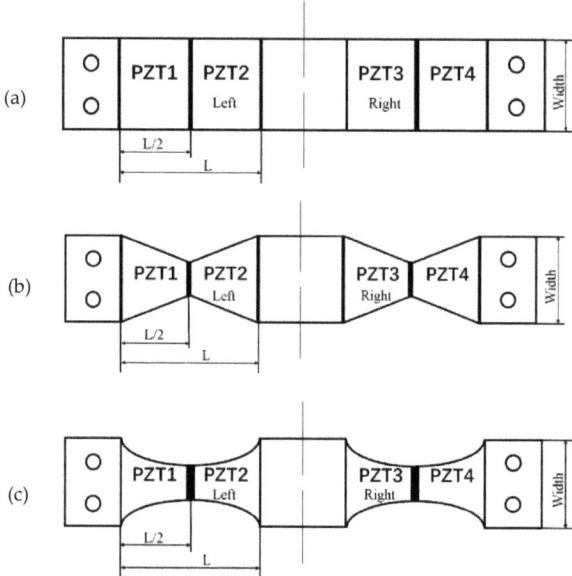

Figure 4. Three kinds of shapes. (**a**) Rectangular. (**b**) Trapezoidal. (**c**) Concave parabolic.

Substituting it into the first governing equation mentioned above, leads to:

$$m\ddot{w}_m + c\dot{w}_m + \left(k + \frac{\Theta^2}{C_p}\right)w_m = ma(t) \tag{19}$$

If $a(t) = A\sin\omega t$ where $\omega^2 = \frac{1}{m}\left(k + \frac{\Theta^2}{C_p}\right)$, the peak value of open voltage is:

$$\overline{V}_{12} = \frac{m\Theta}{2\zeta(kC_p + \Theta^2)}A \tag{20}$$

where ζ is damping ratio.

From this equation it can be seen that the equivalent stiffness $k^* = k + \frac{\Theta^2}{C_p}$ will be affected by the converting factor.

By solving this, when considering the coupling electric field, the open voltage can be obtained.

For a connected closed-circuit system with a load R, the circuit current can be expressed by $I(t) = \frac{V(t)}{R}$. The above equations can be rewritten as:

$$\begin{cases} m\ddot{w}_m + c\dot{w}_m + kw_m - \Theta V = ma(t) \\ \Theta\dot{w}_m + C_p\dot{V} - \frac{V}{R} = 0 \end{cases} \tag{21}$$

In order to analyze the influence of the beam width, we can obtain a simplified quasi-static model by ignoring the inertial part and the damping part of the Equation (8) as:

$$\begin{cases} kw_m - \Theta V = ma(t) \\ Q = \Theta w_m + C_p V \end{cases} \tag{22}$$

In this case, the beam is just an elastic element in the vibration system. The charges are obtained as:

$$Q = \frac{\Theta}{k} ma(t) + \left(\frac{\Theta^2}{k} + C_p\right) V \tag{23}$$

where $k = \frac{K_D}{\psi^2(L)} \int_0^L \varphi(x)[\psi''(x)]^2 dx$, $\Theta = e_{31}\left(\xi + \frac{\delta}{2}\right)\frac{1}{\psi(L)} \int_0^L \varphi(x)\psi''(x)dx$, $C_p = \frac{\varepsilon_{33} S}{\delta}$.

Taking the rectangular shape as an example $\varphi(x) = constant$, For an open circuit, there are:

$$Q = \frac{e_{31}\left(\xi + \frac{\delta}{2}\right)\frac{1}{\psi(L)}\int_0^L \psi''(x)dx}{\frac{K_D}{\psi^2(L)}\int_0^L [\psi''(x)]^2 dx} ma(t) \tag{24}$$

$$V = \frac{\Theta}{C_p} w_m \tag{25}$$

It can be seen roughly that, as the width of the beam becomes narrower, the converting factor and the stiffness of the beam become smaller simultaneously. From the formula, the amount of charge seems to be independent of the width of the beam. The voltage is different. Although the converting factor and the capacitance become smaller simultaneously, the vibration amplitude of the mass will increase so that the voltage will increase. However, this is at the cost of increasing space.

From the perspective of the electrode coverage area, the simplified quasi-static model shows that when the electrode only covers the high-stress area at the root of the beam, whose length is $L'(L' \ll L)$, the converting factor and the capacitance will become small as $\Theta = e_{31}\left(e + \frac{\delta}{2}\right)\frac{1}{\psi(L)}\int_0^{L'} \varphi(x)\psi''(x)dx$ and $C_p = \frac{\varepsilon_{33} S L'}{\delta L'}$ respectively due to the shorting of the integration area. Therefore, the voltage change is not too large. On the contrary, since the stiffness does not change, whereas the converting factor will become smaller due to the shorter integration region, and then the amount of charge will be reduced a lot.

Of course, the above discussion is just a rough analysis.

3. Experiments and Verifications

In order to verify the theoretical model, we conducted two kinds of experiments. One kind was for millimeter-scale structures, the other was for micro-scale structures.

The experimental setup includes a vibrating shaker controlled by a signal generator and a power amplifier in which alternative frequency and amplitude excitations can be provided, a dynamic signal analyzer was used to record output voltage, and an accelerometer was used to record the vibration acceleration of the shaker.

The open voltage outputs are recorded by a dynamic signal analyzer. Its internal resistance is 2 MΩ. This resistance is much higher than the impedance of harvester which is only about 10–20 kΩ.

For millimeter-scale structures, three shape kinds of double-clamped beam structures were designed and manufactured shown in Figure 5. One is rectangular as shown in Figure 5a, one is segmentally trapezoidal as shown in Figure 5b, and the other is concave parabolic as shown in Figure 5c. Two discrete pieces of polarized PZT-5H piezoelectric layer stick symmetrically on the upper surface of each half substructure beam. The substructure is made of copper material, whose Young's Modulus is relatively small to benefit from the lower frequency. A concentrated proof mass is fixed at the center of the double-clamped beam. Corresponding discrete silver layers as electrodes are covered on the upper surface of the PZT-5H piezoelectric layers. A continuous silver layer as a sharing electrode is fully covered on the lower surface of the piezoelectric layer of each half beam to conduct the two piezoelectric parts of a half beam structure.

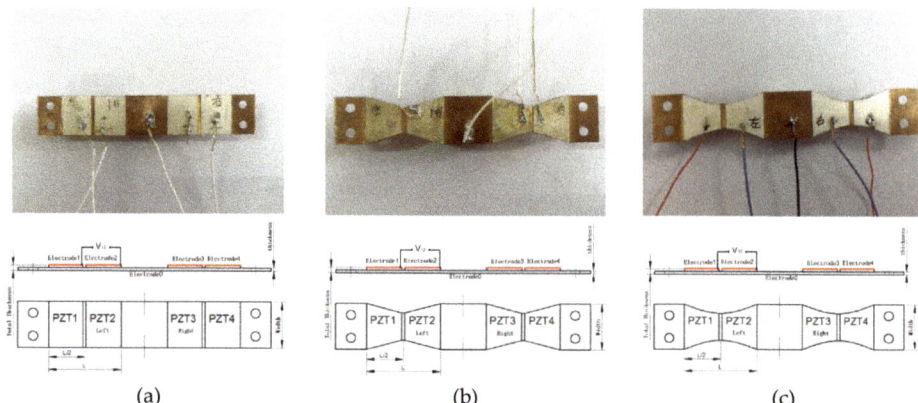

Figure 5. Three shape kinds of designed and manufactured double-clamped beam structures. (**a**) Rectangular, (**b**) Trapezoidal, (**c**) Concave parabolic.

The open voltage V_{12} is the voltage between the upper electrode 1 and electrode 2 corresponding to piezoelectric part 1 and piezoelectric part 2 by series connecting these two parts.

The material parameters of the substructure beam and the piezoelectric layer are listed in Table 1 where ρ_s and are mass densities of the substructure beam and piezoelectric layer, respectively. The geometric parameters of the harvester structure are listed in Table 2 where ζ is damping ratio, A is the amplitude of the excitation acceleration, L is the half length of all these three kinds of beam, a is the base width of all these three kinds of beam, b_{rect}, b_{trapez}, b_{parab} are widths at center ($x = \frac{L}{2}$) of the half beam corresponding to the rectangular, trapezoid, and parabolic shapes, respectively.

Table 1. Material parameters.

Parameters	Value
ρ_s	8300 kg/m^3
ρ_p	7450 kg/m^3
Y_s	131 GPa (10^9 N/m^2)
Y_p	60 GPa (10^9 N/m^2)
μ_s	0.35
ε_{33}	1470 ε_0
ε_0	8.854 × 10^{-12} F/m
e_{31}	11.16 C/m^2

Table 2. Geometric parameters.

Parameters	Value
h	0.3 mm
δ	0.2 mm
$2m_2$	17 g
ζ	0.013
A	2 m/s^2
L	23 mm
a	15 mm
b_{rect}	15 mm
b_{trapez}	12 mm
b_{parab}	12 mm

The damping ratio ζ is obtained by testing based on the principles of vibration mechanics [19]. When the excitation of the beam structure in the resonance state suddenly terminates, the vibration amplitude of the structure will attenuate as a logarithm function. This amplitude (pixel values) is measured by high-speed camera (Photron SA4) as shown in Figure 6. For n cycles apart, logarithmic decrement of amplitude obeys the following relationship $\frac{1}{n} ln \frac{A_i}{A_{i+n}} = \frac{2\pi\zeta}{\sqrt{1-\zeta^2}}$. By use of the measured data A_i and A_{i+n}, the damping ratio ζ can be calculated as 0.013.

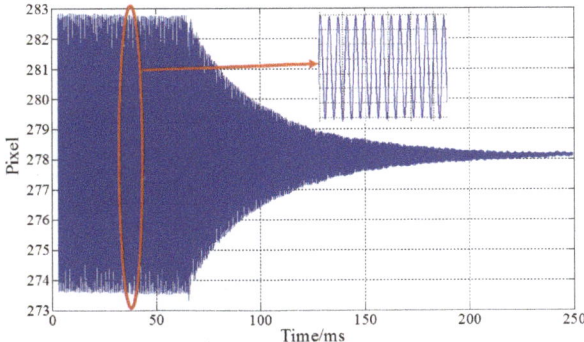

Figure 6. Measured amplitudes after excitation suddenly terminates.

For these three different shapes, by experimental and theoretical analysis, some open voltage outputs are obtained shown as in Figure 7. From the curves it can be found that the voltage amplitude of the concave parabolic shape is the maximum and its resonant frequency is minimum among these three kinds of shape, whereas the rectangular shape has a minimum voltage amplitude and maximum resonant frequency.

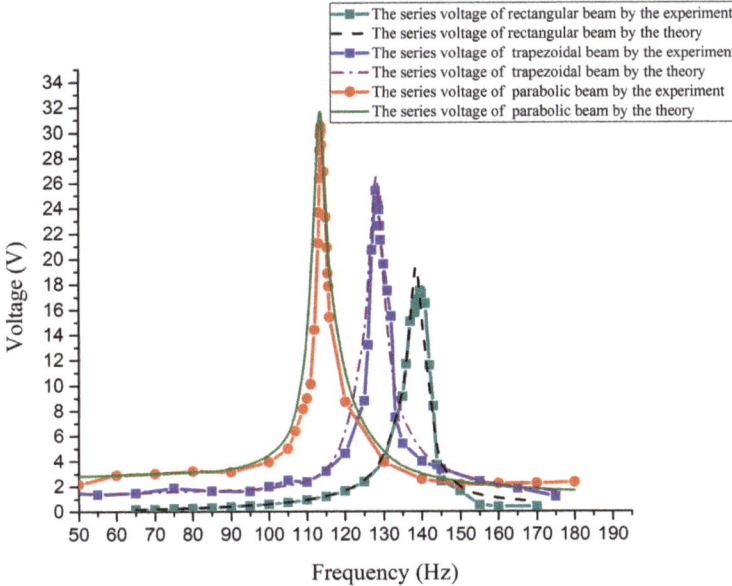

Figure 7. Open circuit voltage outputs versus exciting frequencies.

For micro-scale (MEMS) structures, two shape kinds of double-clamped beam structures were designed and fabricated with the MEMS process [20]. One kind of structure is a double-beam structure as shown in Figure 8, the other is single beam structure, as shown in Figure 9. Every kind of structure includes two kinds of beam shapes. One kind is a rectangular shape, the other is a concave trapezoidal shape. They are all fabricated with the MEMS process. The MEMS process includes: (a) the oxidizing of the silicon wafer into SiO2 and the patterning of the SiO2 layer; (b) the sputtering of the Ti layer and Pt layer; (c) depositing of the PZT thin film on the Pt/Ti/SiO2/Si substrate with the sol-gel method; (d) the fabricating of the top electrode with the sputtering method; (e) the corroding of the PZT thin film; (f) the freeing of beam structures by a dry-etching process.

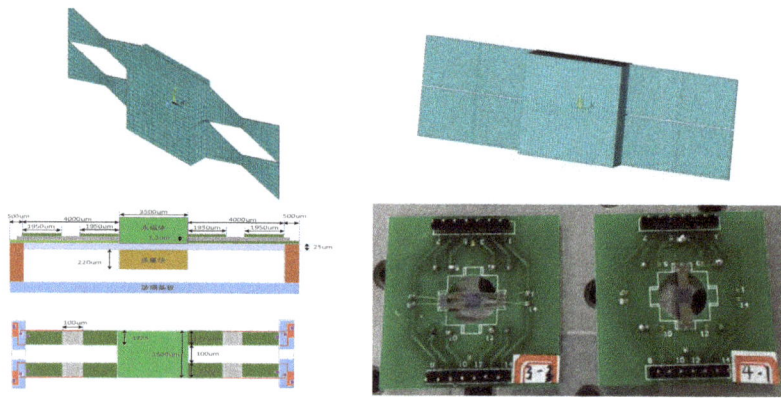

Figure 8. Double beams MEMS structures.

Figure 9. Single beam microelectromechanical system (MEMS) structures.

Because the micro-scale structure is fabricated by MEMS processing technology, the base beam (substructure) is made of silicon (Si) material which is compatible with this process.

The thickness of the substructure beam is 27.7 µm, its Young's modulus is 190 GPa, and its mass density is 2330 kg/m^3. The piezoelectric layer is PZT whose thickness is 1.3 µm, Young's modulus is 60 GPa, mass density is 7720 kg/m^3. The lengths of a half beam for both double beam and single beam are 4000 µm, the width at the fixed end of the single beam for both rectangular and trapezoid shapes is 2300 µm, and the width at the fixed end of the double beam is 1725 µm. Other geometric parameters are shown in Figure 9.

The open voltage and power outputs for a double-beam structure are shown in Figure 10.

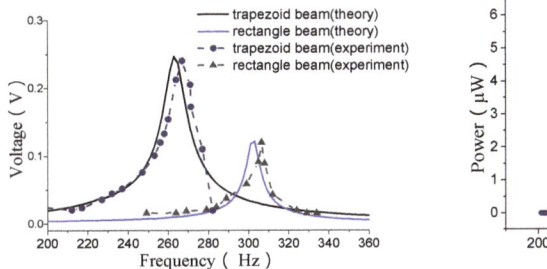

Figure 10. Open voltage and power outputs.

The power is calculated according to the optimal load resistance. The optimal load resistance is obtained through testing. The test is conducted for a loop as shown in Figure 11, where the load resistance is adjustable. A dynamic signal analyzer is used to test the voltage across the load resistance. The internal resistance R_i of the dynamic signal analyzer is very high, which is 2 MΩ, the load resistance R, the resistance of the energy harvester and R_i form a parallel connection. For the resonant frequency, the voltage will increase as the adjustable resistance increases to the open circuit voltage, as shown in Figure 12. At the beginning it increased very quickly. The output power can be calculated according to the voltage and adjustable load resistance by $p = \frac{V^2}{R}$. In the first stage, the output power increases with the increase of the load resistance. But after a peak value, it will decrease. The resistance corresponding the peak value is 12 kΩ which is called the optimal load resistance. This peak value of power is the power for the given frequency we obtain. In the same way, a series of power corresponding different frequencies as shown in Figure 10 can also obtained.

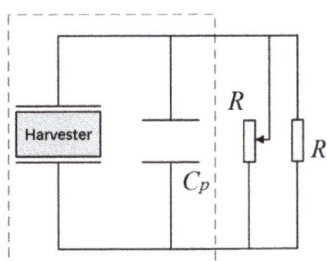

Figure 11. A loop circuit with an adjustable load resistance.

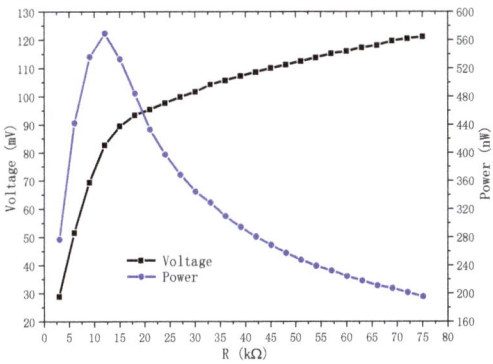

Figure 12. Curves of voltage and power.

The open voltage outputs for single beam structure are shown in Figure 13.

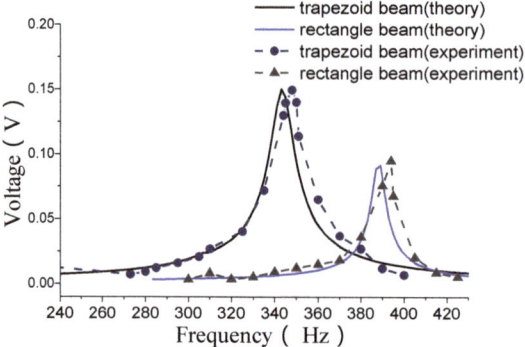

Figure 13. Open voltage outputs.

From these results, it can also be observed that, the energy outputs of a MEMS harvester with a trapezoidal-shaped beam are better than that with a rectangular-shaped beam. At the same time, the resonant frequencies will decrease.

4. Conclusions

In this work, two kinds of double-clamped piezoelectric energy harvester with different width shapes are investigated. In an experimental and analytical way, some electrical outputs are obtained. Not only for millimeter-scale structures but also for micro-scale (MEMS) structures, the theoretical results are in good agreement with those of the experiments. From these results, it can be observed that both the width shapes of the beams and the arrangement of the electrodes have a direct effect on the electrical outputs of piezoelectric energy harvesters.

In future research, on the one hand, through MEMS process design and the processing, we hope to complete the analysis of the micro-scale parabolic beam structure; on the other hand, we hope to explore the effects of thickness shape (such as wedge, tapering, etc.) on the electric output of the energy harvester.

Author Contributions: Conceptualization, L.J. and S.G.; validation, X.Z. and Q.W.; investigation, L.J.; writing—original draft preparation, S.G. All authors have read and agreed to the published version of the manuscript.

Funding: This research received no external funding.

Acknowledgments: We acknowledge Zezhang Li for providing us with some figures.

Conflicts of Interest: The authors declare no conflict of interest.

References

1. Roundy, S.; Leland, E.S.; Baker, J.; Carleton, E.; Reilly, E.; Lai, E.; Otis, B.; Rabaey, J.M.; Wright, P.K.; Sundararajan, V. Improving power output for vibration-based energy scavengers. *IEEE Pervasive Comput.* **2005**, *4*, 28–36. [CrossRef]
2. Ayed, S.B.; Abdelkefi, A.; Najar, F.; Hajj, M.R. Design and performance of variable-shaped piezoelectric energy harvesters. *J. Intell. Mater. Syst. Struct.* **2014**, *25*, 174–186. [CrossRef]
3. Zhang, G.; Gao, S.; Liu, H.; Niu, S. A low frequency piezoelectric energy harvester with trapezoidal cantilever beam: Theory and experiment. *Microsyst. Technol.* **2017**, *23*, 3457–3466. [CrossRef]
4. Zhang, G.; Gao, S.; Liu, H. A utility piezoelectric energy harvester with low frequency and high-output voltage: Theoretical model, experimental verification and energy storage. *AIP Adv.* **2016**, *6*, 095208. [CrossRef]

5. Jackson, N.; O'Keeffe, R.; Waldron, F.; O'Neill, M.; Mathewson, A. Evaluation of low-acceleration MEMS piezoelectric energy harvesting devices. *Microsyst. Technol.* **2014**, *20*, 671–680. [CrossRef]
6. Benasciutti, D.; Moro, L.; Zelenika, S.; Brusa, E. Vibration energy scavenging via piezoelectric bimorphs of optimized shapes. *Microsyst. Technol.* **2010**, *16*, 657–668. [CrossRef]
7. Mateu, L.; Moll, F. Optimum piezoelectric bending beam structures for energy harvesting using shoe inserts. *J. Intell. Mater. Syst. Struct.* **2005**, *16*, 835–845. [CrossRef]
8. Muthalif, A.G.A.; Nordin, N.H.D. Optimal piezoelectric beam shape for single and broadband vibration energy harvesting: Modeling, simulation and experimental results. *Mech. Syst. Signal Process.* **2015**, *54–55*, 417–426. [CrossRef]
9. Tabatabaei, S.M.K.; Behbahani, S.; Rajaeipour, P. Multi-objective shape design optimization of piezoelectric energy harvester using artificial immune system. *Microsyst. Technol.* **2016**, *22*, 2435–2446. [CrossRef]
10. Jin, L.; Gao, S.; Zhou, X.; Zhang, G. The effect of different shapes of cantilever beam in piezoelectric energy harvesters on their electrical output. *Microsyst. Technol.* **2017**, *23*, 4805–4814. [CrossRef]
11. Chun, J.; Kishore, R.A.; Kumar, P.; Kang, M.G.; Kang, H.B.; Sanghadasa, M.; Priya, S. Self-Powered Temperature-Mapping Sensors Based on Thermo-Magneto-Electric Generator. *ACS Appl. Mater. Interfaces* **2018**, *10*, 10796–10803. [CrossRef] [PubMed]
12. Annapureddy, V.; Na, S.-M.; Hwang, G.-T.; Kang, M.-G.; Sriramdas, R.; Palneedi, H.; Yoon, W.-H.; Hahn, B.-D.; Kim, J.-W.; Ahn, C.-W.; et al. Exceeding milli-watt powering magneto-mechano-electric generator for standalone-powered electronics. *Energy Environ. Sci.* **2018**, *11*, 818–829. [CrossRef]
13. Li, P.; Gao, S.; Niu, S.; Liu, H.; Cai, H. An analysis of the coupling effect for a hybrid piezoelectric and electromagnetic energy harvester. *Smart Mater. Struct.* **2014**, *23*, 065016.1–065016.17. [CrossRef]
14. Li, P.; Gao, S.; Cai, H.; Wang, H. Coupling Effect Analysis for Hybrid Piezoelectric and Electromagnetic Energy Harvesting from Random Vibrations. *Int. J. Precis. Eng. Manuf.* **2014**, *15*, 1915–1924. [CrossRef]
15. Zhou, X.; Gao, S.; Liu, H.; Guan, Y. Effects of introducing nonlinear components for a random excited hybrid energy harvester. *Smart Mater. Struct.* **2017**, *26*, 015008. [CrossRef]
16. Lu, Z.; Wen, Q.; He, X.; Wen, Z. A Nonlinear Broadband Electromagnetic Vibration Energy Harvester Based on Double-Clamped Beam. *Energies* **2019**, *12*, 2710. [CrossRef]
17. Gao, C.; Gao, S.; Liu, H.; Jin, L.; Lu, J. Electret Length Optimization of Output Power for Double-End Fixed Beam Out-of-Plane Electret-Based Vibration Energy Harvesters. *Energies* **2017**, *10*, 1122.
18. Meeker, T.R. Publication and proposed revision of ANSI/IEEE standard 176-1987. *IEEE Trans. Ultrason. Ferroelectr. Freq. Control* **1996**, *43*, 717–773.
19. Li, P.; Gao, S.; Cai, H.; Cui, Y. Design, fabrication and performances of MEMS piezoelectric energy harvester. *Int. J. Appl. Electromagn. Mech.* **2015**, *47*, 125–139. [CrossRef]
20. Cui, Y.; Yu, M.; Gao, S.; Kong, X.; Gu, W. Fabrication and characterization of a piezoelectric energy harvester with clampedclamped beams. *AIP Adv.* **2018**, *8*, 055028. [CrossRef]

© 2020 by the authors. Licensee MDPI, Basel, Switzerland. This article is an open access article distributed under the terms and conditions of the Creative Commons Attribution (CC BY) license (http://creativecommons.org/licenses/by/4.0/).

Article

Finite Element Modeling and Performance Evaluation of Piezoelectric Energy Harvesters with Various Piezoelectric Unit Distributions

Cong Du [1], Pengfei Liu [1,*], Hailu Yang [2,3,*], Gengfu Jiang [1], Linbing Wang [4] and Markus Oeser [1]

1. Institute of Highway Engineering (ISAC), RWTH Aachen University, 52074 Aachen, Germany; du@isac.rwth-aachen.de (C.D.); gengfu.jiang@rwth-aachen.de (G.J.); oeser@isac.rwth-aachen.de (M.O.)
2. National Center for Materials Service Safety, University of Science and Technology Beijing (USTB), Beijing 100083, China
3. Research and Development Center of Transport Industry of New Materials, Technologies Application for Highway Construction and Maintenance, Beijing 100088, China
4. Joint USTB Virginia Tech Lab on Multifunctional Materials, University of Science and Technology Beijing (USTB), Beijing 100083, China; wangl@vt.edu
* Correspondence: liu@isac.rwth-aachen.de (P.L.); yanghailu@ustb.edu.cn (H.Y.)

Citation: Du, C.; Liu, P.; Yang, H.; Jiang, G.; Wang, L.; Oeser, M. Finite Element Modeling and Performance Evaluation of Piezoelectric Energy Harvesters with Various Piezoelectric Unit Distributions. *Materials* **2021**, *14*, 1405. https://doi.org/10.3390/ma14061405

Academic Editor: Daniele Davino

Received: 9 February 2021
Accepted: 11 March 2021
Published: 14 March 2021

Publisher's Note: MDPI stays neutral with regard to jurisdictional claims in published maps and institutional affiliations.

Copyright: © 2021 by the authors. Licensee MDPI, Basel, Switzerland. This article is an open access article distributed under the terms and conditions of the Creative Commons Attribution (CC BY) license (https://creativecommons.org/licenses/by/4.0/).

Abstract: The piezoelectric energy harvester (PEH) is a device for recycling wasted mechanical energy from pavements. To evaluate energy collecting efficiency of PEHs with various piezoelectric unit distributions, finite element (FE) models of the PEHs were developed in this study. The PEH was a square of 30 cm × 30 cm with 7 cm in thickness, which was designed according to the contact area between tire and pavement. Within the PEHs, piezoelectric ceramics (PZT-5H) were used as the core piezoelectric units in the PEHs. A total of three distributions of the piezoelectric units were considered, which were 3 × 3, 3 × 4, and 4 × 4, respectively. For each distribution, two diameters of the piezoelectric units were considered to investigate the influence of the cross section area. The electrical potential, total electrical energy and maximum von Mises stress were compared based on the computational results. Due to the non-uniformity of the stress distribution in PEHs, more electrical energy can be generated by more distributions and smaller diameters of the piezoelectric units; meanwhile, more piezoelectric unit distributions cause a higher electrical potential difference between the edge and center positions. For the same distribution, the piezoelectric units with smaller diameter produce higher electrical potential and energy, but also induce higher stress concentration in the piezoelectric units near the edge.

Keywords: piezoelectric energy harvester; finite element simulation; piezoelectric unit distributions; electrical potential and energy; von Mises stress

1. Introduction

With the development of economy and society, the number of traffic loads on asphalt pavements increases in the recent years. During the service life of pavements, millions of axle's loads causes large amounts of wasted mechanical energy. As a remedy, new technologies have been developed and applied to recycle the energy from urban roads and highways by converting them to other types of energy resources. One such example is the energy harvesting technology using piezoelectric and magnetostrictive materials, which can convert the mechanical energy generated by the traffic loads to electrical energy [1–6]. Among others, piezoelectric energy harvester (PEH) shows significant advantages for maintenance of energy output along with traffic flow without being influenced by weather, environmental temperature, and so on [7].

Many researches have been conducted focusing on the piezoelectric materials for the energy harvesting [8,9]. For examples, Anton and Sodano [10] reviewed the piezoelectric materials used for energy saving, including the lead zirconate titanate, also known as piezoelectric ceramic (PZT), poly(vinylidene fluoride) (PVDF) [11], and the macro-fiber composite (MFC) [12]. They found that the PZT materials were the most commonly used piezoelectric materials in the energy harvesting due to its high efficiency. Besides, many scholars [13,14] developed the fiber-based materials in which the PZT fibers with various diameters were consisted. The results showed that a relatively small fiber-based piezoelectric power harvester can supply useable amounts of power from cyclic strain vibration in the local environment. However, the piezoelectric ceramic is very brittle, and its piezoelectric feature only works under undamaged strain conditions. In addition, the stiffness of the piezoelectric ceramic is much higher than pavement materials, which could cause the stress concentration behavior and therefore induce damages.

To address this issue, numerous researches packaged the PZT materials into PEHs using various package materials and shapes to reduce the damages and improve the energy harvesting efficiency [15]. For examples, Yesner et al. [16] developed a bridge transducer based on the cymbal design, which exhibits higher energy generation in horizontal loading condition comparing with the conventional design. Moure et al. [17] tested the electrical energy conversion of piezoelectric cymbals with 29-mm diameter, and the piezoelectric cymbal were integrated into asphalt pavements to evaluate the energy harvesting ability in normal traffic conditions. Zhao et al. [18] developed the multilayer PZT-5 stack configuration for the civil infrastructure application, and the results indicates that the analytical and numerical predications used in their research exhibited very good agreement with the experimental measurements. Xiong et al. [19] developed the PEH prototype that consists of PZT disks sealed in a protective package, and the PEHs were fabricated in the pavement to evaluate their feasibility of energy harvesting. The results showed that the energy harvesters are highly relevant to the axle configuration and magnitude of passing vehicles. Liu et al. [20,21] investigated the influence of PEHs on the structural response of asphalt pavement, which provides the basic information for improving the design of PEHs in application in pavement engineering. Zhang et al. [22] proposed a new packaging method using monomer cast nylon and epoxy resin as the main protective materials for the PEHs. The normalized output power of the PEH system was found to rely on the normalized electrical resistive load and normalized embedded depth. To further improve the efficiency of the piezoelectric energy collection, Yang et al. [23,24] developed the PEH by laboratory and in-situ tests. The PZT-5H was selected to serve as the core piezoelectric units within PEHs. Their researches successfully recycled energy from the pavements, and thus provided a useful guideline for optimization of PEH system in practical roadway applications.

The abovementioned researches provide a general overview on the piezoelectric energy harvesting on asphalt pavements. It is foreseeable that there is a high potential to harvest kinetic energy from pavement using the PEHs. However, with consideration of the laboratory cost and convenience, the design of the PEHs in current researches is still mainly based on empirical approaches, and the PEHs with higher efficiency need to be further developed.

To this end, numerical technologies, like finite element (FE) method, provide possibilities to researchers to massively and comprehensively investigate the mechanical and electrical responses of the PEHs. Zhao et al. [25] designed a cymbal for harvesting energy from asphalt pavement, and the efficiency and coupling effects with pavement of cymbals with various sizes were discussed through FE simulations. As an initial research, the FE models of the cymbals in their study were simplified to some extent, which were not directly applied in realistic pavements. Yang et al. [26] evaluated the efficiencies of PEHs in different locations in asphalt pavement based on the FE simulation, the results can be used to guide the future PEHs applications in pavement engineering. However, the PEH in Yang's simulation was simplified as a homogeneous structure, and the details about the internal structure of the PEH were ignored.

2. Objectives and Outlines

In this study, the piezoelectric energy harvesting efficiency of the PEH is further investigated using FE method. A flowchart as shown in Figure 1 is provided to clearly exhibit the simulation and analysis process in this study.

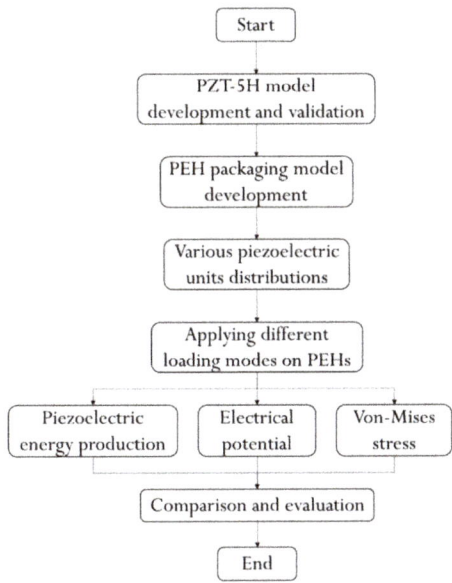

Figure 1. Flowchart of the study.

The internal structure of the PEH was reconstructed based on the authors' previous research [23,24]. Various distributions of the piezoelectric units inside the PEH were modeled in FE software ABAQUS (version 2017). A total of three loading modes were applied on the PEH models to simulate the realistic non-uniform traffic loading conditions. According to the computational results, including the total electrical energy and electrical potential, the energy collecting efficiencies of the PEH were evaluated. In addition, the mechanical performances of the piezoelectric units were analyzed. At the end of this study, recommendations for the future PEH design were proposed according to the computational results.

3. Methodology

3.1. Foundation of Piezoelectric Theory

According to [27,28], the basic equations for the piezoelectric linear medium in this numerical study are defined as below

$$\sigma_{ij} = D^E_{ijkl}\varepsilon_{kl} - e^{\varphi}_{mij}E_m \tag{1}$$

$$q_i = e^{\varphi}_{ijk}\varepsilon_{jk} + D^{\varphi(\varepsilon)}_{ij} \tag{2}$$

where σ_{ij} and ε_{ij} are stress and strain components, Pa and -, respectively; q_i are the electrical flux components, V·m; D_{ijkl} are the material stiffness, Pa; e^{φ}_{mij} are piezoelectric constant, C/m^2; D^{φ}_{ij} are the dielectric constants, C/(V·m); E_m is the electrical fields, V/m. In the above equations, the superscripts E and ε above a particular property indicate that the property is defined at zero electrical gradient and at zero strain, respectively. For the piezoelectric effects, two working modes are defined for the piezoelectric materials, which, respectively, are 3-1 mode and 3-3 mode [26], as shown in Figure 2.

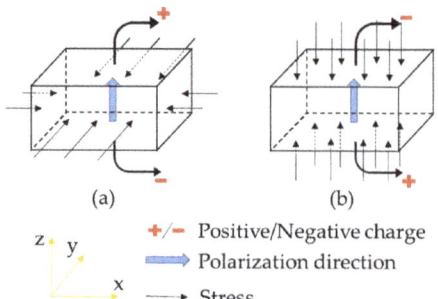

Figure 2. Working modes of the piezoelectric materials: (**a**) 3-1 mode; (**b**) 3-3 mode.

The 3-1 mode refers to that the stress components are perpendicular to the polarization direction of the piezoelectric materials. The 3-3 stands for that the stress component is parallel to the polarization direction.

3.2. Prototype of the PEH

The structure of the PEH used in this study is based on the authors' previous researches [23,24,26]. The PEH was designed as a box and was buried in asphalt pavements, which is shown in Figure 3a. The detail of PEH inside design is shown in Figure 3b. The components within the PEH include the packaging materials, the packaging materials, piezoelectric units and internal circuit board. The packaging material for the PEH was PA66-GF30, which was a type of nylon reinforced with 30% glass fiber. The PA66-GF30 was selected to serve as the upper and lower protective layers, in which the upper layer directly undertook the vehicle load, and the ground reaction force was supported by the lower layer. It was selected for the protective packaging of the PEH owing to its high toughness, load resistance, strength, and resistance to repeated shocks. A rubber gasket was employed between the upper and lower layers, which can prevent water leakage and reset the protective layers after loading. The piezoelectric ceramics are stacked to serve as the piezoelectric units between upper and lower layers, which are the core components of the PEHs. Within this internal circuit board, each power unit is connected to a full bridge rectifier and switched to an output bus after rectifying to reduce the adverse effects of uneven force [23], and these full bridge rectifiers are connected in parallel to output the generated voltage, as presented in Figure 3c.

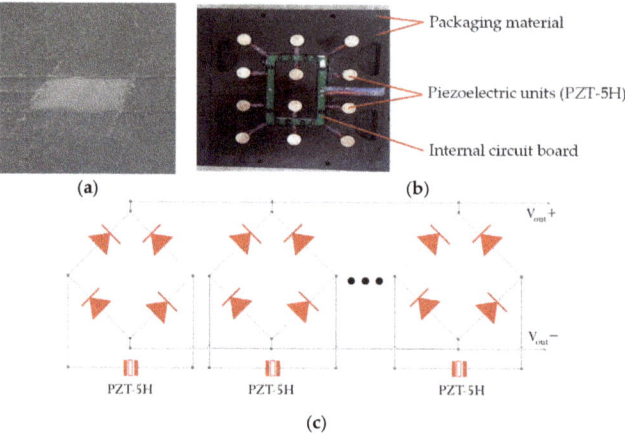

Figure 3. Prototype of the piezoelectric energy harvester (PEH) [23,24]: (**a**) PEH installation in pavement; (**b**) Details of PEH inside design; (**c**) Connection of the piezoelectric units.

As soon as vehicles pass cross on the PEH, the electrical energy could be produced. In terms of the contact patch of the tires and the thickness of asphalt pavements, the PEH was designed to have a square shape with a side length of 30 cm, and its thickness was 7 cm. The piezoelectric units are cylindrical structures with diameter of 2 cm and height of 2.25 cm.

3.3. Numerical Modeling of Piezoelectric Unit and Verification

Based on the research discovery from Cook-Chennault [29], the 3-3 working mode can achieve a higher energy conversion for PZT materials. According to the preliminary researches from Yang [30], the PZT-5H is a polycrystal made by lead titanate, lead zirconate and lead dioxide, which has a relative higher piezoelectric coefficients and compressive strength. Hence, in this study, three plates of PZT-5H with a thickness of 0.75 cm were electrically connected in parallel and the two adjacent contact surfaces have the same polarity, as shown in Figure 4a. Some parameters of the PZT-5H provided by the producer are listed in Table 1.

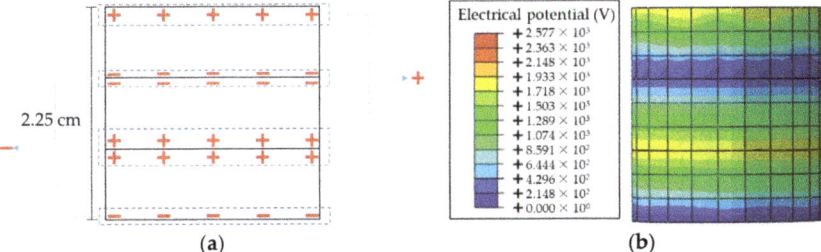

Figure 4. Piezoelectric unit used in this study: (**a**) Structure of piezoelectric unit; (**b**) Example of computational result.

Table 1. Basic parameters for PZT-5H.

Parameters	Value
Density (kg/m^3)	7500
Poisson's ratio	0.3
Electromechanical coupling factor k_p	0.65
Mechanical quality factor Q_m	70
Curie temperature Tc (°C)	200

Before assembling the piezoelectric units into the PEH in the simulation, it is necessary to create the FE model of the piezoelectric unit first and verify the reliability of simulating its piezoelectric performance. To this end, the laboratory compressive loads were applied on the piezoelectric unit. The test was performed by the universal servo hydraulic test device (Cooper HYD25-II), which can randomly set the temperature and provide sinusoidal loads. During the tests, the sinusoidal loadings ranged from 1 to 6.5 kN with the interval of 0.5 kN were applied under loading frequency of 10 Hz and temperature of 20 °C [23]. Meanwhile, the FE model of the piezoelectric units was established in ABAQUS. The loading and boundaries conditions were defined as same as laboratory ones. The material parameters will be introduced in the next section. One example of the computational result is shown in Figure 4b. The distribution of the electrical potential is illustrated. In addition, Figure 5 compares the values of open-circuit current voltage from laboratory and the electrical potential from simulation.

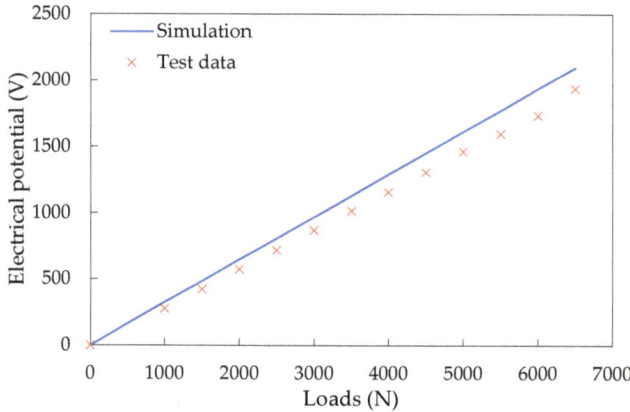

Figure 5. Validation of the piezoelectric unit simulation.

The results show that the numerical results are consistent with the experimental results within this loading range. Therefore, the developed FE model of the piezoelectric unit can effectively predict its piezoelectric performance.

3.4. Development of PEH Finite Element Model

In the FE simulation of the PEH, the packaging and piezoelectric materials were modeled. To clearly exhibit the overview for a PEH FE model, Figure 6 shows the detailed constituents of the PEH model. In this model, tie bonding was assumed between packaging materials and between package and piezoelectric units, and therefore no slips and separations will occur.

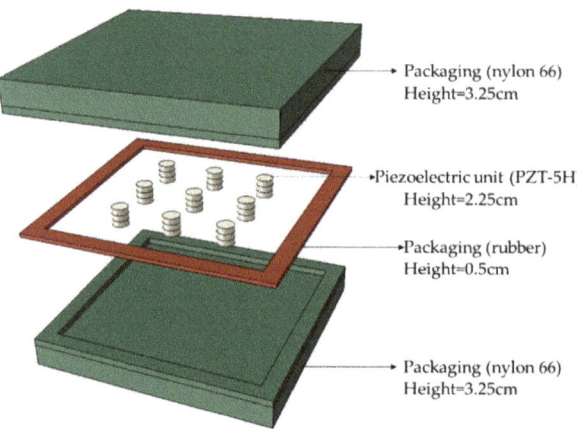

Figure 6. Overview of a PEH finite element model.

To deeply investigate the efficiency of PEHs with different piezoelectric unit distributions, six different PEH FE models were developed as presented in Figure 7. To maximally utilize the space, the distributions of the piezoelectric units were designed in matrixes by 3×3, 3×4 and 4×4, respectively. According to previous researches [23,24,26], the cross section area of the piezoelectric units is related to the electrical potential; therefore, the total cross section area of the piezoelectric units was controlled in this study. For the distribution of 3×3 with diameter of 2 cm, 3×4 with diameter of 1.73 cm and 4×4 with diameter of 1.5 cm, the total cross section area of the units was 28.27 cm^2. For the distribution of 3×3

with diameter of 2.3 cm, 3 × 4 with diameter of 2 cm and 4 × 4 with diameter of 1.73 cm, the total cross section area of the units was 37.7 cm².

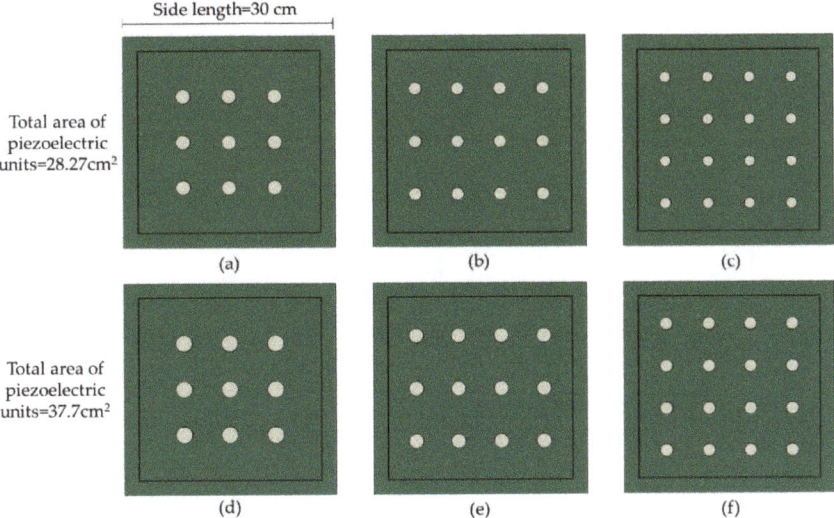

Figure 7. Piezoelectric unit distributions in PEH by: (**a**) 3 × 3 with diameter = 2 cm; (**b**) 3 × 4 with diameter = 1.73 cm; (**c**) 4 × 4 with diameter = 1.5 cm; (**d**) 3 × 3 with diameter = 2.3 cm; (**e**) 3 × 4 with diameter = 2 cm; (**f**) 4 × 4 with diameter = 1.73 cm.

The elastic properties of PA66-GF30 and rubber were defined as typical values [31,32]. The piezoelectric properties for piezoelectric units (PZT-5H) were defined according to [27]. The parameters used in the simulation are listed in Table 2.

Table 2. Model parameters.

	PZT-5H	PA66-GF30	Rubber
Elastic constants (Pa)	$C_{11} = 12.6 \times 10^{10}$ $C_{12} = 5.50 \times 10^{10}$ $C_{13} = 5.30 \times 10^{10}$ $C_{33} = 11.7 \times 10^{10}$ $C_{44} = 3.53 \times 10^{10}$	–	–
Piezoelectric constants (C/m²)	$e_{31} = -6.5$ $e_{33} = 23.3$ $e_{15} = 17.0$	–	–
Dielectric constants (C/(V·m))	$\varepsilon_{11} = 1.511 \times 10^{-8}$ $\varepsilon_{33} = 1.301 \times 10^{-8}$	–	–
Elastic modulus (Pa)	–	5.9×10^9	8×10^6
Poisson's ratio	–	0.35	0.47

Based on a comprehensive mesh study, the element types for the packaging materials and piezoelectric units were C3D8 with size of 2 cm and C3D8E with size of 0.25 cm, respectively. To simulate the realistic loading conditions of the PEH in pavements, the bottom and sides of the PEH were restricted in vertical and horizontal directions, respectively. The uniform pressure was applied on the top of the PEH. To consider different traffic loading conditions from the moving vehicles, Figure 8 exhibits three modes of pressure loadings, including the full loading, half loading, and quarter loading.

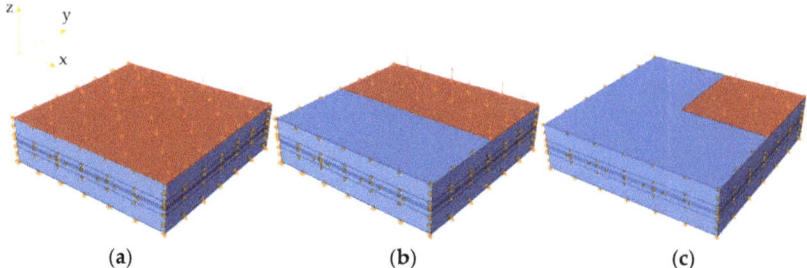

Figure 8. Loading modes on the PEH: (**a**) full loading; (**b**) half loading along y-direction; (**c**) quarter loading.

For the distribution of 3 × 4, the half loading modes were, respectively, applied along x-direction (the loading area can fully cover 6 piezoelectric units) and along y-direction (the loading area can fully cover 4 piezoelectric units and partially cover another 4 piezoelectric units). The loading amplitude was 0.7 MPa.

4. Results and Discussion
4.1. Comparison of Piezoelectric Energy Production

According to [23], the electrical energy produced by the piezoelectric units can be calculated by

$$E_i = \frac{1}{2}\frac{d_z^2 \sigma_z^2 A_i h}{\varepsilon} \qquad (3)$$

where d_z is the piezoelectric coefficient in z-direction, C/N; σ_z is the stress in z-direction, Pa; ε is the dielectric constant, F/m; A_i and h are the cross section area and height of i-th piezoelectric unit, m^2 and m, respectively.

To quantitatively exhibit the piezoelectric energy harvesting, Figure 9 exhibits the extracted total electrical energy production in the PEHs under the three loading modes.

It can be observed that higher electrical energy can be produced by piezoelectric units with smaller diameters. The total electrical energy production is linearly related to the loading conditions, i.e., the PEHs under half and quarter loading modes produced around half and quarter energy of that produced under full loading mode. For the piezoelectric units in 3 × 4 distribution, the total energy productions under half loading along x-direction and y-direction are equivalent. In addition, even the total cross section areas of the piezoelectric units are equivalent, the electrical energy production still shows large difference when the PEHs are under the same loading conditions. For instance, when the total cross section areas of the piezoelectric units are 28.27 cm^2 (3 × 3 distribution with 2 cm diameter, 3 × 4 distribution with 1.73 cm diameter and 4 × 4 distribution with 1.5 cm diameter), however, they produced 2.15, 3.69, and 4.15 J electrical energy under full loading condition, respectively. This phenomenon can be explained by the stress distribution variations. According to Equation (3), the electrical energy is dependent on the stress response of the piezoelectric unit. Therefore, although the total cross section areas of the piezoelectric units in the three PEHs are equivalent, the stress conditions on units are different due to various distributions, and thus generate difference electrical energy.

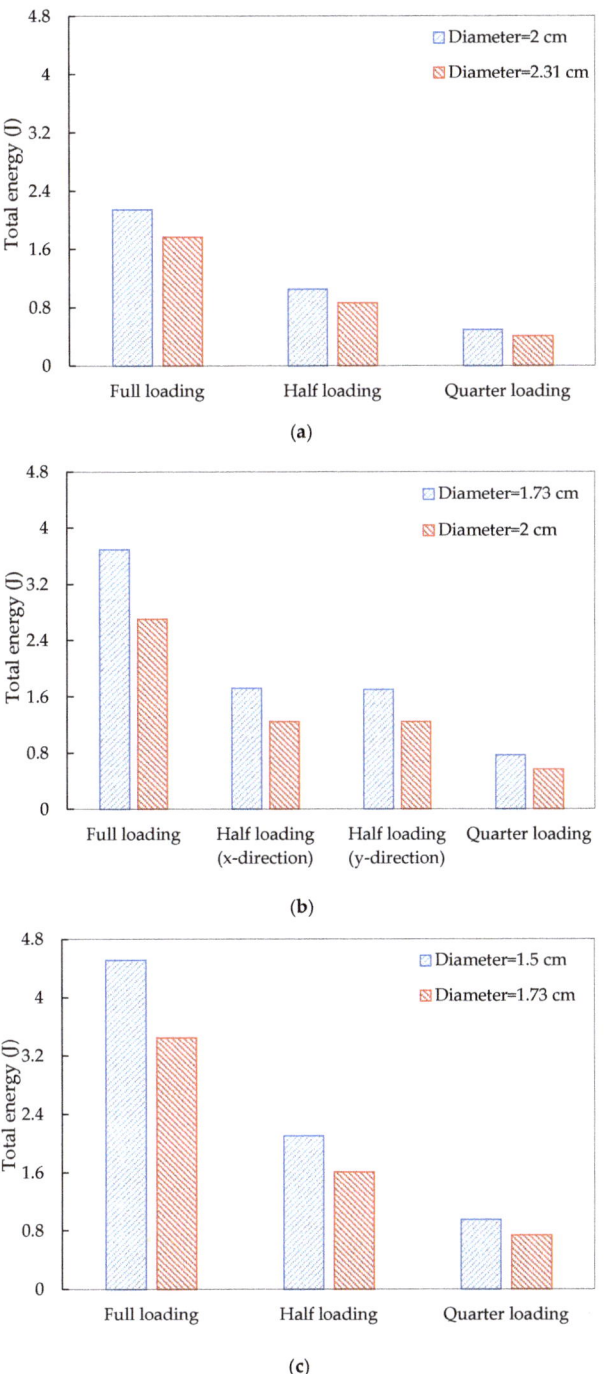

Figure 9. Total electrical energy production in the PEHs with different piezoelectric unit distribution: (**a**) 3 × 3; (**b**) 3 × 4; (**c**) 4 × 4.

4.2. Comparison of Electrical Potential

The electrical potential distributions within the piezoelectric units are, respectively, presented in Figures 10–12. Only the models with smaller diameters (2, 1.73, and 1.5 cm) are compared here under the three loading modes. It can be observed that the electrical potential exhibits extremely non-uniform distribution in the PEHs. Piezoelectric units near the edges show relatively higher electrical potential than those in the centers of the PEHs. The aforementioned phenomenon can be explained by the stress concentration in the piezoelectric units near the edge. The stiffness of the piezoelectric units (PZT-5H) is much higher than that of the packaging and sealing materials (PA66-GF30 and rubber), and therefore, higher stress concentrations mostly exist in the units near the edge. Under the half and quarter loading conditions, even though the electrical potential in the piezoelectric units beyond the loading area is almost zero, very high electrical potential still appears near the edge of PEHs. These results indicate that the current PEH design will cause large difference in the electrical potential between different piezoelectric units, especially under the non-uniform loading conditions. The similar distributions of the electrical potential can be found in the other three models with larger diameters (2.31, 2, and 1.73 cm).

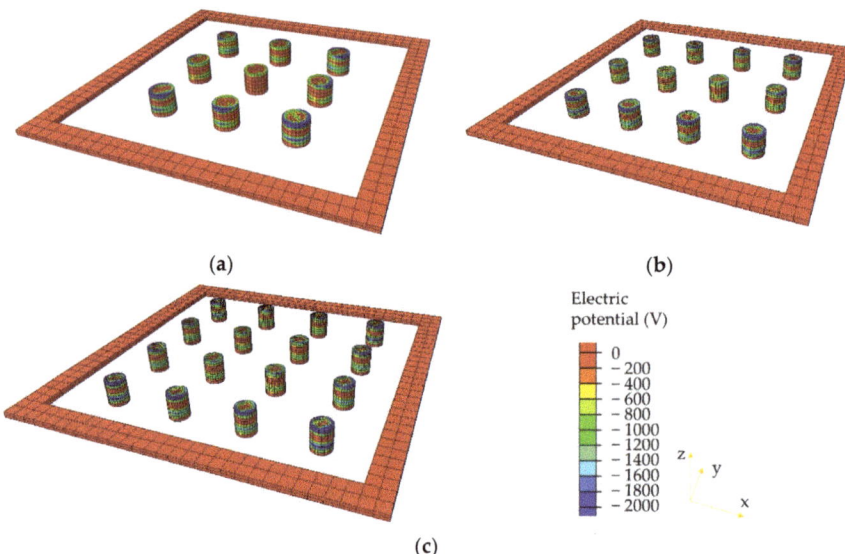

Figure 10. Electrical potential distributions in piezoelectric units under full loading condition: (**a**) 3 × 3 with diameter = 2 cm; (**b**) 3 × 4 with diameter = 1.73 cm; (**c**) 4 × 4 with diameter = 1.5 cm.

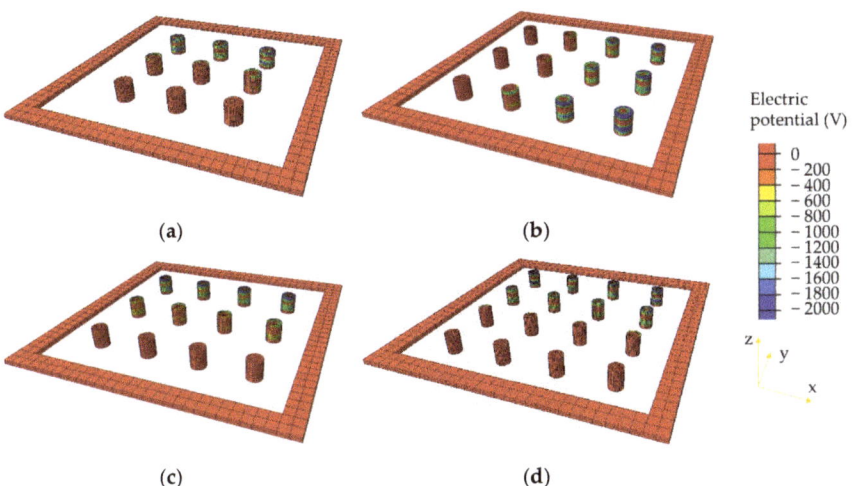

Figure 11. Electrical potential distributions in piezoelectric units under half loading condition: (**a**) 3 × 3 with diameter = 2 cm; (**b**) 3 × 4 with diameter = 1.73 cm, along x-direction; (**c**) 3 × 4 with diameter = 1.5 cm, along y-direction; (**d**) 4 × 4 with diameter = 1.5 cm.

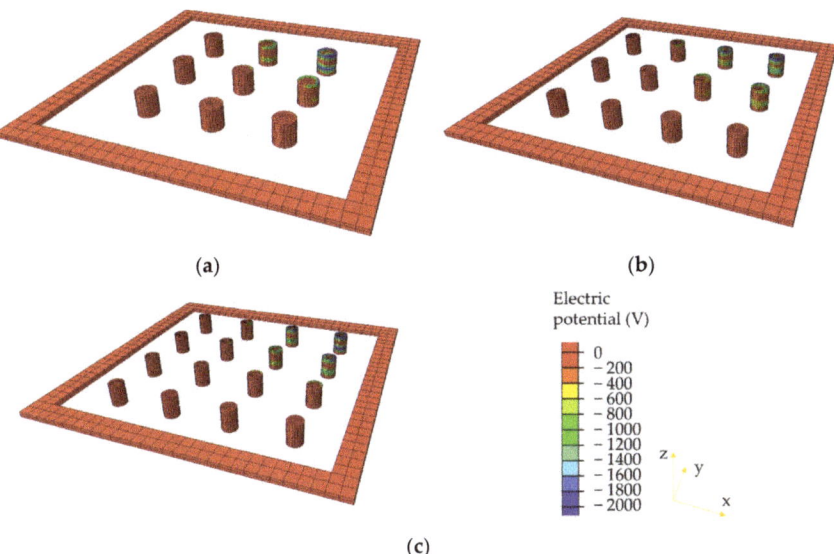

Figure 12. Electrical potential distributions in piezoelectric units under quarter loading condition: (**a**) 3 × 3 with diameter = 2 cm; (**b**) 3 × 4 with diameter = 1.73 cm; (**c**) 4 × 4 with diameter = 1.5 cm.

As mentioned above, high difference in the electrical potential between piezoelectric units will reduce the efficiency of the energy harvesting. To evaluate the efficiency, average electrical potentials in the PEHs under the three loading modes are shown in Figure 13. It can be observed that the 4 × 4 distribution has the largest electrical potential, followed by 3 × 4 and 3 × 3 distributions. The results indicate that the electrical potential is highly related to the number of the piezoelectric units in PEHs.

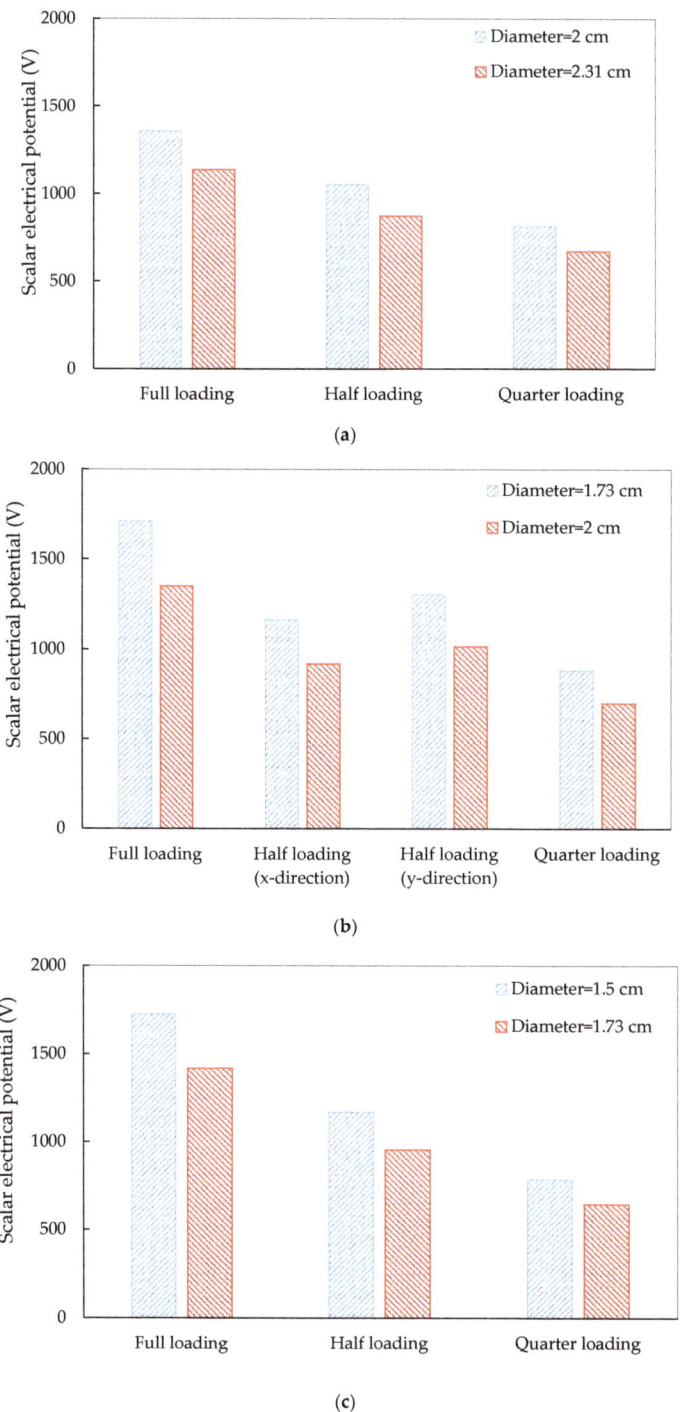

Figure 13. Electrical potential in the PEHs with different piezoelectric unit distributions: (**a**) 3 × 3; (**b**) 3 × 4; (**c**) 4 × 4.

In addition, the electrical potential difference between the piezoelectric units within one PEH is also an essential factor impacting the piezoelectric harvesting efficiency. In this study, the maximum and minimum electrical potentials are, respectively, produced by the piezoelectric units at the edge and center positions of the PEHs. The differences are listed in Table 3. Despite that the PEH with 4 × 4 distribution produced the highest average electrical potential in the piezoelectric units, largest potential differences can be observed in the electrical potentials under half and quarter loading modes. The PEHs with 3 × 4 piezoelectric unit distributions gain a better balance between the average electrical potential and the potential difference. In addition, the PEHs with 3 × 4 distribution under half loading mode along y-direction have higher electrical potential and smaller potential difference. This can be explained by the stress distributions. When the PEH bears the half loading along y-direction, four piezoelectric units near edge positions stand the stress concentration, which can generate higher electrical potential than that along x-direction. In addition, the difference between maximum and minimum electrical potential can be reduced by more piezoelectric units under the loading area. For the half loading along x-direction, six units were under the loading area; for the half loading along y-direction, eight units were under (or partially under) the loading area.

Table 3. Electrical potential difference in PEHs.

Piezoelectric Units	Loading Modes	Maximum Electrical Potential (V)	Minimum Electrical Potential (V)	Difference (V)
3 × 3 Diameter = 2 cm	Full loading	1728	589	1139
	Half loading	1777	296	1481
	Quarter loading	1821	144	1677
3 × 3 Diameter = 2.31 cm	Full loading	1487	421	1066
	Half loading	1528	206	1322
	Quarter loading	1562	99	1463
3 × 4 Diameter = 1.73 cm	Full loading	2038	1207	831
	Half loading (x-direction)	2108	180	1928
	Half loading (y-direction)	2079	598	1481
	Quarter loading	2130	86	2044
3 × 4 Diameter = 2 cm	Full loading	1599	976	623
	Half loading (x-direction)	1643	150	1493
	Half loading (y-direction)	1628	488	1140
	Quarter loading	1666	73	1593
4 × 4 Diameter = 1.5 cm	Full loading	2253	1209	1044
	Half loading	2308	188	2120
	Quarter loading	2346	20	2326
4 × 4 Diameter = 1.71 cm	Full loading	1859	980	879
	Half loading	1904	139	1765
	Quarter loading	1934	11	1923

4.3. Comparison of Von Mises Stress

In engineer practice of PEH design, the mechanical responses of the piezoelectric units should also be considered to prevent or reduce the damages. To this end, the maximum von Mises stress in the piezoelectric units are derived and shown in Tables 4 and 5. The two tables, respectively, include the piezoelectric units with smaller and larger diameters, i.e., the total cross section areas of the piezoelectric units, respectively, are 28.37 cm^2 for Table 4 and 37.7 cm^2 for Table 5. The maximum von Mises stress always exists near the edge of the PEHs. From these results, it can be observed that the highest von Mises stress occurs in the

PEH with piezoelectric units in 3 × 3 distribution, which indicates that less piezoelectric units would induce higher von Mises stress. However, the piezoelectric units with 3 × 4 distributions have slightly lower von Mises values than that with 4 × 4 distributions. This can be explained by the detailed spatial arrangement of the piezoelectric units in the two PEHs. As one can see in Figure 7, the piezoelectric units with 4 × 4 distribution were closer to the edge than that with 3 × 4 distribution, and hence higher stress concentration exists in the 4 × 4 distribution. For the same piezoelectric unit distribution, the larger cross section area can effectively reduce the von Mises stress.

Table 4. Maximum von Mises stress in PEHs. (total cross section area = 28.37 cm^2).

Distribution	Diameter (cm)	Von Mises Stress (MPa)		
		Full Loading	Half Loading	Quarter Loading
3 × 3	2	295	288	260
3 × 4	1.73	158	136 (x-direction) 133 (y-direction)	134
4 × 4	1.5	175	173	173

Table 5. Maximum von Mises stress in PEHs. (total cross section area = 37.7 cm^2).

Distribution	Diameter (cm)	Von Mises Stress (MPa)		
		Full Loading	Half Loading	Quarter Loading
3 × 3	2.31	212	211	150
3 × 4	2	126	137 (x-direction) 124 (y-direction)	113
4 × 4	1.73	125	117	119

4.4. Evaluation of the Piezoelectric Effect

To evaluate the piezoelectric effects of different PEHs, a radar chart was provided in Figure 14, in which the output electrical energy, electrical potential, potential difference, and von Mises stress of the PEHs under the full loading mode were exhibited.

According to previous researches [23,24], higher electrical energy and potential are required for piezoelectric energy harvesting. In addition, larger von Mises stress and potential difference can, respectively, increase the damage behavior of the piezoelectric units and decrease the energy harvesting efficiency. It can be observed that the PEHs with more piezoelectric units can produce higher electrical energy and potential, and meanwhile reduce the von Mises stress concentration and potential difference. However, the PEHs with 4 × 4 distributions experience higher von Mises stress and potential difference than that with 3 × 4 distributions, which could be caused by the spatial locations of the piezoelectric units. Amongst the six PEHs, the 3 × 4 distribution with smaller cross section of the piezoelectric units achieves a better balance between electrical energy harvesting and stress concentration.

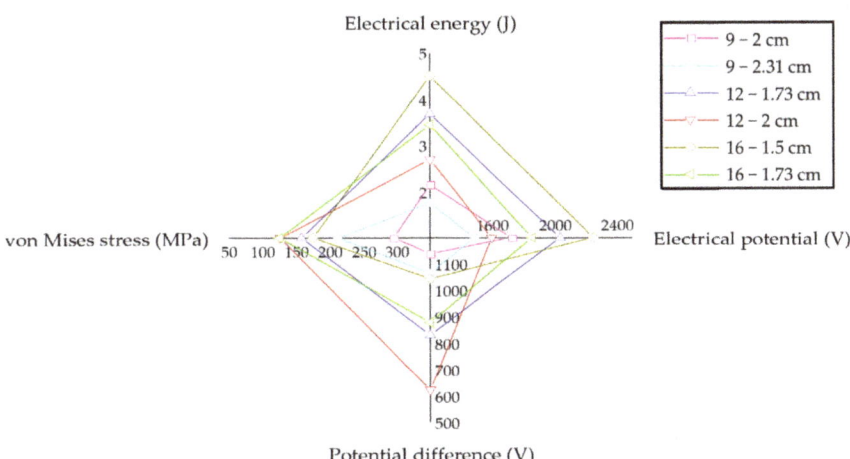

Figure 14. Radar chart of the output value of PEHs under full loading model.

5. Conclusions and Outlook

In this study, different PEHs with different piezoelectric unit distributions were modeled using FE method to evaluate their efficiency of energy harvesting. The PEH model in the simulation includes piezoelectric units and packaging materials. In total, three different piezoelectric unit distributions (3×3, 3×4 and 4×4) were developed. In addition, different cross section areas (28.27 and 37.7 cm^2) of the piezoelectric units were considered. To simulate the loading and boundary condition of PEH in asphalt pavements, three loading modes were applied on the PEH models (full, half, and quarter loading modes).

To sum up, the PEH with more piezoelectric units can increase the non-uniformity of the stress distribution, and produce more electrical power. Furthermore, higher electrical potential can be produced by piezoelectric units with smaller cross section area. The PEH with piezoelectric units in 4×4 distribution can generate more power from the traffic loads. However, remarkable electrical potential difference can be observed in the PEH with 4×4 distribution especially under non-uniform loading conditions. The highest electrical potential occurs near the edge of PEHs whilst the piezoelectric units in the center position produced the lowest electrical potential. In addition, larger cross section area could significantly reduce the electrical potential in the piezoelectric units. The stress results indicate that less piezoelectric units normally induce higher von Mises stress in PEHs. For the same distribution, the von Mises stress can be reduced by increasing the cross section area of the piezoelectric units.

Based on the aforementioned conclusions, for the PEH design in the future, some recommendations are proposed. When the total cross section areas of the piezoelectric units are the same, in order to produce higher energy, more piezoelectric units are suggested to be used. In addition, the diameter of the piezoelectric units near the edge could be larger than those at the center of the PEH, which would not only reduce the difference in the electrical potential between piezoelectric units at edge and center positions, but also effectively prevent or reduce the high stress concentration at edge positions.

Author Contributions: Conceptualization, P.L. and H.Y.; methodology, P.L., H.Y., and C.D.; software, C.D. and G.J.; validation, P.L., H.Y., and L.W.; formal analysis, C.D.; investigation, C.D. and P.L.; resources, P.L., H.Y., and M.O.; data curation, C.D.; writing—original draft preparation, C.D.; writing—review and editing, C.D., P.L., and H.Y.; visualization, C.D.; supervision, P.L. and H.Y.; project administration, P.L. and H.Y.; funding acquisition, P.L., H.Y., L.W., and M.O. All authors have read and agreed to the published version of the manuscript.

Funding: This research is funded by German Research Foundation (Grant Project No. FOR 2089/2, OE 514/1-2), Fundamental Research Funds for the Central Universities (Grant Project No. FRF-TP-18-048A1), and Excellence Strategy of the German Federal and State Governments (Grant Project No. StUpPD373-20). The authors gratefully acknowledge their financial support.

Institutional Review Board Statement: Not applicable.

Informed Consent Statement: Not applicable.

Data Availability Statement: The data presented in this study are available on request from the corresponding author. The data are not publicly available due to project requirement.

Conflicts of Interest: The authors declare no conflict of interest.

References

1. Ferreira, A.; Duarte, F.; Champalimaud, J. Waynergy Vehicles—An Innovative Pavement Energy Harvest System. *ICE Proc. Munic. Eng.* **2015**, *169*, 343–348.
2. Ferreira, A. Briefing: Recent developments in pavement energy harvest systems. *ICE Proc. Munic. Eng.* **2012**, *165*, 189–192. [CrossRef]
3. Song, G.; Cai, S.; Li, H.-N. Energy dissipation and vibration control: Modeling, algorithm, and devices. *Appl. Sci.* **2017**, *7*, 801. [CrossRef]
4. Davino, D.; Giustiniani, A.; Visone, C. Capacitive Load Effects on a Magnetostrictive Fully Coupled Energy Harvesting Device. *IEEE Trans. Magn.* **2009**, *45*, 4108–4111. [CrossRef]
5. Davino, D.; Giustiniani, A.; Visone, C.; Adly, A. Experimental analysis of vibrations damping due to magnetostrictive based energy harvesting. *J. Appl. Phys.* **2011**, *109*, 07E509. [CrossRef]
6. Alam, M.M.; Ghosh, S.K.; Sultana, A.; Mandal, D. Lead-free ZnSnO3/MWCNTs-based self-poled flexible hybrid nanogenerator for piezoelectric power generation. *Nanotechnology* **2015**, *26*, 165403. [CrossRef]
7. Wang, C.H.; Chen, S.; Li, Y.W.; Shi, X.; Li, Q. Design of piezoelectric elements' protection measures and energy output of intelligent power pavement. *China J. Highw. Transp.* **2016**, *29*, 41–49.
8. Wang, H.; Jasim, A.; Chen, X. Energy harvesting technologies in roadway and bridge for different applications—A comprehensive review. *Appl. Energy* **2018**, *212*, 1083–1094. [CrossRef]
9. Di Graziano, A.; Marchetta, V.; Cafiso, S. Structural health monitoring of asphalt pavements using smart sensor networks: A comprehensive review. *J. Traffic Transp. Eng. (Engl. Ed.)* **2020**, *7*, 639–651. [CrossRef]
10. Anton, S.R.; Sodano, H.A. A review of power harvesting using piezoelectric materials (2003–2006). *Smart Mater. Struct.* **2007**, *16*, R1–R21. [CrossRef]
11. Lee, C.S.; Joo, J.; Han, S.; Koh, S.K. Multifunctional transducer using poly (vinylidene fluoride) active layerand highly conducting poly (3,4-ethylenedioxythiophene) electrode: Actuator and generator. *Appl. Phys. Lett.* **2004**, *85*, 1841–1843. [CrossRef]
12. Sodano, H.; Lloyd, J.; Inman, D. *An Experimental Comparison between Several Active Composite Actuators for Power Generation*; SPIE: San Francisco, CA, USA, 2004; Volume 5390.
13. Mohammadi, F.; Khan, A.; Cass, R.B. Power Generation from Piezoelectric Lead Zirconate Titanate Fiber Composites. *MRS Proc.* **2002**, *736*, 55. [CrossRef]
14. Churchill, D.L.; Hamel, M.J.; Townsend, C.P.; Arms, S.W. Strain Energy Harvesting for Wireless Sensor Networks. In Proceedings of the Smart Structures and Materials 2003: Smart Electronics, MEMS, BioMEMS, and Nanotechnology, San Diego, CA, USA, 2–6 March 2003; International Society for Optics and Photonics: Bellingham, WA, USA, 2003; pp. 319–327.
15. Safaei, M.; Sodano, H.A.; Anton, S.R. A review of energy harvesting using piezoelectric materials: State-of-the-art a decade later (2008–2018). *Smart Mater. Struct.* **2019**, *28*, 113001. [CrossRef]
16. Yesner, G.; Kuciej, M.; Safari, A.; Jasim, A.; Wang, H.; Maher, A. Piezoelectric Energy Harvesting Using a Novel Cymbal Transducer Design. In Proceedings of the 2016 Joint Ieee International Symposium on the Applications of Ferroelectrics, European Conference on Application of Polar Dielectrics, and Piezoelectric Force Microscopy Workshop (Isaf/Ecapd/Pfm), Darmstadt, Germany, 21–25 August 2016.
17. Moure, A.; Izquierdo Rodríguez, M.A.; Rueda, S.H.; Gonzalo, A.; Rubio-Marcos, F.; Cuadros, D.U.; Pérez-Lepe, A.; Fernández, J.F. Feasible integration in asphalt of piezoelectric cymbals for vibration energy harvesting. *Energy Convers. Manag.* **2016**, *112*, 246–253. [CrossRef]
18. Sodano, H.; Zhao, S.; Erturk, A. Energy harvesting from harmonic and noise excitation of multilayer piezoelectric stacks: Modeling and experiment. In *Active and Passive Smart Structures and Integrated Systems 2013*; SPIE Digital Library: San Diego, CA, USA, 2013.
19. Xiong, H.; Wang, L. Piezoelectric energy harvester for public roadway: On-site installation and evaluation. *Appl. Energy* **2016**, *174*, 101–107. [CrossRef]
20. Liu, P.; Zhao, Q.; Yang, H.; Wang, D.; Oeser, M.; Wang, L.; Tan, Y. Numerical Study on Influence of Piezoelectric Energy Harvester on Asphalt Pavement Structural Responses. *J. Mater. Civ. Eng.* **2019**, *31*, 04019008. [CrossRef]

21. China Journal of Highway and Transport. Review on China's Pavement Engineering Research·2020. *China J. Highw. Transp.* **2020**, *33*, 1–66.
22. Zhang, H.; Huang, K.; Zhang, Z.; Xiang, T.; Quan, L. Piezoelectric Energy Harvesting From Roadways Based on Pavement Compatible Package. *J. Appl. Mech.* **2019**, *86*, 1–17. [CrossRef]
23. Yang, H.; Guo, M.; Wang, L.; Hou, Y.; Zhao, Q.; Cao, D.; Zhou, B.; Wang, D. Investigation on the factors influencing the performance of piezoelectric energy harvester. *Road Mater. Pavement Des.* **2017**, *18*, 180–189. [CrossRef]
24. Yang, H.; Wang, L.; Hou, Y.; Guo, M.; Ye, Z.; Tong, X.; Wang, D. Development in Stacked-Array-Type Piezoelectric Energy Harvester in Asphalt Pavement. *J. Mater. Civ. Eng.* **2017**, *29*, 04017224. [CrossRef]
25. Zhao, H.D.; Yu, J.A.; Ling, J.M. Finite element analysis of Cymbal piezoelectric transducers for harvesting energy from asphalt pavement. *J. Ceram. Soc. Jpn.* **2010**, *118*, 909–915. [CrossRef]
26. Yang, H.; Zhao, Q.; Guo, X.; Zhang, W.; Liu, P.; Wang, L. Numerical Analysis of Signal Response Characteristic of Piezoelectric Energy Harvesters Embedded in Pavement. *Materials (Basel)* **2020**, *13*, 2770. [CrossRef] [PubMed]
27. Ding, H.J.; Liang, J.A. The fundamental solutions for transversely isotropic piezoelectricity and boundary element method. *Comput. Struct.* **1999**, *71*, 447–455. [CrossRef]
28. Arnau, A.; Soares, D. Fundamentals of Piezoelectricity. In *Piezoelectric Transducers and Applications*; Vives, A.A., Ed.; Springer: Berlin/Heidelberg, Germany, 2008; pp. 1–38.
29. Cook-Chennault, K.A.; Thambi, N.; Sastry, A.M. Powering MEMS portable devices—A review of non-regenerative and regenerative power supply systems with special emphasis on piezoelectric energy harvesting systems. *Smart Mater. Struct.* **2008**, *17*, 17. [CrossRef]
30. Yang, H. *Development of a Piezoelectric Energy Harvesting System for Applications in Collecting Pavement Deformation Energy*; University of Science and Technology Beijing: Beijing, China, 2018.
31. Palmer, R.J. Polyamides, Plastics. In *Encyclopedia of Polymer Science and Technology*; Wiley Online Library: Hoboken, NJ, USA, 2000.
32. Greve, H.-H. Rubber, 2. Natural. In *Ullmann's Encyclopedia of Industrial Chemistry*; VCH Publishers: Hoboken, NJ, USA, 1985.

Article

A New Prospect in Road Traffic Energy Harvesting Using Lead-Free Piezoceramics

Manuel Vázquez-Rodríguez [1,2,*], Francisco J. Jiménez [2,3], Lorena Pardo [4], Pilar Ochoa [2,3], Amador M. González [2,3] and José de Frutos [2,3]

1. DTE-ETSIST, Universidad Politécnica de Madrid, 28031 Madrid, Spain
2. CEMDATIC-POEMMA R & D Group, Universidad Politécnica de Madrid, 28040 Madrid, Spain; franciscojavier.jimenez@upm.es (F.J.J.); ochoa@etsist.upm.es (P.O.); amador.m.gonzalez@upm.es (A.M.G.); jose.defrutos@upm.es (J.d.F.)
3. Departamento de Electrónica Física, Ingeniería Eléctrica y Física Aplicada, Universidad Politécnica de Madrid, 28031 Madrid, Spain
4. Instituto de Ciencia de Materiales de Madrid (ICMM), Consejo Superior de investigaciones Científicas (CSIC), C/Sor Juana Inés de la Cruz, 3. Cantoblanco 28049 Madrid, Spain; lpardo@icmm.csic.es
* Correspondence: m.vazquez@upm.es; Tel.: +34-91-0673289

Received: 30 September 2019; Accepted: 6 November 2019; Published: 11 November 2019

Abstract: In this paper, a new prospect using lead-free piezoelectric ceramics is presented in order to determine their behavior in piezoelectric-based road traffic energy harvesting applications. This paper will describe the low-cost and fully programmable novel test bench developed. The test bench includes a traffic simulator and acquires the electrical signals of the piezoelectric materials and the energy harvested when stress is produced by analogous mechanical stimuli to road traffic effects. This new computer-controlled laboratory instrument is able to obtain the active electrical model of the piezoelectric materials and the generalized linear equivalent electrical model of the energy storage and harvesting circuits in an accurate and automatized empirical process. The models are originals and predict the extracted maximum power. The methodology presented allows the use of only two load resistor values to empirically verify the value of the output impedance of the harvester previously determined by simulations. This parameter is unknown a priori and is very relevant for optimizing the energy harvesting process based on maximum power point algorithms. The relative error achieved between the theoretical analysis by applying the models and the practical tests with real harvesting systems is under 3%. The environmental concerns are explored, highlighting the main differences between lead-containing (lead zirconate titanate, PZT) and lead-free commercial piezoelectric ceramics in road traffic energy harvesting applications.

Keywords: piezoelectric ceramics; lead-free piezoceramics; energy harvesting; virtual instrument

1. Introduction

Nowadays, climate change is one of the most extended concerned topics worldwide. Classical electrical energy generation models have opened toward clean energies, reducing their carbon footprint by gradually increasing the power produced in hydroelectric, wind, and solar power plants. However, this trend is still far from achieving that as most of the electrical production comes from energy with low CO_2 emission to the atmosphere. For context, the 2018 annual report [1] about the Spanish electrical system shows that 19.8% was wind production, 13.8% was hydraulic, and 4.8% was solar (thermal, 1.8%, and photovoltaic, 3%).

Other subjects related to the environmental concerns are the reduction of harmful chemical waste, i.e., electrochemical accumulators or other electronic components that use lead (Pb) in their composition [2].

New techniques have been developed in micro-renewable energy generation, namely energy harvesting applications. Energy harvesting can be defined as electrical energy generation from natural and clean primary energy sources or from human activity to power electronic devices of low consumption. Some examples are wearable electronics, IoT (Internet of Things) devices, or wireless sensor networks. The source energies [3] are the well-known wind, solar, and mechanical energy from vibration, stress, or impacts generated from ambient or in residential or industrial human activities. Other primary energy sources are thermal energy and the RF (radio frequency) spectrum produced by human broadcast and telecommunication networks.

Applications in piezoelectric energy harvesting have been published since the beginning of the 21st century. The mechanical source is vibrational and the prevalent shape of the electromechanical transducer is the cantilever. Several enhancements were built-in such as magnetic elements, springs, L-shapes, and connections between them [4] to broaden the frequencies where maximum power generation is achieved.

The framework of the applied research presented in this article is electrical energy generation using ceramic piezoelectric transducers that optimizing the energy conversion from mechanical road traffic stimuli. A comparison is done between the behavior of lead-containing lead zirconate titanate (PZT) and lead-free commercial piezoelectric ceramics.

Table 1 shows a review of road traffic piezoelectric energy harvesting publications from 2010.

Table 1. Summary of road traffic energy harvesting publications. Review from 2010.

Published [Reference]	Contribution
2010 [5]	Finite elements theoretical and simulation study of the application of cymbal-type housing for piezoelectric materials. 1.2 mW generated at 20 Hz
2012 [6]	Several piezoelectric packages are studied using the finite elements technique for asphalt inlay highlighting cymbal and bridge for its efficiency in energy conversion
2015 [7]	Three encapsulation options for bridge-type housing are studied to minimize the fracture of the piezoelectric material by fatigue. It is concluded that the arch bridge is optimal for burying on asphalt. An applied pressure of 0.7 MPa generated 286 V
2016 [8]	A prototype consisting of 4, 8, or 16 piezoelectric disks sandwiched between two copper plates was assembled in-between asphalt mixtures. A uniaxial compression test was performed to measure the output power directly on a resistor
2016 [9]	Based on the Ph. D. thesis of the first author, piezoelectric degradation measurements in an USA real road installation are presented. Over 14% of the asphalt stress produced by the vehicles is transmitted to the road-embedded prototypes producing 3.106 mW of harvested power
2016 [10]	Two prototypes formed by stacked prismatic or cylindrical piezoelectric elements are tested in the laboratory. Assuming daily moderately busy USA Interstate highway traffic of 30,000 vehicles/day, the first prototype will produce 9.66 Wh per year and the second one 240.95 Wh
2016 [11]	A cymbal structure is modified in seven piezoelectric parallelized sections. In a laboratory test over a 400 kΩ resistor, 2.1 mW of power is produced
2016 [12]	An association of piezoelectric cantilevers produces 184 µW over an empirically optimized resistor of 70 kΩ. A Universal Test Machine (UTM) performs the laboratory tests
2016 [13]	Wheel tracking tests are performed assuming a continuous rate of traffic. Several recommendations are obtained to adjust the geometry and composition of the piezoelectric material in order to maximize the extracted power in response to variable speed and distance between vehicles
2017 [14]	Up to 60 PVDF layers are associated in parallel to generate 200 mW of peak power. Viability of using flexible material is shown
2017 [15]	A new structure formed by a layer of piezoelectric material embedded between two layers of conductive asphalt generates 1.2 mW in UTM tests
2018 [16]	A stacked array type of piezoelectric energy harvester is field-tested, generating a voltage between 250 and 400 V when a test vehicle is passes. The obtained piezoelectric energy lights LED signs
2018 [17]	A new prototype of 11 stacked piezoelectric elements is presented and compared to the prototype results presented in [8]. The energy output estimated per prototypes I and II was 360 and 171 Wh annually

The main things lacking that has been appreciated in the previous review are summarized in the following. There is a reduced number of piezoelectric harvesters in roadway installations; instead, laboratory tests mostly apply uniaxial stress by means of Universal Test Machine (UTM) equipment. There is a low number of models of piezoelectric elements in road traffic environments. The influence of the instruments in the experimental measurements is generally not considered. The scalability of the power generated by harvesters is often not demonstrated.

The 2014 report [18] for the California Energy Commission estimated a high cost, at $600,000–$1,000,000, of a demonstration project that included laboratory, acceleration, and field tests.

This paper will describe the low-cost original and fully programmable instrument developed by some of the authors at the Universidad Politécnica de Madrid [19]. This test bench is able to obtain accurate models of piezoelectric-based energy harvesters and carry out the accelerated tests in a much more economically affordable way. The test bench includes a traffic simulator and acquires the electrical signals of the piezoelectric materials and the energy harvested when the stress is produced by mechanical stimuli, analogous to the road traffic effect.

The parameters of those models, as well as the harvested power, will be empirically verified by performing a reduced set of practical tests.

Finally, the main differences in energy harvesting applications between PZT and lead-free commercial piezoelectric ceramics will be highlighted.

2. The New Piezoelectric Characterization System

Figure 1 shows a block diagram of the complete harvesting and piezoelectric test system. The test bench is made up of a Road Traffic Simulator driven by an AC geared motor. The angular speed ω (expressed in rpm) of the upper rotating platform shaft is fully programmable. The mechanical topology of this platform is built in an open way. Their wheels may be disposed in several locations to configure the angle between the simulated axes of the vehicles β (°). The static platform, below the rotating upper platform, includes, in the track way, the piezoelectric devices under test (PDUTs). Equation (1) calculates the simulated speed v (km/h) of the tests for each vehicle type. The data acquisition card (DAQ) sends the control signals to the driver control electronic card, which commands the AC motor driver.

$$v = \frac{21.6 \cdot b \cdot \omega}{\beta}. \tag{1}$$

The simulated speed in the test bench for a sedan-style car, which has a wheelbase, b (m), of 2.64 m, is between 14 km/h (8 mph) and 180 km/h (112 mph), as a maximum value for laboratory test purposes only.

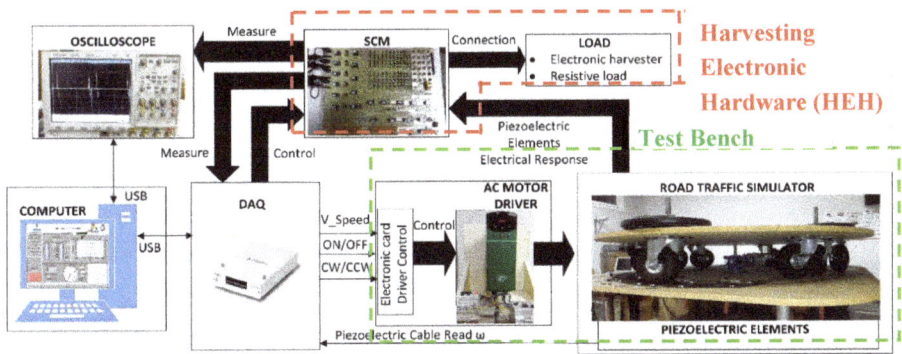

Figure 1. Piezoelectric characterization system block diagram.

A picture of the system performing the laboratory test is shown in Figure 2.

The harvesting electronic hardware (HEH) in Figure 1 performs the automatized electrical measurement. The relay-based switched circuit's matrix (SCM) is electronically controlled by the DAQ. The SCM can control up to six PDUTs. The first routing stage of this matrix connects the PDUTs to a USB-controlled oscilloscope or to the selected diode rectifier topologies. The SCM second routing stage selects the rectifier topology and the series or parallel associations between them. The last stage connects the automatized load, selecting the cyclic or single test. Our developed control software is programmed in the National Instruments LABVIEW™ graphical language. The software commands the acquisition of the measured piezoelectric signal of the PDUT at the first routing stage to obtain the active piezoelectric simulation model, according to the periodical operation of the test bench. The software obtains the transient and the steady state of the energy harvesting measured voltage. The power and load regulation graphs are obtained by applying different loads. The open load voltage and the output equivalent impedance of the energy harvesting capacitor filtered rectifier circuit are computed. An example of the acquisition of four electrical signals from a PZT PDUT using the oscilloscope is presented in Figure 3a. In Figure 3b, our developed software user interface acquires channel number 1 of the piezoelectric response shown in Figure 3a.

Figure 2. Piezoelectric characterization system in action.

This piezoelectric characterization system configures a new virtual instrument (VI). A virtual instrument performs the functions of the traditional measurement instruments but engineers and scientists can build automated measurement systems that suit their needs exactly instead of being conditioned and limited by standard instruments.

measurement impedance Z_meas with C_p and R_p being the capacitive and resistive values of the probe connected to the oscilloscope, respectively.

$$V_{pz}(f_i) = \frac{V_o(f_i)}{Z\_meas} \cdot (Z_{pz} + Z\_meas),\qquad(2)$$

$$Z\_meas = \frac{R_p - jR_p^2\omega_i C_p}{1 + R_p^2\omega_i^2 C_p^2},\qquad(3)$$

$$\omega_i = 2\pi f_i.\qquad(4)$$

Table 2 shows the values of the properties of the lead-containing and lead-free materials. The values in Table 2 show that the lead-containing material is more piezoelectric, polarizable, and lossy, as well as more compliant, than the lead-free material.

Table 2. Piezoelectric (g_{33} and d_{33}) and elastic ($s_{33}{}^D$; or $Y_{33} = 1/s_{33}$) coefficients, dielectric permittivity and losses ($K_{33}{}^T$ and tan δ), and electromechanical coupling factors (k_{33}) of the lead-containing, hard lead titanate zirconate (Navy I-type PZT; APC International, Ltd., Mackeyville, PA, USA) and lead-free, tetragonal bismuth sodium barium titanate (BNBT) (PIC700; PI Ceramic GmbH, Lederhose, Germany) commercial ceramic materials (longitudinally poled cylinders of 6 mm diameter and 15 mm length). The catalog values are shown for PZT, and PIC700 was characterized using the resonance method (f_s = 148.3 kHz, f_p = 160.1 kHz).

Material	g_{33} (10^{-3} Vm/N)	d_{33} (10^{-12} C/N)	$s_{33}{}^D$ (10^{-12} m²/N)	$K_{33}{}^T$	tan δ (%)	k_{33}
PZT	26	>260	12.5	1280	0.6	>0.68
BNBT	16	98	7.5	710	0.4	0.40

In Figure 7, the detailed housing of the PDUTs and their location in the test bench are depicted. These are two cylinders connected electrically in parallel, but mechanically in series. The piezoelectric elements are placed in a mechanically amplified (lever) holder (see Figure 7a,b in exploded view), and disposed in very shallow cavities (lever projects only 2 mm from the nonrotating platform) in diametric positions in the test bench inner path (see Figure 7c,d).

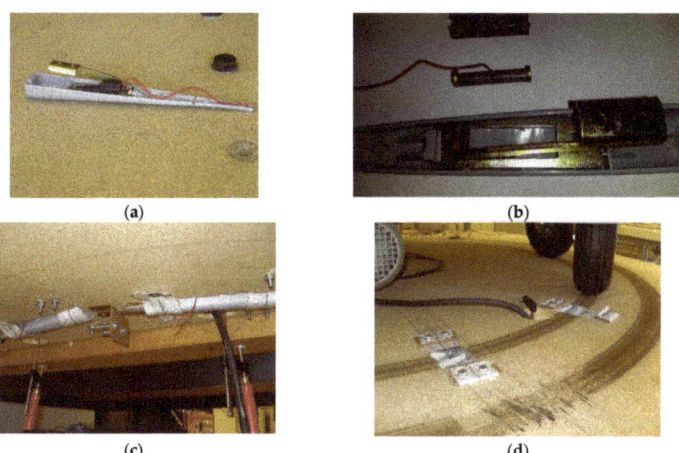

Figure 7. (a) Open view of the commercial piezoelectric housing; (b) exploded view of the commercial piezoelectric showing the lever mechanical amplifier and the piezoelectric material outside the holder; (c) bottom view of the commercial piezoelectric placement in the test bench; (d) top view of the PDUTs in the inner path of the road traffic simulator.

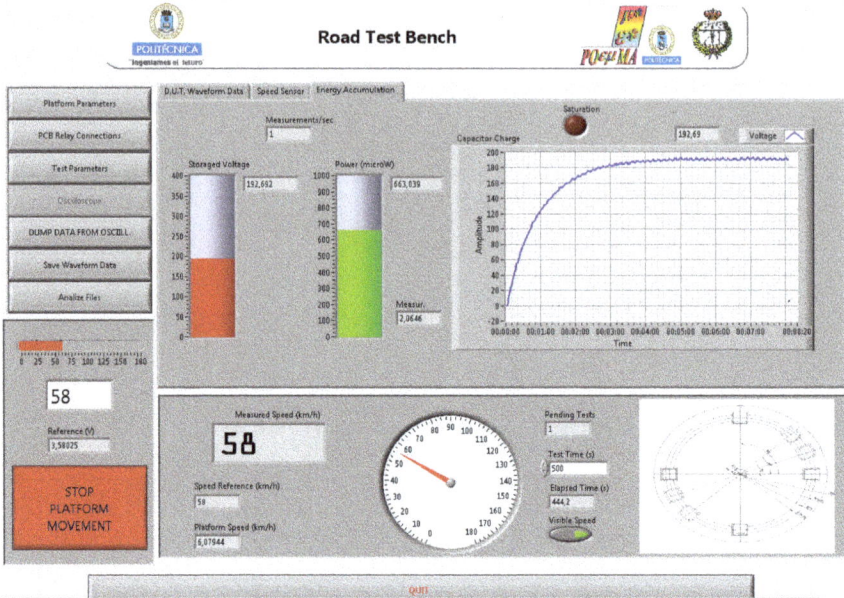

Figure 5. Practical test results.

2.1. Piezoelectric Ceramic Material Characterization under Harvesting Conditions

Our methodology computes, at the first stage, an active electrical model of the piezoelectric material mechanically excited by the road traffic. The model is the series association of the impedance of the material with active inner Thévenin voltage generators. This is calculated with Fourier analysis of the measured piezoelectric voltage (Figure 3a), the equivalent input impedance of the oscilloscope, and the impedance of the piezoelectric elements. Figure 6 shows the electrical circuit needed to solve the active electrical model of the piezoelectric ceramic. The Fourier generator V_{pz} and the piezoelectric impedance Z_{pz} are the elements of the active electrical model of the piezoelectric ceramic materials. The impedance of the measurement equipment is a key factor to calculate the active Fourier electrical model that predicts its behavior in whatever energy harvesting application. In this case, the measurement oscilloscope probe ($Z_{\_meas}$ in Figure 6) has an equivalent input impedance of 10 MΩ in parallel with a capacitance of 4 pF when it is connected to the input impedance of the oscilloscope (which is of 1 MΩ in parallel with a capacitance of 11 pF).

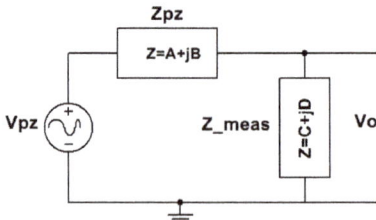

Figure 6. Electrical equivalent circuit needed to obtain the active piezoelectric model in energy harvesting road traffic applications.

Equation (2) calculates the component values of the Fourier active generator V_{pz}, when the spectrum of frequencies of the measured voltage V_o is computed by the VI. Equation (3) calculates the

(2) The test bench is set in action. The piezoelectric voltage is acquired and its active Fourier model is calculated. The active Fourier model is obtained by calculating each Fourier component of the inner piezoelectric generators, taking into account the input impedance of the measurement equipment and the impedance of the PDUTs.
(3) The active Fourier model is sent to the LabVIEW® PSpice-based software module. An iterative process is started. The harvesting circuit formed by a capacitor-filtered rectifier stage is simulated for n different load resistance values. The high accuracy of the active Fourier models achieves a low simulation error.
(4) The VI computes the voltage–current and power graphs. A first estimation of the open circuit voltage (V_{oc}) and the equivalent output resistance (R_o) of the harvester in the maximum power zone is obtained.
(5) The next step is to verify the accuracy of the first estimation obtained for the key parameters V_{oc} and R_o. Analyzing the simulation results, a pair of appropriate values for the load resistance (R_{load1} and R_{load2}) are chosen. These resistor values are connected in the HEH module.
(6) The test bench is set in action. The voltage, current, and power are registered for both load resistance values.
(7) The practical values of output resistance (R_o), open circuit voltage (V_{oc}), and maximum power point (Po_{max}) are obtained and empirically verified.

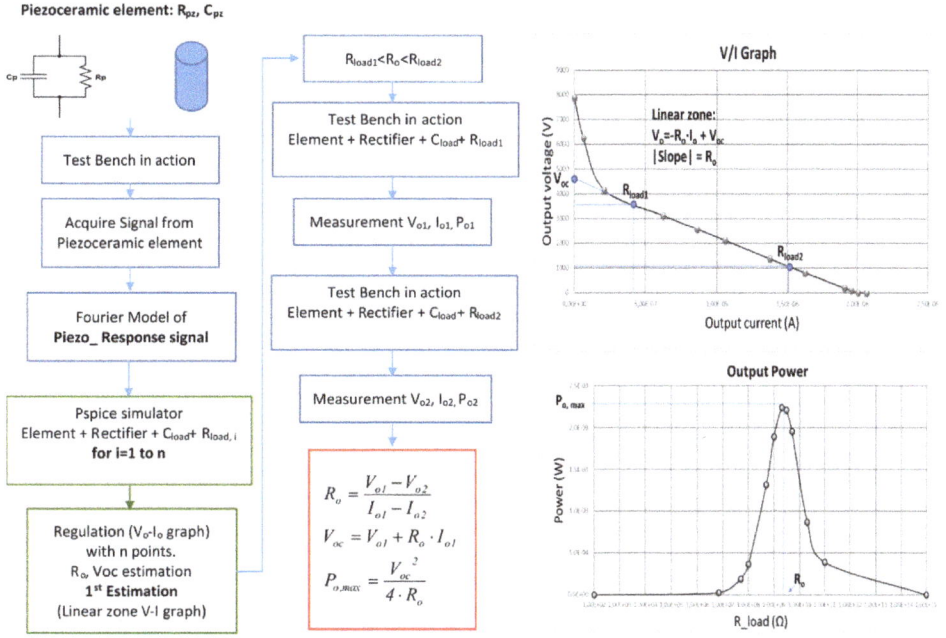

Figure 4. Methodology applied in the original piezoelectric characterization system.

The actual measurements on the harvesting electronic hardware (Figure 1) module validate the methodology. In Figure 5, the VI screen of the accumulated voltage measured in the energy harvesting module of the Test Bench is presented.

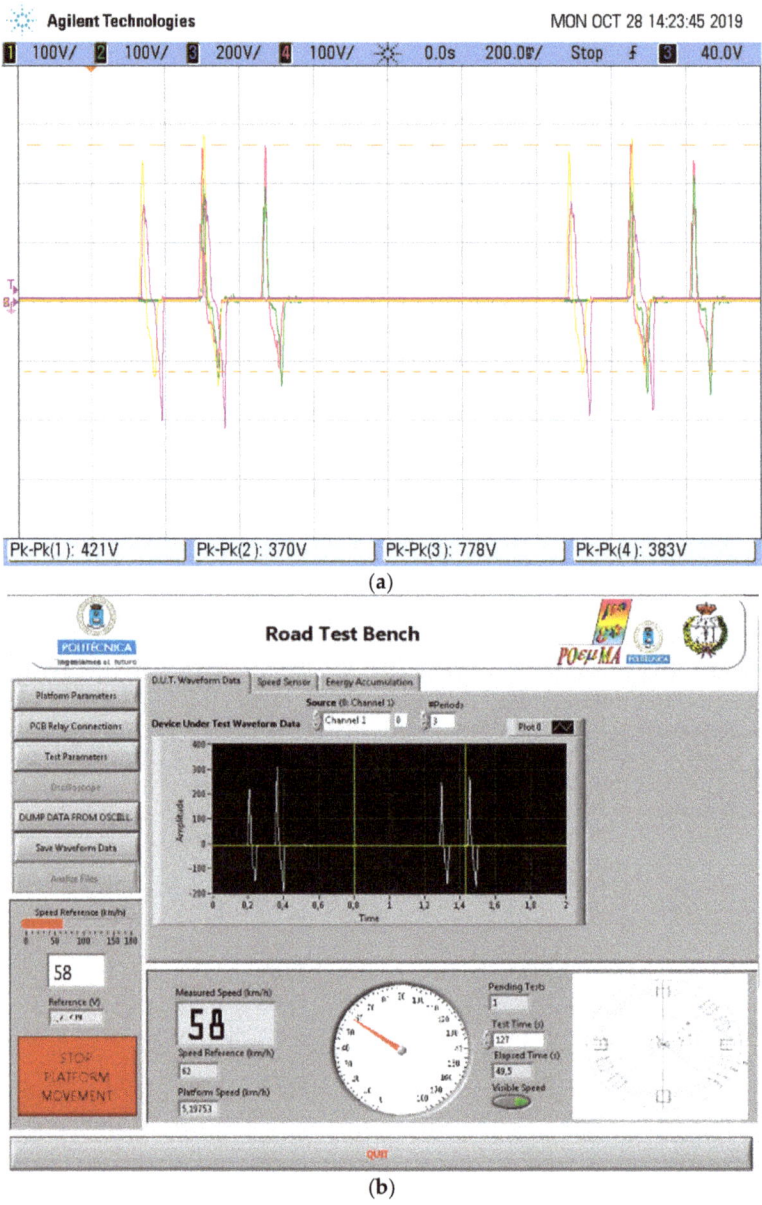

Figure 3. (a) Piezoelectric electrical signals from four lead-containing lead zirconate titanate (PZT) piezoelectric devices under test (PDUTs). (b) Software interface acquiring one channel of electrical PDUT response to obtain the active electrical model.

The methodology to obtain and validate the models is presented in Figure 4. The steps that cover the process are as follows:

(1) The PDUTs are electrically characterized. Their impedance is measured with an impedance meter. The piezoelectric elements are placed in the test bench.

2.1.1. Impedance of the PDUTs

The impedance of the PDUTs was determined with the impedance analyzer Solartron 1260 from AMETEK Scientific Instruments. The impedance analyzer provides the real and imaginary parts of the impedance (Z'(a) and Z"(b)). Equations (5)–(10) obtain the modulus and phase of the impedance, the admittance, the capacitance, and the resistance of the material. The results for the impedance module of PZT and PIC700 are shown in Figure 8.

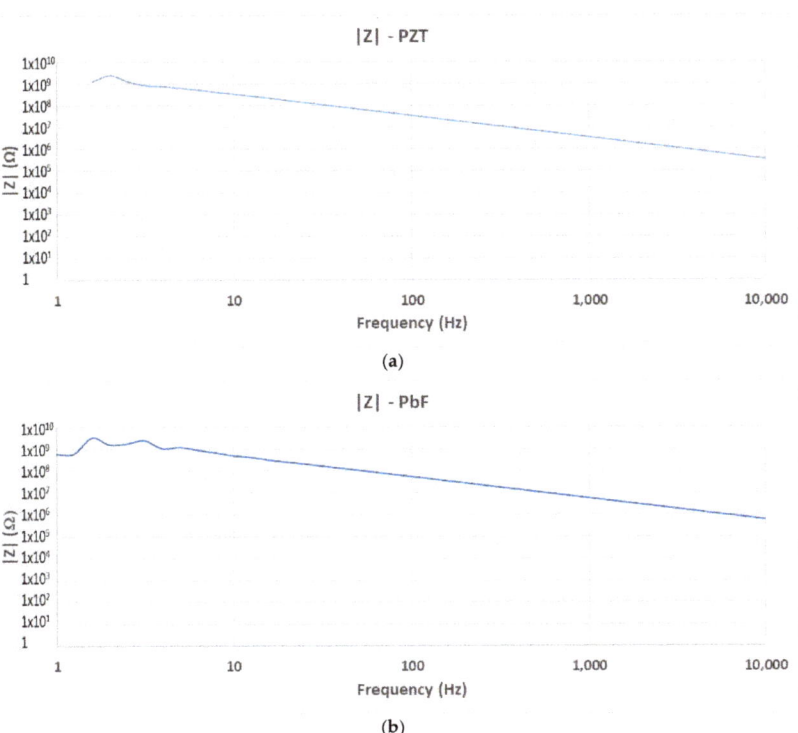

Figure 8. Impedance measurement of the PDUTs: (**a**) Lead-containing PZT; (**b**) lead (Pb)-free.

$$\varnothing_Z = tan^{-1}\frac{Z''(b)}{Z'(a)}, \tag{5}$$

$$|Z| = \sqrt{(Z'(a))^2 + (Z''(b))^2}, \tag{6}$$

$$|Y| = \frac{1}{|Z|}, \tag{7}$$

$$\varnothing_Y = -\varnothing_Z, \tag{8}$$

$$R_{pz} = \frac{|Z|}{cos\varnothing_Z} = \frac{1}{|Y|cos\varnothing_Y}, \tag{9}$$

$$C_{pz} = \frac{-sen\varnothing_Z}{|Z|2\pi f} = \frac{|Y|sen\varnothing_Y}{2\pi f}. \tag{10}$$

The capacitive effect is relevant in both piezoelectric materials on the impedance of the PDUTs.

2.1.2. Piezoelectrically Active Electrical Model

The Test Bench, programmed to perform the road test at 58 km/h of simulated car speed, stresses both piezoelectric materials in the same way to the consecutive tests. The generated voltage (V_o in Figure 6) was recorded in the VI to compute their Fourier spectrum. The modulus of the PZT Fourier analysis is shown in the Figure 9. The voltage V_o measured with the oscilloscope and the modulus of the active generator from the spectral Fourier analysis, $|V_{pz}|$, calculated by the VI are presented in Figure 10 for the PZT and the lead-free piezoceramics.

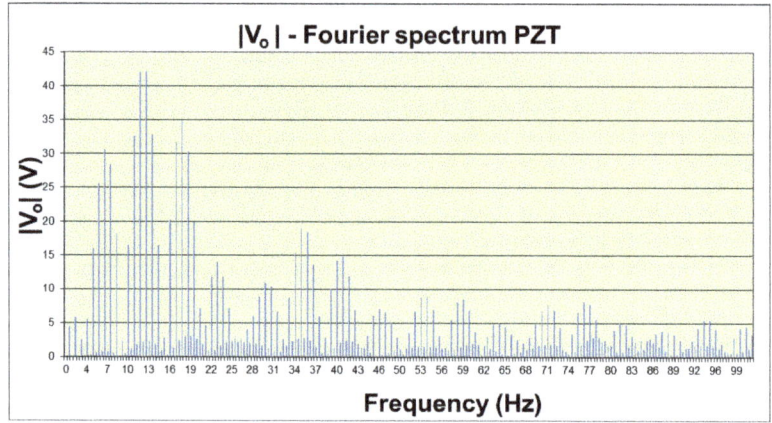

Figure 9. Fourier spectrum modulus of the measured voltage (V_o) in the PZT ceramics.

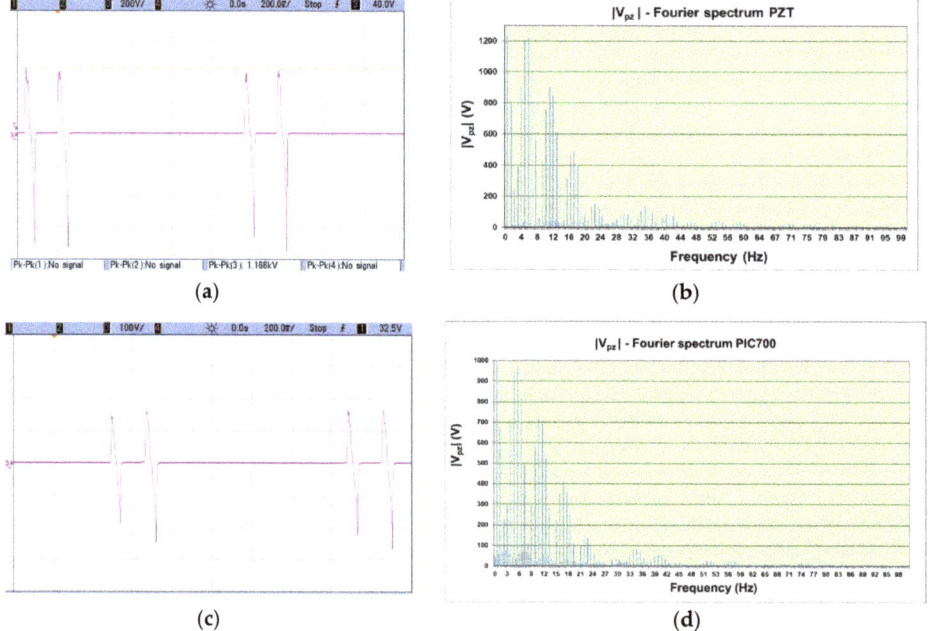

Figure 10. (a) PZT material measured voltage; (b) Fourier spectrum (modulus) of the active piezoelectrical generator for the PZT ceramic material; (c) lead-free ceramics measured voltage; (d) Fourier spectrum (modulus) of the active piezoelectrical generator for the PIC700 ceramic material.

The amplitude of the spectral components of the measured V_o voltage is on the tens of volts range; meanwhile, the amplitude of the components in the inner active piezoelectric generator (V_{pz}, see Figure 6) is on the order of magnitude of a thousand volts. The effect of the load impedance and the high impedance of the PDUTs explains this behavior in practical energy harvesting applications.

In energy harvesting road traffic environmental applications, the working conditions are in the very low frequency band. The frequencies of interest are always below 100 Hz because the Fourier spectral analysis of the piezoelectric response shows a bandwidth up to 100 Hz at the Test Bench maximum speed. This practical conclusion points to the main difference of this work with respect to other research works that show interest in working with piezoelectric elements in the resonance points of the material (here at ~150 kHz, see Table 2).

The recorded voltages show that lead-free piezo-ceramics generates a lower peak-to-peak voltage than the PZT material, in agreement with the values in Table 2.

Once the active electrical model is computed, it is possible to start the next stage of harvesting simulations to conclude with energy harvesting application results.

3. Energy Harvesting Results

The VI computes the piezoelectric active model. The model is different for each value of simulated speed. The active electrical model is exported to perform the electrical simulations in PSpice-based software connecting the piezoelectric model to the diode rectifier circuit filtered by the capacitor. The capacitor accumulates the extracted charge. The load resistance (R_{load} in Figure 11) is varied in successive simulations from 100 Ω (practical zone of short circuit) to 1000 GΩ (practical zone of open load) to obtain the voltage and current load graph. The practical graphic results are presented in Figure 12 for the PZT and lead-free PIC700 ceramic.

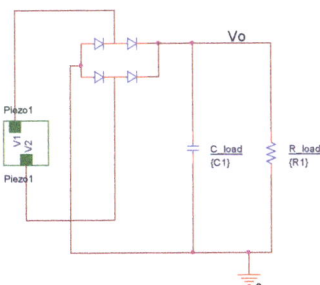

Figure 11. Harvesting circuit.

In Figure 12a, the regulation graph of voltage V_o (see Figure 11) versus load current (Io) in resistor R_{load} is presented for tests at 58 km/h of simulated speed using PZT and PIC700 lead-free ceramics. The parameters R_o (output resistance, calculated as the slope of the linear zone where maximum power is achieved) and V^*_{oc} (open circuit voltage: Intersection of the ordinate axis with the extended line of the linear maximum power zone) are the key factors to estimate the maximum power point of the harvesting power.

The maximum extracted power point verifies Equation (11), when the R_{load} applied equals the output equivalent (R_o) resistance of the piezoelectric harvesting circuit. The parameter R_o is previously unknown and is of significant relevance to design energy harvesting systems that achieve the maximum energetic efficiency. Our methodology calculates R_o and estimates V^*_{oc} with high precision.

$$Po_{max} = \frac{V^{*2}_{oc}}{4 \cdot R_o}. \tag{11}$$

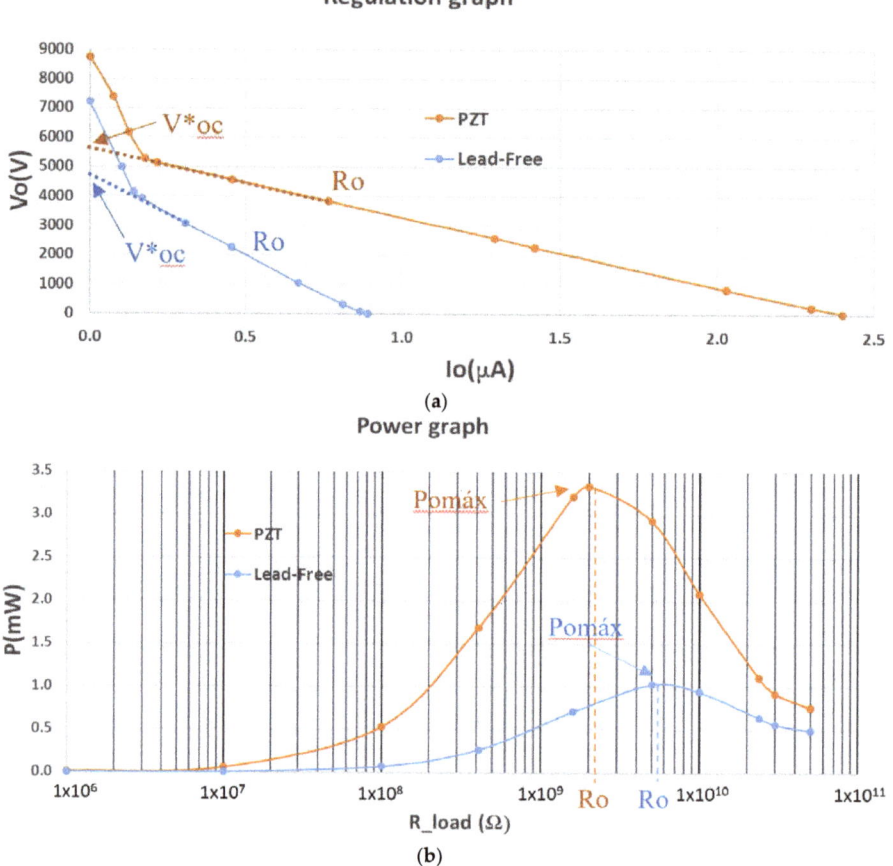

Figure 12. Comparative results: PZT vs. lead-free piezoceramics: (**a**) Regulation graph; (**b**) power generated graph.

The practical results of the simulation stage are summarized in Table 3.

The data in Table 3 show that the impedance of the ceramic set (piezo + accumulator circuit) of maximum power delivery is approximately three times higher in the lead-free piezoelectric ceramic.

It is also observed that the deliverable power for the optimum R_load is approximately three times lower in the lead-free ceramic.

The results of the experiments verify that the tested materials are different from the point of view of electric power generation. However, the differences are not so distant. To equalize the maximum power capability, the lead-free material should be excited to provide a piezoelectric amplitude (V^*_{oc}) of approximately $\sqrt{3}$ times greater. This conclusion opens the way to the ecological materials in alternative energy generation.

Table 3. Parameters of the piezoelectric energy harvesting application system.

Parameter	PZT	PIC700
R_o (GΩ)	2.36	5.57
V^*_{oc} (V)	5640	4800
Po_{max} (mW)	3.4	1.03

The validation procedure stage was performed next in the Test Bench. A couple of R_load values were selected to be in the linear zone of maximum harvesting power. The practical values of the accumulated voltage V_o in the energy harvesting circuit are presented in Figure 13 for the PZT material. Table 4 calculates the practical parameter R_o and the relative error (Er) between empirically validated data and previous results from simulations.

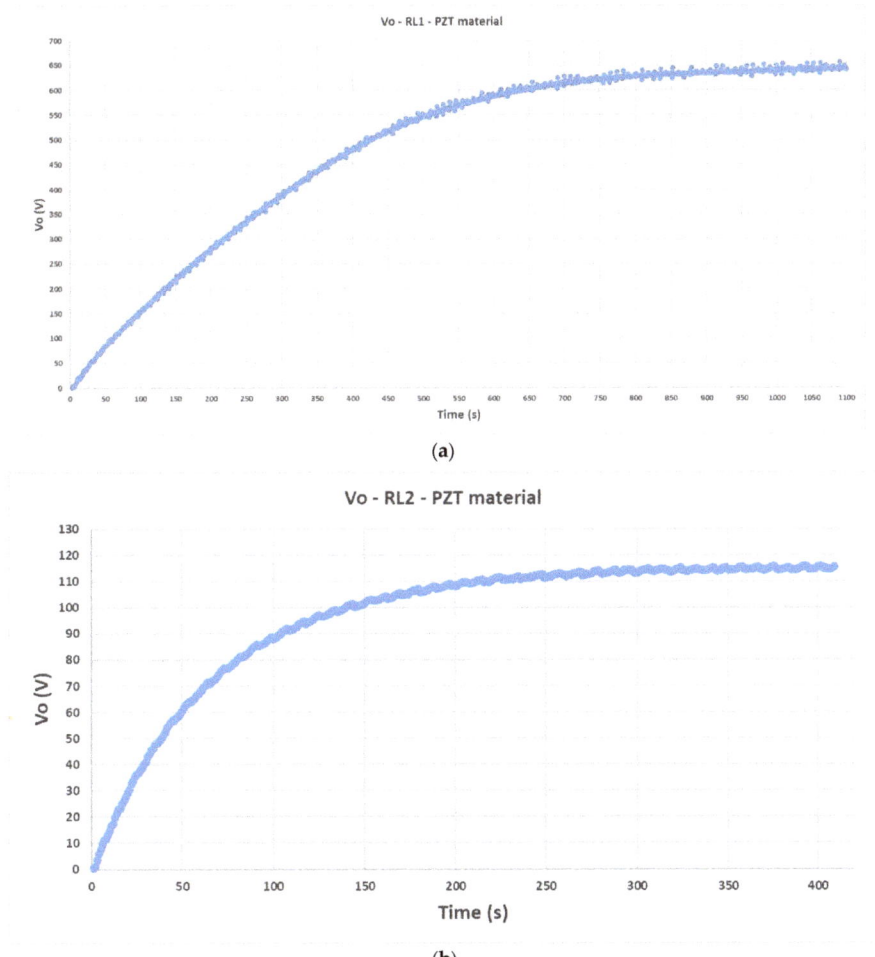

Figure 13. Transient response and steady state of the accumulated voltage in the capacitor ($C\_\text{load} = 1$ µF) of the harvesting circuit when the PZT piezoelectric material is utilized in the Test Bench: (**a**) Output voltage recorded by virtual instrument (VI) when using a set of resistors of equivalent $R\_\text{load1} = 300$ MΩ; (**b**) output voltage when $R\_\text{load2} = 50$ MΩ.

Table 4. Empirical verification of the methodology. PZT material.

Measurements		Simulations	Er %
$R_o = \frac{|V_{o1}-V_{o2}|}{|I_{o1}-I_{o2}|} =$	$\frac{645-115}{|2.12-2.35|\cdot 10^{-6}} = 2.30$ GΩ	2.36 GΩ	−2.54

The measurements of the accumulated voltage in the harvesting circuit when PIC700 is utilized are presented in the Figure 14.

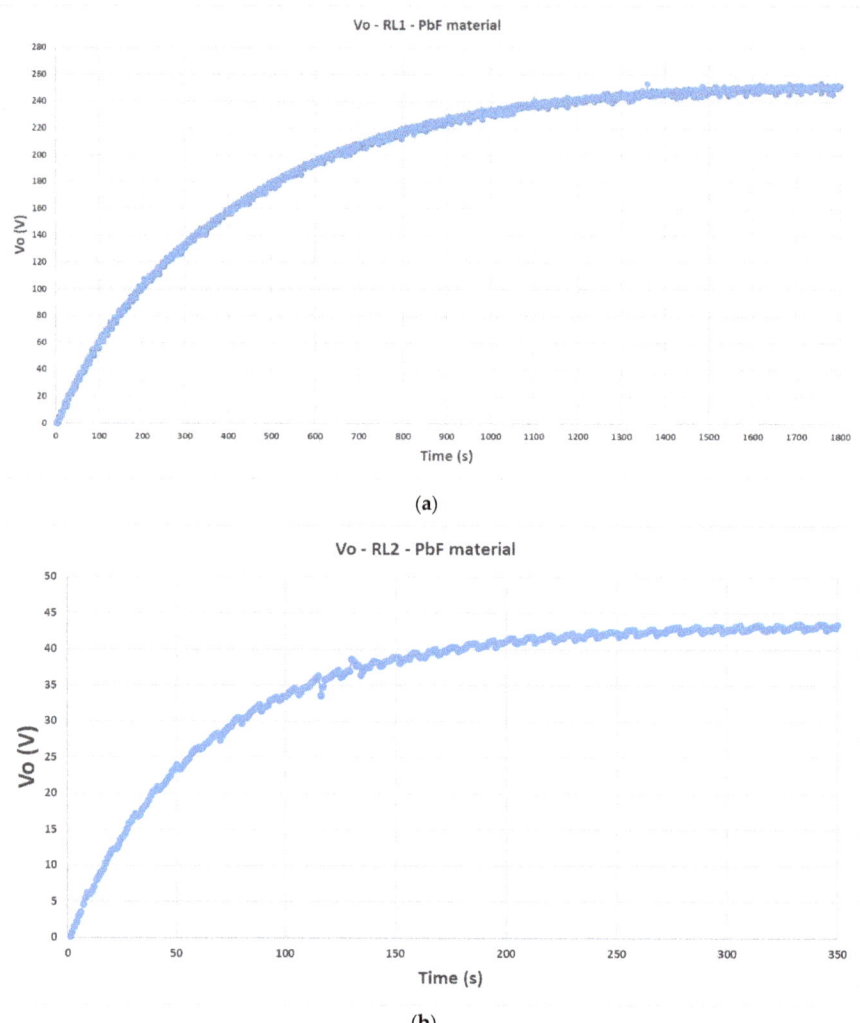

Figure 14. Transient response and steady state of the accumulated voltage in the capacitor ($C\_load = 1~\mu F$) of the harvesting circuit when PIC700 lead-free piezoelectric material is utilized in the Test Bench: (a) Output voltage recorded by VI when using a set of resistors of equivalent $R\_load1 = 300~M\Omega$; (b) output voltage when $R\_load2 = 50~M\Omega$.

Table 5 presents the empirically determined R_o and the relative error achieved between previous results from simulations and test validated data.

Table 5. Empirical verification of the methodology. PIC700 lead-free material.

Measurements		Simulations	Er %
$R_o = \frac{|V_{o1}-V_{o2}|}{|I_{o1}-I_{o2}|} =$	$\frac{250.7-43.6}{|0.835-0.873|\cdot 10^{-6}} = 5.45~G\Omega$	5.57 GΩ	−2.15

The methodology presented allows the use of only two load resistor values to empirically verify the value of output impedance of the harvester previously determined by simulations. This value is relevant for optimizing the energy harvesting process in maximum power point algorithms.

The originality of the new instrument developed and adapted to perform road traffic tests in a laboratory environment achieves practical results with low error in the modeling characterization process of piezoelectric materials and energy harvesting systems.

The influence of the measurement equipment is considered in the development of the practical methodology exposed.

The results obtained in a single device under test can be generalized to topological associations between harvesters, as it was previously published [20]. The influence of the rate of traffic (vehicles/minute) and of peak-to-peak piezoelectric voltage on the harvested power was discussed in [21]. The topologies of associated harvesters verify the modeling process described in References [19–21].

Those previous results have opened the prospects of using lead-free piezoelectric materials in clean electrical energy generation.

4. Conclusions

The models used here to analyze and predict the energy generation of harvesters based on piezoelectric ceramics are original. With this original methodology, we were able to compare the performance in piezoelectric energy harvesting in road traffic of lead-containing (PZT) and lead-free (PIC700) piezoelectric ceramics. Classical research about energy harvesting using piezoelectric materials is based on vibrational behavior, at which the two materials present differences, particularly at resonance (see Table 2). The vibrational component in the stress applied by road traffic is not relevant in the presented analysis. The low relative error achieved between the theoretical analysis of applying the models and the practical tests with real harvesting systems is under 3% both for the lead-containing and lead-free material.

The data in Table 3 show that the impedance of the ceramic set (Piezo + accumulator circuit) of maximum power delivery is approximately three times higher in the lead-free piezoelectric ceramic. The results of the experiments verify that the tested materials are different from the point of view of electric power generation. However, the differences are not so distant. To equalize the maximum power capability, the lead-free material should be excited to provide a piezoelectric amplitude (V^*_{oc}) of approximately $\sqrt{3}$ times greater. This conclusion opens the way to the ecological materials in alternative clean energy generation.

Author Contributions: M.V.-R., F.J.J., P.O., A.M.G., and J.d.F. conceived and designed the experiments, compiled the literature data, and discussed the experimental results; M.V.-R. performed the data acquisition and analysis, elaborated the figures, and wrote the manuscript; L.P. provided and characterized the lead-free material and discussed the experimental results. All authors revised and discussed the manuscript.

Funding: This work was supported in part by the Spanish R&D Project No. MAT 2017 86168R.

Acknowledgments: Authors are indebted to PI Ceramic GmbH for marketing a small batch of lead-free PIC700 cylinders.

Conflicts of Interest: The authors declare no conflict of interest.

References

1. Red Eléctrica de España. Available online: https://www.ree.es/sites/default/files/11_PUBLICACIONES/Documentos/InformesSistemaElectrico/2018/inf_sis_elec_ree_2018.pdf (accessed on 29 September 2019).
2. Villafuerte-Castrejón, M.E.; Morán, E.; Reyes-Montero, A.; Vivar-Ocampo, R.; Peña-Jiménez, J.-A.; Rea-López, S.-O.; Pardo, L. Towards Lead-Free Piezoceramics: Facing a Synthesis Challenge. *Materials* **2016**, *9*, 21. [CrossRef] [PubMed]
3. Calio, R.; Rongala, U.B.; Camboni, D.; Milazzo, M.; Stefanini, C.; de Petris, G.; Oddo, C.M. Piezoelectric Energy Harvesting Solutions. *Sensors* **2014**, *14*, 4755–4790. [CrossRef] [PubMed]

4. Abramovich, H.; Har-nes, I. Analysis and Experimental Validation of a Piezoelectric Harvester with Enhanced Frequency Bandwidth. *Materials* **2018**, *11*, 1243. [CrossRef] [PubMed]
5. Zhao, H.; Yu, J.; Ling, J. Finite element analysis of Cymbal piezoelectric transducers for harvesting energy from asphalt. *J. Ceram. Soc. Jpn.* **2010**, *118*, 909–915. [CrossRef]
6. Zhao, H.; Ling, J.; Yu, J. A comparative analysis of piezoelectric transducers for harvesting energy from asphalt pavement. *J. Ceram. Soc. Jpn.* **2012**, *120*, 317–323. [CrossRef]
7. Zhao, H.; Qin, L.; Ling, J. Test and Analysis of Bridge Transducers for Harvesting Energy from Asphalt Pavement. *Int. J. Transp. Sci. Technol.* **2015**, *4*, 17–28. [CrossRef]
8. Roshani, H.; Dessouky, S.; Montoya, A.; Papagiannakis, A.T. Energy harvesting from asphalt pavement roadways vehicle-induced stresses: A feasibility study. *Appl. Energy* **2016**, *30*, 210–218. [CrossRef]
9. Xiong, H.; Wang, L. Piezoelectric energy harvester for public roadway: On-site installation and evaluation. *Appl. Energy* **2016**, *174*, 101–107. [CrossRef]
10. Papagiannakis, A.G.; Dessouky, S.; Montoya, A.; Roshani, H. Energy Harvesting from Roadways. *Procedia Comput. Sci.* **2016**, *83*, 758–765. [CrossRef]
11. Yesner, G.; Kuciej, M.; Safari, A.; Jasim, A.; Wang, H.; Maher, A. Piezoelectric Energy Harvesting Using a Novel Cymbal Transducer Design. In Proceedings of the 2016 Joint IEEE International Symposium on the Applications of Ferroelectrics, European Conference on Application of Polar Dielectrics, and Piezoelectric Force Microscopy Workshop (ISAF/ECAPD/PFM), Darmstadt, Germany, 21–25 August 2016.
12. Song, Y.; Yang, C.H.; Hong, S.K.; Hwang, S.J.; Kim, J.H.; Choi, J.Y.; Ryu, S.K.; Sung, T.H. Road energy harvester designed as a macro-power source using the piezoelectric effect. *Int. J. Hydrogen Energy* **2016**, *41*, 12563–12568. [CrossRef]
13. Chen, Y.; Zhang, Y.; Li, C.; Yang, Q.; Zheng, H.; Lü, C. Mechanical Energy Harvesting from Road Pavements Under Vehicular Load Using Embedded Piezoelectric Elements. *J. Appl. Mech.* **2016**, *83*, 081001. [CrossRef]
14. Jung, I.; Shin, Y.-H.; Kim, S.; Choi, J.; Kang, C.-Y. Flexible piezoelectric polymer-based energy harvesting system for roadway applications. *Appl. Energy* **2017**, *197*, 222–229. [CrossRef]
15. Guo, L.; Lu, Q. Modeling a new energy harvesting pavement system with experimental verification. *Appl. Energy* **2017**, *208*, 1071–1082. [CrossRef]
16. Yang, H.; Wang, L.; Zhou, B.; Wei, Y.; Zhao, Q. A preliminary study on the highway piezoelectric power supply system. *Int. J. Pavement Res. Technol.* **2018**, *11*, 168–175. [CrossRef]
17. Roshani, H.; Jagtap, P.; Dessouky, S.; Montoya, A.; Papagiannakis, A.T. Theoretical and Experimental Evaluation of Two Roadway Piezoelectric-Based Energy Harvesting Prototypes. *J. Mater. Civ. Eng.* **2018**, *30*, 04017264. [CrossRef]
18. DNV KEMA Energy & Sustainability. *Final Project Report: Assesment of Piezoelectric Materials for Roadway Energy Harvesting: Cost of Energy and Demonstration Roadmap*; California Energy Commission: Sacramento, CA, USA, 2014.
19. Vázquez Rodríguez, M. Contribución al Estudio de la Generación de Energía Eléctrica a Partir de Materiales Piezoeléctricos. Ph.D. Thesis, Universidad Politécnica de Madrid, Madrid, Spain, 2019. [CrossRef]
20. Vázquez-Rodríguez, M.; Jiménez, F.J.; de Frutos, J.; Alonso, D. Piezoelectric energy harvesting computer controlled test bench. *Rev. Sci. Instrum.* **2016**, *87*, 095004. [CrossRef]
21. Vázquez-Rodríguez, M.; Jiménez, F.J.; de Frutos, J. Virtual instrument to obtain electrical models of piezoelectric elements used in energy harvesting. *Adv. Appl. Ceram.* **2018**, *117*, 201–211. [CrossRef]

© 2019 by the authors. Licensee MDPI, Basel, Switzerland. This article is an open access article distributed under the terms and conditions of the Creative Commons Attribution (CC BY) license (http://creativecommons.org/licenses/by/4.0/).

Article

Numerical Analysis of Signal Response Characteristic of Piezoelectric Energy Harvesters Embedded in Pavement

Hailu Yang [1,2], Qian Zhao [1], Xueli Guo [1], Weidong Zhang [1], Pengfei Liu [3,*] and Linbing Wang [4,*]

1. National Center for Materials Service Safety, University of Science and Technology Beijing (USTB), Beijing 100083, China; yanghailu@ustb.edu.cn (H.Y.); zhaoqian928@126.com (Q.Z.); 18291923897@163.com (X.G.); zwd@ustb.edu.cn (W.Z.)
2. Research and Development Center of Transport Industry of New Materials, Technologies Application for Highway Construction and Maintenance, Beijing 100088, China
3. Institute of Highway Engineering (ISAC), RWTH Aachen University, 52074 Aachen, Germany
4. Joint USTB Virginia Tech Lab on Multifunctional Materials, USTB, Virginia Tech, Department Civil & Environmental Engineering, Blacksburg, VA 24061, USA
* Correspondence: liu@isac.rwth-aachen.de (P.L.); wangl@vt.edu (L.W.)

Received: 31 May 2020; Accepted: 16 June 2020; Published: 18 June 2020

Abstract: Piezoelectric pavement energy harvesting is a technological approach to transform mechanical energy into electrical energy. When a piezoelectric energy harvester (PEH) is embedded in asphalt pavements or concrete pavements, it is subjected to traffic loads and generates electricity. The wander of the tire load and the positioning of the PEH affect the power generation; however, they were seldom comprehensively investigated until now. In this paper, a numerical study on the influence of embedding depth of the PEH and the horizontal distance between a tire load and the PEH on piezoelectric power generation is presented. The result shows that the relative position between the PEH and the load influences the voltage magnitude, and different modes of stress state change voltage polarity. Two mathematic correlations between the embedding depth, the horizontal distance, and the generated voltage were fitted based on the computational results. This study can be used to estimate the power generation efficiency, and thus offer basic information for further development to improve the practical design of PEHs in an asphalt pavement.

Keywords: piezoelectric energy harvester; finite element method; open circuit voltage; moving load

1. Introduction

With the continuous development of society, more and more new technologies are applied in road engineering to meet the needs of energy saving, environmental protection, and intelligent infrastructure [1–3]. The rapid digitalization of society has greatly promoted the rise and progress of intelligent road transport systems. Sensors are embedded in infrastructure and serve as intelligent nodes in a communication network. These are limited by the requirement of traditional centralized power supply or battery power supply. The development of low-cost, decentralized, and sustainable energy is a necessity to facilitate a wide application of embedded, off-grid sensors [4]. Environmental energy harvesting is the process of transforming the energy that exists in the environment (such as light, heat, mechanical, electromagnetic, biological, and wind energy, among others) into electricity that can be used; it is a potential way to extend the service life of embedded, off-grid sensors [5]. Pavements are designed for millions of axle's loads during service, which absorbs large amount of wasted mechanical energy. In recent years, researchers have paid more attention to pavement energy harvesting [6,7].

On the basis of the analysis of vehicle-induced vibrations on bridges and pavements, Ashebo et al. addressed the feasibility of vibration energy harvesting from transportation infrastructures to power wireless sensors with a peak power output of 1 mW [8]. It has been found that piezoelectric technology exhibited the highest power density for vibration energy harvesting [9]. The piezoelectric effect is an electromechanical coupling effect, which includes electrical and mechanical boundary conditions [10]. Piezoelectric materials are anisotropic and this property affects their working modes. The working mode of piezoelectric energy harvester (PEH) is determined by its stress state when it is embedded in the pavement. The working modes corresponding to different stress states are shown in Figure 1. In this case, the polarization direction of piezoelectric materials is fixed along the Z axis. When the stress direction is perpendicular to the polarization direction, the working mode is 3-1. When the stress direction is parallel to the polarization direction, the working mode is 3-3. Particularly, at the stress state shown in Figure 1a, the positive charge is generated on the upper surface of the PEH, while the negative one is generated on the lower surface in the 3-1 mode. The sign of the charge will be opposite when the stress state is converse. The relationship between the stress state and the sign of the charge in the 3-3 mode is shown in Figure 1b.

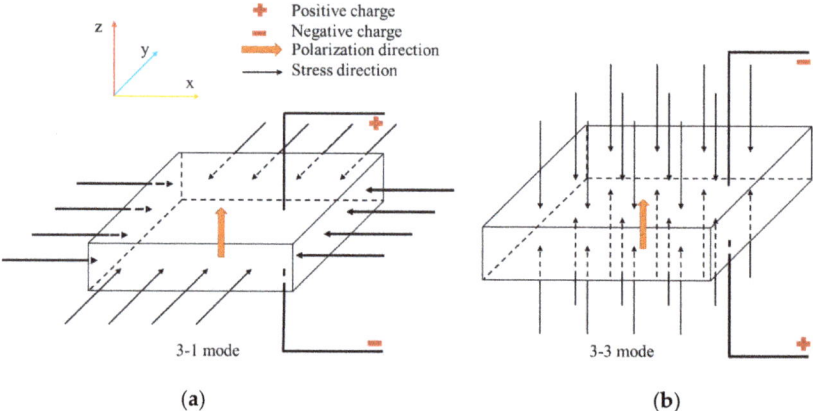

Figure 1. Working modes of the piezoelectric energy harvester (PEH): (**a**) 3-1 mode; (**b**) 3-3 mode.

Numerous piezoelectric energy harvesters (PEHs) have been designed and analyzed in different fields [11–14]. Zhao et al. designed a cymbal-shaped energy harvester for asphalt pavement. Its performance was analyzed for different sizes. The result suggested that the thicker piezoelectric transducers (PZTs) and smaller total cymbal size could increase its efficiency [15]. On the basis of the cymbal structure, Yesner et al. provided a novel electrode design to promote bridge transducer [16]. The exact output energy amount was calculated by Moure et al. considering different kinds of traffic loading by cymbal harvesters [17]. Vázquez-Rodríguez et al. presented a new prospect using lead-free piezoelectric ceramics to determine their behavior in piezoelectric-based road traffic energy harvesting applications [18]. The PEHs of the stack structure have been studied as well [19]. Two numerical methods of the electromechanical model of multilayer PZT stacks were proposed by Zhao and Erturk [20]. Ling et al. designed a kind of piezoelectric bridge transducer that produced a peak voltage of 154 V under 0.7 MPa with 5 Hz half-sine loads [21]. When applied in field experiments and pilot projects, the piezoelectric technology showed potential in pavement energy harvesting, and more analyses on the factors affecting the efficiency have been conducted [22]. Some researchers have figured out that the ambient temperature has a great impact on output power of PEHs and there is a linear relationship between the loading frequency and the output open circuit voltage [23]. Studies have found that the axle load and the axle number of passing vehicles have a significant effect on the power generation of the PEHs [24]. Furthermore, researchers have made numerous finite element (FE) studies on pavement materials and structures with PEHs [25–27]. Liu et al. studied the mechanical responses of the asphalt pavement embedded with PEHs and provided the basis for optimization design of PEHs [28].

A review of the literature shows that the wander of the tire load and the embedding location of the PEH in the pavement may significantly affect the power generation of the PEH. However, among the existing simulations and experimental analyses, their influence on the power generation efficiency has not been sufficiently analyzed. Therefore, in this paper, FE pavement models with a moving tire load were developed and verified according to a demonstration pavement embedded with PEHs. On the basis of the numerical model, the influence of different depths of PEH on the output energy was analyzed. The influence of the distance between the load and the PEH on piezoelectric power generation performance was also analyzed. The conclusions can be used to estimate the power generation efficiency of the PEH, and thus offer basic information for further development to improve the practical design of PEHs in asphalt pavement.

2. Demonstration Pavement Embedded with PEHs

The pre-fabricated PEHs were embedded in the Highway G320, a highway near the City of Kunming in China. Taking into consideration the contact patch of the tires, it was decided to design the PEH in a square shape with a side length of 0.3 m. The PEH was 0.07 m thick. The PEH was composed of four component parts: the packaging materials, the piezoelectric units, the full bridge rectifiers, and other components for sealing and fastening, as shown in Figure 2. Piezoelectric material was the core component of the PEHs. The piezoelectric unit was a cylindrical structure with a diameter of 20 mm and a height of 23 mm. The type of piezoelectric ceramics was PZT-5H. PA66 nylon with 30% glass fiber was selected for the protective packaging of the PEH owing to its high toughness, load resistance, strength, and resistance to repeated shocks. There were upper, middle, lower layers, and a middle frame, where the upper layer directly undertook the vehicle load, while the ground reaction force was supported by the lower layer, and the middle layer was left to facilitate the holes for nine piezoelectric units and nine full bridge rectifiers.

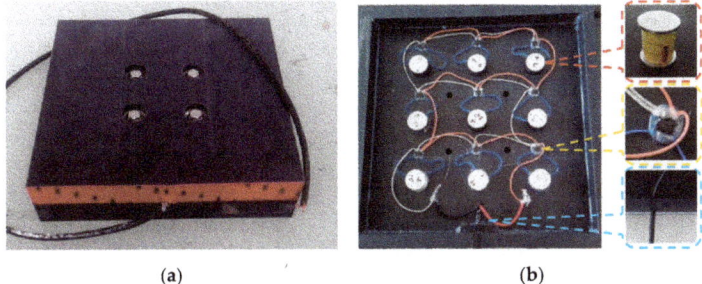

Figure 2. Illustration of the PEH: (**a**) outward appearance; (**b**) internal structure.

The output power exited via the cable once the piezoelectric units were connected to the circuit board. Water leakage was prevented by the rectifier bridge being sealed with electronic glue. This was achieved via the application of a silicone gasket between the upper and the lower encapsulation structure. To avoid stress concentration, a 0.03 m diameter stainless steel gasket was inserted between the piezoelectric materials, and then wrapped using a protection package. Therefore, the PEH had a good performance level in terms of compression, resistance to fatigue, and water resistance. Light-emitting diodes (LEDs) were connected to the PEH to receive the piezoelectric power.

Figure 3 shows the PEH installation process and the final state of the demonstration pavement. More information and results about this demonstration pavement can be found in the previous study [29].

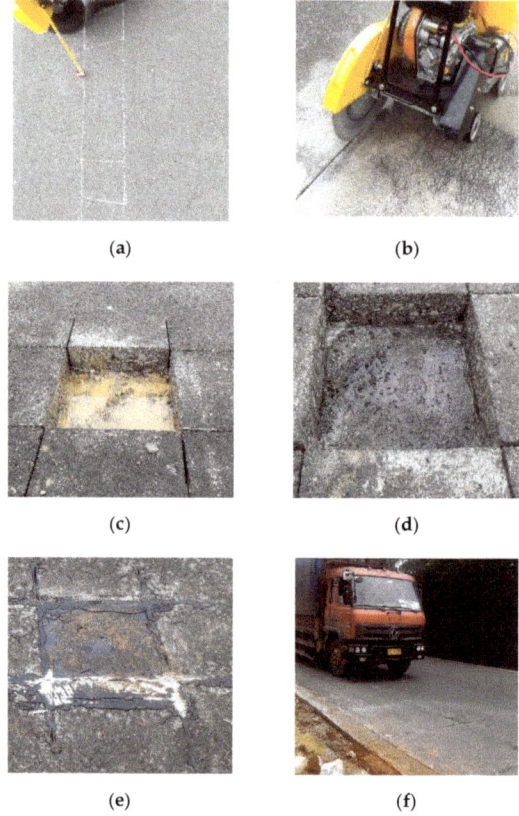

Figure 3. The flow chart of PEH installation: (**a**) positioning and scribing; (**b**) slotting; (**c**) coring for installation; (**d**) cleaning and levering the holes' bottom; (**e**) putting PEHs in the pavement and laying the cable; (**f**) final state of the demonstration pavement.

3. Development and Validation of Finite Element Model

In pavement engineering, full-scale tests are one of the most important methods for material testing and structural analysis. However, the cost of field tests has always been high; furthermore, the conditions cannot always be controlled completely in field tests [30]. Therefore, numerical simulations have been widely applied in research and have been accepted as an effective and efficient substitute [31].

In this study, the simulation and analysis of the signal response characteristic of the PEH embedded in the pavement was conducted with the FE analysis software ANSYS. The FE model was generated based on the aforementioned demonstration pavement structures and the respective material parameters. The pavement structure was simplified as a multi-layer elastic model, which was divided into three layers: surface layer, base layer, and subgrade. The specific parameters are given in Table 1. The model dimensions are as follows: length × width × thickness = 10 m × 6 m × 2.5 m.

Table 1. Geometry and material parameters of the pavement model.

Layer	Material	Thickness/m	Young's Modulus/MPa	Poisson's Ratio	Density/ kg·m^{-3}	Damping Coefficient α
Surface	AC-20	0.18	1300	0.35	2400	0.25
Base	cement stabilized macadam	0.72	1600	0.30	2100	0.25
Subgrade	Soil	1.60	50	0.40	1900	0.25

The full-scale pavement model is shown in Figure 4, in which the different colors represent the different layers. In ANSYS, the element type of SOLID185 was selected to model the three-dimensional (3D) pavement structures. In order to reduce calculation cost and speed up the simulation, the mesh density gradually becomes sparse from the middle to both sides. The surface mesh density is the largest and the mesh becomes coarser as the depth increases; the total number of elements is 700,000.

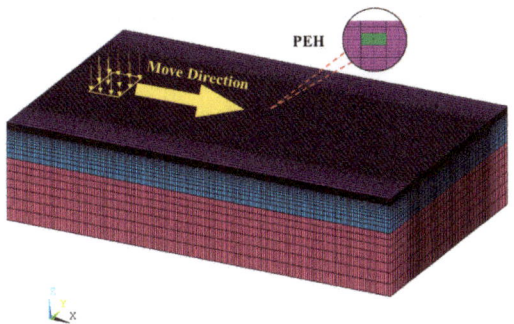

Figure 4. Full scale pavement model with ANSYS.

In the surface layer, 24 elements with element type of SOLID5 were set to be piezoelectric materials representing a PEH. These elements have a 3D piezoelectric and structural field capability with coupling between the fields. In this research, the PEH was set to be anisotropic and its input parameters include density ρ, dielectric constant ε, stiffness **C**, and piezoelectric stress constant **e**. These parameters were determined based on laboratory test and the values are listed in Table 2. For the sake of brevity, the details about the definition of the material properties can refer to other research [32].

Table 2. Material properties for the piezoelectric energy harvester (PEH).

Density ρ (kg·m^{-3})	Dielectric Constant ε (C/Vm)	Stiffness C (MPa)	Piezoelectric Stress Constant e (C/m^2)
7500	$\varepsilon_{11} = 3.27 \times 10^{-9}$	$C_{11} = 13.9 \times 10^{10}$	$e_{31} = -5.2$
-	$\varepsilon_{33} = 5.16 \times 10^{-9}$	$C_{12} = 7.78 \times 10^{10}$	$e_{33} = 15.1$
-	-	$C_{13} = 7.43 \times 10^{10}$	$e_{15} = 12.7$
-	-	$C_{33} = 11.5 \times 10^{10}$	-
-	-	$C_{44} = 2.56 \times 10^{10}$	-
-	-	$C_{66} = 3.06 \times 10^{10}$	-

The horizontal position of the PEH was in the center of the pavement. The buried depth of the PEH was defined as the distance from the pavement surface to the upper surface of the PEH. In order to study the influence of the buried depth on the piezoelectric voltage, five pavement models with different buried depths of the PEH were created, and the corresponding depths are 2 cm, 4 cm, 6 cm, 8 cm, and 10 cm, respectively.

The displacements of nodes on the four boundary sides of the model were constrained in the X and Y directions, while the displacements of the bottom surface nodes were constrained in the X, Y, and Z directions. The other nodes were not constrained. Because of the existence of electrodes, the upper and lower surfaces of the PEH were considered as voltage equipotential surfaces. In the FE model, the electrode was simulated by coupling the voltage degrees of freedom of the nodes on the upper and lower surfaces, respectively. The nodes coupled with voltage degrees of freedom would keep the same voltage value in the simulation process. Moreover, the voltage of the bottom surface of the PEH was constrained to 0 V.

A truck tire load was applied in the model, which was assumed as rectangular with 30 cm wide and 20 cm long, as shown in Figure 4. It moved along the longitudinal direction with a contact pressure of 0.7 MPa. Five different loading paths were considered in the simulation to investigate the influence of the tire load on the piezoelectric voltage, that is, the horizontal distances between the center of the load and the center of the PEH were set as 0, 0.2, 0.4, 0.6, and 0.8 m, respectively. The moving speed of the tire was equivalent to a velocity of 40 km/h, which was determined by setting up a step function in a different duration on any of the passing elements [33]. During the loading process, the upper surface voltage of the piezoelectric structures was monitored.

The numerical model was verified by comparing the computational value with measured ones derived from the demonstration pavement. In the comparison, the tire load directly passes the PEH and the buried depth of the PEH is 2 cm. The peak voltage derived from the simulation is 626 V, whereas those obtained from the two measurements are 704 V and 608 V, respectively. A range of error of 20% is considered to allow for uncertainties and fluctuations. As the absolute values of relative error are 11% and 3% between the simulation and measurement, based on this criterion, the reliability of the developed FE model is validated.

4. Results and Discussion

4.1. Open Circuit Voltage Response of PEH

The FE simulations were carried out based on the validated models. Different horizontal distances between the tire load and the PEH as well as the different buried depths of the PEH were considered. The numerical piezoelectric voltages of the open circuit structure are shown in Figure 5.

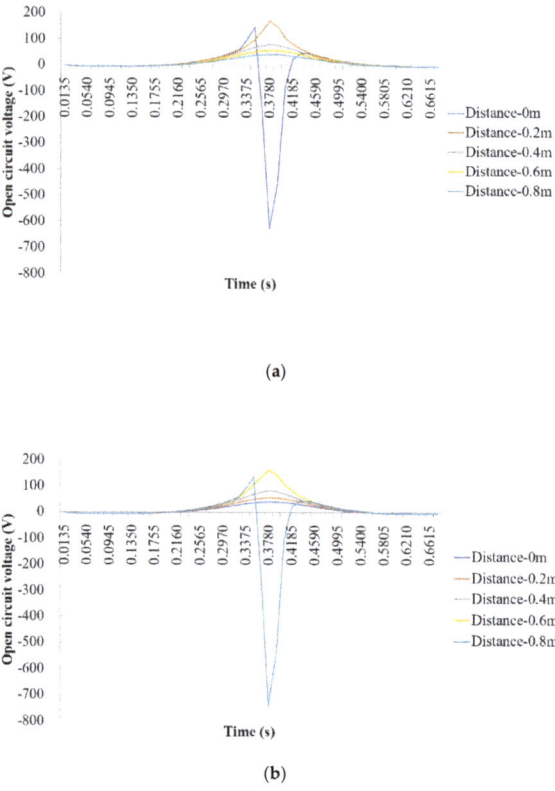

Figure 5. Cont.

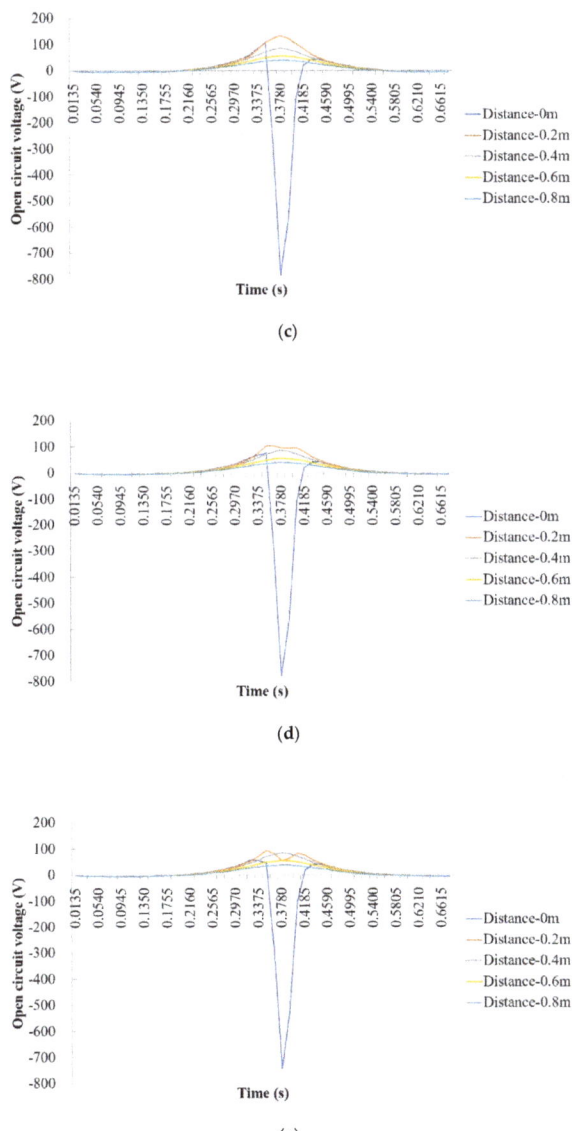

Figure 5. Open circuit voltage response of PEH at different horizontal loading distances and buried depths: (**a**) buried depth of 2 cm; (**b**) buried depth of 4 cm; (**c**) buried depth of 6 cm; (**d**) buried depth of 8 cm; (**e**) buried depth of 10 cm.

In general, the responses of the open circuit voltage of PEH at different buried depths are similar, that is, the changes over time are similar, while the values are different. Particularly, when the tire load passes directly above the PEH (the horizontal distance equals to 0 m), the stress state of the PEH is first 3-1 mode, then 3-3 mode, and finally 3-1 mode. The output from the 3-1 mode and 3-3 mode produces positive and negative voltage, respectively. Therefore, the values of the open circuit voltage are first positive, then become negative, and finally return to be positive. When the loading path does not pass directly above the PEH, it is not subjected to vertical compression, the main stress is horizontal stress, and the stress state is shifted from 3-3 mode to 3-1 mode. The power generation is a combination of the two modes. By comparison, 3-3 mode has high power generation capacity, but its corresponding range is small, while 3-1 mode has the opposite effect. As a result, in the application of pavement energy harvesting, the design of embedding location of the PEH should consider the spatial distribution of the traffic vehicle load.

4.2. Maximum Positive Voltage of PEH

The positive voltage peaks at different buried depths of the PEH and horizontal distances are concluded and listed in Table 3. In order to see the changes, the data are drawn in Figure 6.

Table 3. Maximum positive voltage of PEH.

Buried Depth	Peak Voltage at Different Horizontal Distances (V)				
	0 cm	20 cm	40 cm	60 cm	80 cm
2 cm	146.99	170.40	78.36	56.58	41.56
4 cm	135.15	162.534	82.73	56.76	41.18
6 cm	108.14	133.49	85.50	56.44	40.46
8 cm	76.44	103.68	86.51	55.82	39.60
10 cm	45.72	94.08	85.36	54.80	38.54

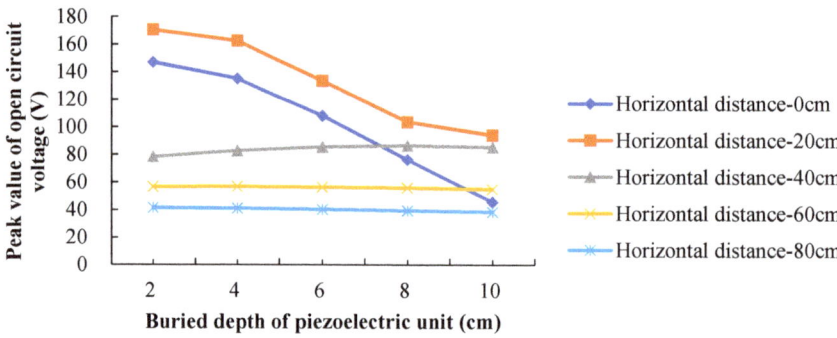

Figure 6. Response of PEH at different buried depths with different horizontal loading distances.

It can be seen that, when the load passes directly above the PEH, the peak voltage decreases sharply with the increase of the buried depth. When the horizontal distance between the load and the PEH is 20 cm, the buried depth of the PEH still has a great influence on the response, that is, the piezoelectric voltage decreases significantly as the depth increases. When the horizontal distance reaches and exceeds 40 cm, the influence of the buried depth of the PEH on the open circuit voltage is sharply reduced. It is worth noting that the peak voltage produced by the PEH with horizontal loading distance of 20 cm is larger than the one in case the load passes directly above the PEH. Complex stress state within the pavement structure is generated when the pavement bears the moving tire load. As the PEH is embedded in the pavement, it withstands not only the vertical stress, but also the horizontal stresses. The voltage polarity produced by the vertical stress is opposite to the polarity of the voltage produced by the horizontal stresses. Therefore, the piezoelectric output is not necessarily better the closer it is to the load, as it depends on its stress state.

To comprehensively quantify the correlation among horizontal distance, buried depth, and peak voltage, the data were fitted with the aid of the software TableCurve 3D. A 3D curve fitting plot is shown in Figure 7. The influence of the horizontal distance (D) and buried depth (H) on the peak voltage (U) is made apparent in the 3D curve fitting image. The fitting equation is shown below:

$$U = 159.53 + 3.19D + 0.88H - 2.67D^2 - 0.11H^2 + 0.34DH + 0.12D^3 \\ + 0.0010H^3 - 0.0032DH^2 + 0.0075D^2H \quad (1)$$

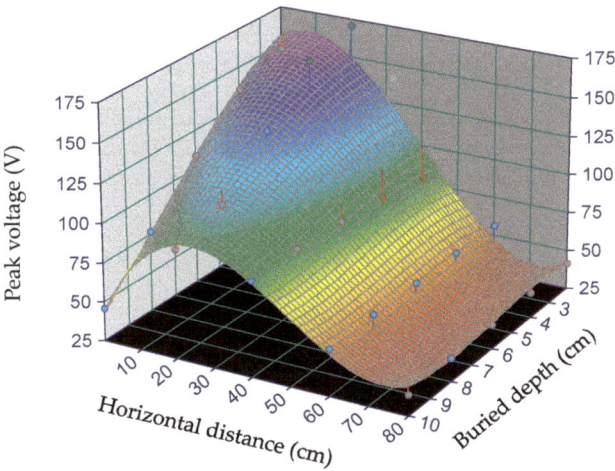

Figure 7. The 3D curve fitting image of horizontal distance D, buried depth H, and peak voltage U.

The goodness-of-fit of the fitting equation is plotted in Figure 8. The correlation coefficient R^2 of the equation is 0.95, which is high enough to prove the good applicability of the fitting equation. With this equation, the peak voltage produced by the PEH can be estimated based on the loading distance and buried depth of the PEH, and thus promote optimization of the pavement design with PEHs.

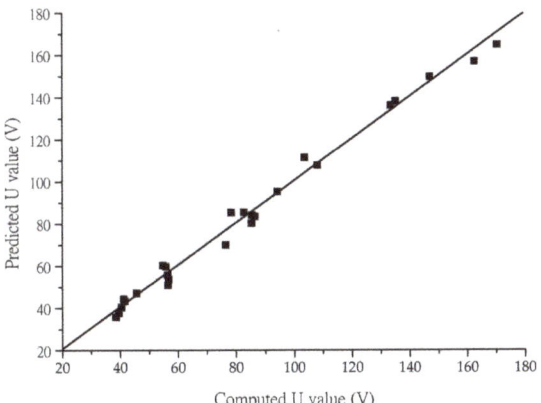

Figure 8. The goodness-of-fit plot for the fitting equation.

4.3. Maximum Negative Voltage of PEH

As aforementioned, when the tire load passes directly above the PEH (the horizontal distance equals to 0 m), the stress state of the PEH is first 3-1 mode, then 3-3 mode, and finally 3-1 mode. The output from the 3-3 mode produces negative voltage. The peak negative voltages derived from this process are presented in Figure 8. As can be seen from Figure 9, the absolute value of the voltage produced by the PEH increases first and then decreases with the increase of the buried depth of the PEH. The reason can be that the polarity and magnitude of the voltage are the result of the entire stress state born within the PEH. The stress evolution along the pavement depth does not change linearly, which has been proven in previous research [34]. The maximum open circuit voltage is the highest when the PEH is buried at the depth of 6 cm and has an output of −779.2 V. The polynomial fitting equation for the relationship between open circuit peak voltage (U) and buried depth (H) is as follows:

$$U = -0.3846H^3 + 13.215H^2 - 125.38H - 424.99 \tag{2}$$

The correlation coefficient R^2 of the equation is 1, which can prove the good applicability of the fitting equation. According to Equation (2), the buried depth H corresponding to the maximum open circuit voltage is 6.7 cm, and the maximum open circuit voltage is −787.49V.

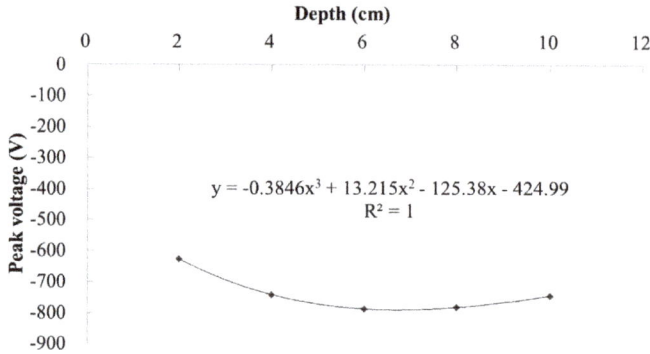

Figure 9. Curves of maximum negative voltages with buried depth when load passes directly above the PEH.

5. Conclusions and Future Work

Owing to the change of stress distributions in the pavement under moving load, the stress state of PEH is not simple, which in turn affects the piezoelectric response. A full scale pavement model with a moving tire load was developed based on the FE method. The influence of buried depth of the PEH and the horizontal distance between the tire load and the PEH on piezoelectric power generation was investigated. The following results can be concluded:

(1) The responses of the open circuit voltage of PEH at different buried depths are similar. When the tire load passes directly above the PEH, the values of the open circuit voltage are first positive, then become negativem and finally return to being positive. When the loading path does not pass directly above the PEH, the value of the open circuit voltage is always positive.

(2) When the load passes directly above the PEH, the peak voltage decreases sharply with the increase of the buried depth. When the horizontal distance between the load and the PEH is 20 cm, the buried depth of the PEH still has a great influence on the response. When the horizontal distance reaches and exceeds 40 cm, the influence of the buried depth of the PEH on the open circuit voltage is sharply reduced.

(3) The power generation is a combination of the 3-1 mode and 3-3 mode. By comparison, 3-3 mode has high power generation capacity, but its corresponding range is small, while 3-1 mode has the opposite effect. The piezoelectric output is not necessarily better the closer it is to the load; it depends on its stress state.

(4) Two mathematic correlations between the buried depth, the horizontal distance, and the generated voltage were fitted based on the computational results. The correlation coefficients of the equations are high enough to prove their good applicability.

The result derived from this study is feasible to be used to estimate the power generation efficiency of the PEHs under different working situations, and thus provide an opportunity to improve the practical design of the pavement with the PEHs. For the future work, a series of PEH with different arrangement methods will be considered. Viscoelasticity of the asphalt materials will be taken into account to compute the recoverable and irrecoverable deformations of asphalt pavements with PEH [35]. A real tire-pavement interaction FE model will be created by simulating an explicit passing tire [36]. The influence of the temperature on the mechanical responses of the asphalt pavement with PEH will be studied as well.

Author Contributions: Conceptualization, H.Y. and L.W.; Data curation, Q.Z.; Formal analysis, P.L.; Funding acquisition, H.Y., W.Z., and P.L.; Methodology, H.Y.; Writing—original draft, Q.Z. and X.G.; Writing—review & editing, H.Y., P.L., and L.W. All authors have read and agreed to the published version of the manuscript.

Funding: This research was funded by China Postdoctoral Science Foundation: 2019M660467, Fundamental Research Funds for the Central Universities: FRF-TP-18-048A1, and Excellence Strategy of the German Federal and State Governments: StUpPD373-20.

Conflicts of Interest: The authors declare no conflict of interest.

References

1. Yang, Q.; Li, X.; Zhang, L.; Qian, Y.; Qi, Y.Z.; Kouhestani, H.S.; Shi, X.; Gui, X.; Wang, D.; Zhong, J. Performance evaluation of bitumen with a homogeneous dispersion of carbon nanotubes. *Carbon* **2020**, *158*, 465–471. [CrossRef]
2. Yang, Q.; Hong, B.; Lin, J.; Wang, D.; Zhong, J.; Oeser, M. Study on the reinforcement effect and the underlying mechanisms of a bitumen reinforced with recycled glass fiber chips. *J. Clean Prod.* **2020**, *251*, 119768. [CrossRef]
3. Lu, G.; Fan, Z.; Sun, Z.; Liu, P.; Leng, Z.; Wang, D.; Oeser, M. Improving the polishing resistance of cement mortar by using recycled ceramic. *Resour. Conserv. Recycl.* **2020**, *158*, 104796. [CrossRef]
4. Su, Y.; Lin, X.; Huang, R.; Zhou, Z. Analytical Electromechanical Modeling of Nanoscale Flexoelectric Energy Harvesting. *Appl. Sci.* **2019**, *9*, 2273. [CrossRef]
5. Clemente, C.S.; Davino, D. Modeling and Characterization of a Kinetic Energy Harvesting Device Based on Galfenol. *Materials* **2019**, *12*, 3199. [CrossRef]
6. Wang, H.; Jasim, A.; Chen, X. Energy harvesting technologies in roadway and bridge for different applications—A comprehensive review. *Appl. Energy* **2018**, *212*, 1083–1094. [CrossRef]
7. Ambrożkiewicz, B.; Litak, G.; Wolszczak, P. Modelling of Electromagnetic Energy Harvester with Rotational Pendulum Using Mechanical Vibrations to Scavenge Electrical Energy. *Appl. Sci.* **2020**, *10*, 671. [CrossRef]
8. Ashebo, D.B.; Tan, C.A.; Wang, J.; Li, G. Feasibility of energy harvesting for powering wireless sensors in transportation infrastructure applications. In *Nondestructive Characterization for Composite Materials, Aerospace Engineering, Civil Infrastructure, and Homeland Security 2008, 69340Y; Proceedings of the SPIE 6934*; Society of Photo Optical: Bellingham, WA, USA, 2008. [CrossRef]
9. Torres, E.O.; Rincón-Mora, G.A. Long-lasting, self-sustaining, and energy-harvesting system-in-package (SIP) wireless micro-sensor solution. In Proceedings of the International Conference on Energy, Environment and Disasters (INCEED 2005), Charlotte, NC, USA, 24–30 July 2005.
10. Butcher, M.; Davino, D.; Giustiniani, A.; Masi, A. An experimental evaluation of the fully coupled hysteretic electro-mechanical behaviour of piezoelectric actuators. *Phys. B Condens. Matter* **2016**, *486*, 116–120. [CrossRef]
11. Wang, C.; Chen, S.; Li, Y.; Shi, X.; Li, Q. Design of piezoelectric elements' protection measures and energy output of intelligent power pavement. *China J. Highw. Transp.* **2016**, *29*, 41–49.
12. Duarte, F.; Champalimaud, J.P.; Ferreira, A. Waynergy Vehicles: An innovative pavement energy harvest system. *Proc. Inst. Civ. Eng.-Munic.* **2016**, *169*, 13–18. [CrossRef]
13. Ferreira, A.J.L. Briefing: Recent developments in pavement energy harvest systems. *Proc. Inst. Civ Eng.-Munic.* **2012**, *165*, 189–192. [CrossRef]
14. Song, G.; Li, H.; Cai, S.C. Editorial for Special Issue "Energy Dissipation and Vibration Control: Materials, Modeling, Algorithm, and Devices". *Appl. Sci.* **2020**, *10*, 572. [CrossRef]
15. Zhao, H.; Yu, J.; Ling, J. Finite element analysis of Cymbal piezoelectric transducers for harvesting energy from asphalt pavement. *J. Ceram. Soc. Jpn.* **2010**, *118*, 909–915. [CrossRef]

16. Yesner, G.; Kuciej, M.; Safari, A.; Jasim, A.; Wang, H.; Maher, A. Piezoelectric Energy Harvesting Using a Novel Cymbal Transducer Design. In Proceedings of the Joint IEEE International Symposium on The Applications of Ferroelectrics, European Conference on Application of Polar Dielectrics, And Piezoelectric Force Microscopy Workshop (ISAF/ECAPD/PFM), Darmstadt, Germany, 21–25 August 2016.
17. Moure, A.; Rodríguez, M.A.I.; Rueda, S.; Hernández, G.A.; Rubio-Marcos, F.; Cuadros, D.U.; Pérez-Lepe, A.; Fernández, J.F. Feasible integration in asphalt of piezoelectric cymbals for vibration energy harvesting. *Energy Convers. Manag.* **2016**, *112*, 246–253. [CrossRef]
18. Vázquez-Rodríguez, M.; Jiménez, F.J.; Pardo, L.; Ochoa, P.; González, A.M.; de Frutos, J. A New Prospect in Road Traffic Energy Harvesting Using Lead-Free Piezoceramics. *Materials* **2019**, *12*, 3725. [CrossRef]
19. Yang, H.; Wang, L.; Hou, Y.; Guo, M.; Ye, Z.; Tong, X.; Wang, D. Development in Stacked-Array-Type Piezoelectric Energy Harvester in Asphalt Pavement. *J. Mater. Civil Eng.* **2017**, *29*, 04017224. [CrossRef]
20. Zhao, S.; Erturk, A. Energy harvesting from harmonic and noise excitation of multilayer piezoelectric stacks: Modeling and experiment. In *Proceedings of Spie. Bellingham: Spie-Int Soc Optical Engineering*; Sodano, H.A., Ed.; SPIE: Bellingham, WA, USA, 2013.
21. Yao, L.; Zhao, H.; Dong, Z.; Sun, Y.; Gao, Y. Laboratory testing of piezoelectric bridge transducers for asphalt pavement energy harvesting. *Key Eng. Mater.* **2012**, *492*, 172–175. [CrossRef]
22. Yang, H.; Cao, D. An investigation on stress distribution effect on multi-piezoelectric energy harvesters. *Front. Struct. Civ. Eng.* **2017**, *11*, 1–7. [CrossRef]
23. Yang, H.; Guo, M.; Wang, L.; Hou, Y.; Zhao, Q.; Cao, D.; Zhou, B.; Wang, D. Investigation on the factors influencing the performance of piezoelectric energy harvester. *Road Mater. Pavement Des.* **2017**, *18* (Suppl. S3), 180–189. [CrossRef]
24. Xiong, H.; Wang, L. Piezoelectric energy harvester for public roadway: On-site installation and evaluation. *Appl. Energy* **2016**, *174*, 101–107. [CrossRef]
25. Liu, P.; Chen, J.; Lu, G.; Wang, D.; Oeser, M.; Leischner, S. Numerical simulation of crack propagation in flexible asphalt pavements based on cohesive zone model developed from asphalt mixtures. *Materials* **2019**, *12*, 1278. [CrossRef] [PubMed]
26. Zhao, H.; Ling, J.; Yu, J. A comparative analysis of piezoelectric transducers for harvesting energy from asphalt pavement. *J. Ceram. Soc. Jpn.* **2012**, *120*, 317–323. [CrossRef]
27. Kollmann, J.; Liu, P.; Lu, G.; Wang, D.; Oeser, M.; Leischner, S. Investigation of the microstructural fracture behaviour of asphalt mixtures using the finite element method. *Constr. Build. Mater.* **2019**, *227*, 117078. [CrossRef]
28. Liu, P.; Zhao, Q.; Yang, H.; Wang, D.; Oeser, M.; Wang, L.; Tan, Y. Numerical study on influence of piezoelectric energy harvester on asphalt pavement structural responses. *J. Mater. Civ. Eng.* **2019**, *31*, 04019008. [CrossRef]
29. Yang, H.; Wang, L.; Zhou, B.; Wei, Y.; Zhao, Q. A preliminary study on the highway piezoelectric power supply system. *Int. J. Pavement Res. Technol.* **2018**, *2*, 168–175. [CrossRef]
30. Otto, F.; Liu, P.; Zhang, Z.; Wang, D.; Oeser, M. Influence of temperature on the cracking behavior of asphalt base courses with structural weaknesses. *Int. J. Transp. Sci. Technol.* **2018**, *7*, 208–216. [CrossRef]
31. Liu, P.; Xing, Q.; Dong, Y.; Wang, D.; Oeser, M.; Yuan, S. Application of finite layer method in pavement structural analysis. *Appl. Sci.* **2017**, *7*, 611. [CrossRef]
32. Ding, H.; Liang, J. The fundamental solutions for transversely isotropic piezoelectricity and boundary element method. *Comput. Struct.* **1999**, *71*, 447–455. [CrossRef]
33. Liu, P.; Wang, D.; Otto, F.; Oeser, M. Application of semi-analytical finite element method to analyze the bearing capacity of asphalt pavements under moving loads. *Front. Struct. Civ. Eng.* **2018**, *12*, 215–221. [CrossRef]
34. Liu, P.; Wang, D.; Oeser, M. Application of semi-analytical finite element method to analyze asphalt pavement response under heavy traffic loads. *J. Traffic Transp. Eng. (Engl. Ed.)* **2017**, *4*, 206–214. [CrossRef]
35. Blasl, A.; Khalili, M.; Falla, G.C.; Oeser, M.; Liu, P.; Wellner, F. Rheological characterisation and modelling of bitumen containing reclaimed components. *Int. J. Pavement Eng.* **2019**, *20*, 638–648. [CrossRef]
36. Liu, P.; Ravee, V.; Wang, D.; Oeser, M. Study of the influence of pavement unevenness on the mechanical response of asphalt pavement by means of the finite element method. *J. Traffic Transp. Eng. (Engl. Ed.)* **2018**, *5*, 169–180. [CrossRef]

© 2020 by the authors. Licensee MDPI, Basel, Switzerland. This article is an open access article distributed under the terms and conditions of the Creative Commons Attribution (CC BY) license (http://creativecommons.org/licenses/by/4.0/).

Article

Harvesting Variable-Speed Wind Energy with a Dynamic Multi-Stable Configuration

Yuansheng Wang [1], Zhiyong Zhou [1,2], Qi Liu [1], Weiyang Qin [1,*] and Pei Zhu [1]

[1] Department of Engineering Mechanics, Northwestern Polytechnical University, Xi'an 710072, China; wangyuansheng@nwpu.edu.cn (Y.W.); 10160091@vip.henu.edu.cn (Z.Z.); liuqiarc@163.com (Q.L.); 10160096@vip.henu.edu.cn (P.Z.)

[2] School of Civil Engineering and Architecture, Henan University, Kaifeng 475004, China

* Correspondence: mengg@nwpu.edu.cn

Received: 26 February 2020; Accepted: 17 March 2020; Published: 19 March 2020

Abstract: To harvest the energy of variable-speed wind, we proposed a dynamic multi-stable configuration composed of a piezoelectric beam and a rectangular plate. At low wind speeds, the system exhibits bi-stability, whereas, at high wind speeds, the system exhibits a dynamic tri-stability, which is beneficial for harvesting variable-speed wind energy. The theoretical analysis was carried out. For validation, the prototype was fabricated, and a piezoelectric material was bonded to the beam. The corresponding experiment was conducted, with the wind speed increasing from 1.5 to 7.5 m/s. The experiment results prove that the proposed harvester could generate a large output over the speed range. The dynamic stability is helpful to maintain snap-through motion for variable-speed wind. In particular, the snap-through motion could reach coherence resonance in a range of wind speed. Thus, the system could keep large output in the environment of variable-speed wind.

Keywords: wind energy harvesting; snap-through motion; dynamic stability; variable-speed

---

1. Introduction

With the development of wireless sensor networks (WSNs), the problem of sensor power is becoming more and more prominent. Thus, harvesting energy from ambient nature and storing it for sensor consumption has received extensive attention in recent decades. In the future, this promising technology may replace the traditional mode of supplying power for sensors with chemical batteries in many extreme environments.

Wind, as a clean, abundant and renewable energy source, is regarded as one of the most promising alternatives to traditional energy. Wind energy harvesting has been studied and developed in both microscale and macroscale [1]. Harvesting wind energy by aerodynamic instability has received increasing attention due to its high power density and easy implementation in practice [2]. Some kinds of wind harvesters have been proposed by exploiting flow-induced vibrations, e.g., flutter, vortex-induced vibration (VIV) and galloping [3,4].

Flutter is a typical flow-induced instability which results from dynamic fluid-structure interaction [5]. According to the mechanism of flutter, the energy harvesters by flutter could be divided into two categories: the extraneously induced energy harvester (EIEH) and the movement-induced energy harvester (MIEH). The flutter motion of EIEH is excited and sustained by a gradient flow pressure caused by von Kármán vortices. One of the primary studies of EIEH was conducted by Allen and Smits [6]. It was demonstrated that the vortex shedding behind a flat plate could be used to excite a piezoelectric membrane to generate electrical power. Akaydin et al. [7,8] placed the flexible piezoelectric cantilever beams in the wake of a circular cylinder to scavenge flow energy. Goushcha et al. [9] immersed a piezoelectric beam in the wake of bluff bodies and employed interaction between the vortex and the flexible beam to harvest flow energy. Hobbs and Hu [10] presented a new piezoelectric harvester that applied tree-like swaying to harvest wind energy. By adjusting the center-to-center distance of the cylinder, the optimal configuration could be obtained. Lahooti and Kim [11] investigated the multi-body interactions in a hydrofoil and its influence on the power generation efficiency. Movement-induced excitation (MIE) is another type of flutter which results from resonantly bending instability caused by the interplay of fluid and elastic forces. Some harvesters were designed based on this mechanism (MIEH). The instability may result in a self-excited motion when the wind speed exceeds the threshold. Tang et al. [12] presented the concept of a flutter-mill based harvester and compared its performance with a representative Horizontal Axis Wind Turbine. Dunnmon et al. [13] put the piezoelectric laminates onto a flexible beam, which could achieve an excellent performance by executing a nonlinear limit cycle motion. The experimental results agree well with the theoretical predictions. The output power per unit length could reach up to 870 W/m at the flow speed of 27 m/s. The response of flexible beam fluttering in viscous flow was studied by Akcabay [14]. The results show that the performance of the harvester depends on the viscous force involved. Moreover, the flutter of a flexible flag was utilized to convert flow energy to electricity [15]. For self-sustained oscillations, the high fluid load could improve the efficiency of energy harvesting. Giacomello et al. [16] explored the flutter instability of a heavy flag hosted ionic-polymer-metal composite. They optimized the parameters of the system so as to maximize the output power. Li et al. [17] proposed a cross-flow stalk-leaf wind energy harvester and tried increasing output power through fluttering motion. The influences of parameters on a harvester from aeroelastic flutter vibration were studied by Bryant [18]. It was found that modifying the parameters, e.g., the flap mass location, the flap mass moment of inertia or the flap mass, could change the cut-in speed and obtain the maximum power. Li et al. [19] compared three operation modes of the polymer piezoelectric energy harvesters and found an optimum value. Zhao et al. [20] designed and experimentally tested three rectangular wings with different aspect ratios. Aquino et al. [21] proposed a Wind-Induced Flutter Energy Harvester (WIFEH). The experimental investigation of the WIFEH was carried out in a wind tunnel. The results examined the WIFEH under various wind tunnel wind speeds varying from 2.3 up to 10 m/s. The WIFEH could generate a RMS voltage of 3 V at v = 2.3 m/s. At v = 5 m/s, the RMS voltage could reach 4.88 V. Shan et al. [22] presented a novel flutter-induced vibration energy harvester with a curved panel for harvesting energy. The experimental results show that the harvesting performance with the segmented piezoelectric patches is better than that of the continuous ones. A sustained output power density of 0.032 mW/cm^3 is obtained at the airflow velocity of 25 m/s. Orrego et al. [23] reported the flutter of an inverted piezoelectric flag fixed in an orientation in ambient wind conditions. Generally, the wind energy harvesters based on linear mechanism may exhibit a good performance for a flow speed, but will be inefficient if the wind speed fluctuates. On the other hand, the aero-dynamic instability often gives rise to a divergent response, which is not desired for keeping a stable electrical output [24].

To overcome this drawback, some researchers tried introducing nonlinear forces and multi-stability so as to enhance the power output performance [25]. Some investigations focused on multi-stability and its nonlinear characteristics. The results suggest that the multi-stable nonlinear elements are helpful to improve harvesting performance of flow energy. Alhadidi and Daqaq [26] designed a bi-stable wake-galloping energy harvester to improve the response bandwidth for varying wind speed. Zhang et al. [27] used two small magnets to form a bi-stable VIV harvester. Huynh and Tjahjowidodo [28] introduced bi-stable springs to broaden the working range of VIV energy harvester. Nasser et al. [29] performed a comparative study on mono-stable and bi-stable VIV harvesters. The results show that the mono-stable harvester exhibits a hardening behavior, while the bi-stable one exhibits a softening behavior. It is found that both harvesters can widen the synchronization region. Valipour et al. [30] considered a hollow cylindrical tube with flowing fluid, which is supported on a Pasternak-type elastic medium. The parameterized perturbation method was used to solve the nonlinear dynamical equation. The results show that the nonlinear flow-induced frequency will change greatly when the amplitude, flow velocity and nonlocal parameter are large. Inspired by the concept of nonlinear multi-stability, a dynamic-stable flutter energy harvester (DFEH) was proposed. It exhibits different types of stability for different wind speeds. The experimental results proved that it could exhibit bi-stability for the low wind speed and tri-stability for the high wind speed [31]. However, the underlying mechanism was not elucidated fully. In this paper, we carried out the theoretical analysis on the DFEH, and proved that with the increase in flow speed, the system could exhibit tri-stability; then, with the wind speed varying from v = 1.5 to 7.5 m/s, we carried out corresponding experimental research; the results proved the transition from the bi-stability to the dynamic tri-stability.

In the natural environment, the wind usually is weak and its speed is variable with time. For WSNs, the node sensors are generally scattered in a large area, and the traditional wind energy harvesting and transmission mode cannot achieve the goal of supplying power to each sensor. Therefore, it is of great significance to design a micro wind energy harvesting structure which could integrate with the sensor to construct a system in the environment of weak and variable-speed wind. The proposed multi-stable harvester could realize snap-through motion, even coherence resonance, over a wide range of wind speeds, thus it could keep a large output for variable-speed wind. The experiment results prove this superiority.

The remainder of the paper is organized as follows: the dynamical model of the DFEH is established first; then, the potential energy analysis is carried out; subsequently, the experiment studies are described for the wind speed ranging from 1.5 to 7.5 m/s; the strain response and the output voltage are obtained for each wind speed, some are magnified to clearly show the jumping characteristics in the bi-stable and dynamic tri-stable states. Finally, some useful conclusions are drawn.

2. Dynamic-Stable Flutter Energy Harvester

The schematic of the proposed DFEH is illustrated in Figure 1, which consists of a piezoelectric cantilever, a rectangular wing and three magnets. The tip and fixed magnets are designed to be attractive, in such a way that the piezoelectric beam can possess a bi-stable characteristic. The rectangular wing is designed such as to induce flutter instability in the air field, which may lead the system to approach one static equilibrium position or jump between two equilibrium positions. Furthermore, as the incoming flow passes through, the combined effect of elastic force, magnetic force and aerodynamic force could make the system have different equilibrium positions for different wind speeds. This characteristic is helpful for exciting snap-through motion for variable-speed wind.

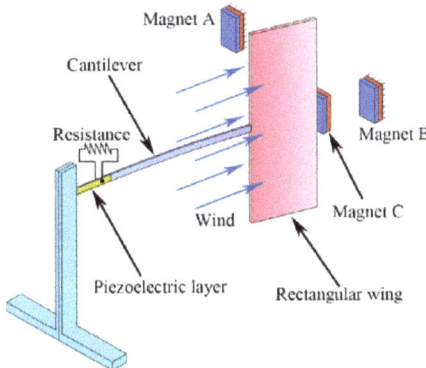

Figure 1. Schematic of the proposed dynamic-stable flutter energy harvester (DFEH).

The dynamical characteristics of the DFEH may be reflected by its potential energy feature. Therefore, for the first step, we derived its potential energy. The total potential energy of the DFEH includes two parts, i.e, the elastic part and the magnetic part.

The elastic potential energy can be calculated from Euler–Bernoulli theory and Hooke's law. For an elastic beam, the strain and stress can be given by

$$\varepsilon_{xx} = -z \frac{\partial w(x,t)}{\partial x^2} \quad (1)$$

$$\sigma_{xx} = E_s \varepsilon_{xx} \quad (2)$$

where ε_{xx} and σ_{xx} represent the normal stress and the normal strain, respectively; E_s is the Young's modulus of the beam; $w(x,t)$ represents the transverse displacement at distance x from the clamped side and at instant t. Then, the elastic potential energy of the beam can be given by:

$$U_s = \frac{1}{2} \int_0^{L_s} \int_A \sigma_{xx} \varepsilon_{xx} dA dx \quad (3)$$

where A is the area of cross section of the beam. Substituting Equations (1) and (2) into Equation (3) yields

$$U_s = \frac{1}{2} E_s I_s \int_0^{L_s} [w''(x,t)]^2 dx \quad (4)$$

where I_s is the inertia moment and can be formulated as $I_s = \frac{1}{12} b_s h_s^3$.

Similarly, the elastic potential energy of piezoelectric laminate can be given by

$$U_p = \frac{1}{2} E_p I_p \int_0^{L_p} [w''(x,t)]^2 dx \quad (5)$$

where I_p is the inertia moment of piezoelectric laminate and can be given by $\frac{1}{12} b_p h_p (4h_p^2 + 6 b_p h_s + 3 h_s^2)$; E_p is the Young's modulus of piezoelectric laminate.

Moreover, the elastic potential energy of rectangular wing can be given by

$$U_{Fp} = \frac{1}{2} E_{Fp} I_{Fp} \int_{L_s - L_{Fp}}^{L_s} [w''(x,t)]^2 dx \quad (6)$$

where the E_{Fp} is Young's modulus of rectangular wing; I_{Fp} is the inertia moment of the rectangular wing.

As a result, the total elastic potential energy of system can be given by

$$U_e = U_s + U_p + U_{Fp}$$
$$= \frac{1}{2} E_s I_s \int_0^{L_s} [w''(x,t)]^2 dx + \frac{1}{2} E_p I_p \int_0^{L_p} [w''(x,t)]^2 dx \quad (7)$$
$$+ \frac{1}{2} E_{Fp} I_{Fp} \int_{L_s - L_{Fp}}^{L_s} [w''(x,t)]^2 dx$$

In deriving magnetic potential energy, the three magnets could be modeled as point dipoles. The geometric configuration of the tip magnet and two external magnets is illustrated in Figure 2. The total magnetic potential energy generated by magnets A and B upon magnet C could be given by [32]

$$U_m = -\frac{\mu_0}{4\pi} \left(\nabla \frac{m_A \cdot r_{AC}}{\|r_{AC}\|_2^3} \right) \cdot m_c - \frac{\mu_0}{4\pi} \left(\nabla \frac{m_B \cdot r_{BC}}{\|r_{BC}\|_2^3} \right) \cdot m_c \quad (8)$$

where m_A (m_B or m_c) denotes the magnetic moment vectors of magnet A (B or C); $\alpha = \arctan[w'(L,t)]$ is the rotation angle of beam tip; μ_0 is the magnetic permeability constant.

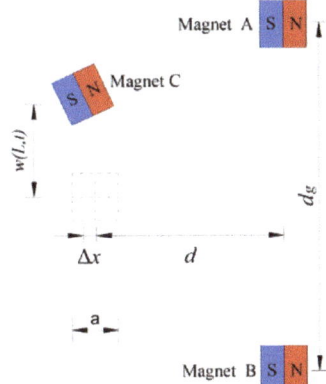

Figure 2. Geometric configuration of the tip magnet and two external magnets.

Considering the beam's deflection and the tip magnet's size, the vertical displacement of magnet C can be represented by $w(L,t) + \frac{a\sin\alpha}{2}$, where a is the side length of magnet C. Then, vectors r_{AC} and r_{BC}, which are directed from the magnetic moment source of magnet A and B to that of magnet C, will become

$$r_{AC} = \begin{bmatrix} d + \Delta x & w(L,t) + \frac{a}{2}(1-\cos\alpha) - \frac{1}{2}d_g \end{bmatrix} \quad (9)$$

$$r_{BC} = \begin{bmatrix} d + \Delta x & w(L,t) + \frac{a}{2}(1-\cos\alpha) + \frac{1}{2}d_g \end{bmatrix} \quad (10)$$

m_A (m_B or m_c) can be estimated from magnetization intensity M_A (M_B or M_C) and material volume V_A (V_B or V_c). They can be expressed by the following formulas:

$$m_A = \begin{bmatrix} 0 & -M_A V_A \end{bmatrix} \quad (11)$$

$$m_B = \begin{bmatrix} 0 & -M_B V_B \end{bmatrix} \quad (12)$$

$$m_C = \begin{bmatrix} M_C V_C \cos\alpha & M_C V_C \sin\alpha \end{bmatrix} \quad (13)$$

It should be noted that the horizontal displacement of magnet A is $\Delta x = \frac{a}{2}(1-\cos\alpha)$. Since $\alpha \approx 0$, then $\Delta x \approx 0$. As a result, the magnetic potential energy expression can be written as

$$U_m = \frac{\mu_0 M_A M_C V_A V_C \left(-\left(w(L,t)+\frac{1}{2}d_g\right)^2 + 2d^2 - 3d\left(w(L,t)+\frac{1}{2}d_g\right)w'(L,t)\right)}{4\pi \sqrt{w'(L,t)^2+1}\left(\left(w(L,t)+\frac{1}{2}d_g\right)^2+d^2\right)^{5/2}}$$
$$+ \frac{\mu_0 M_B M_C V_B V_C \left(-\left(w(L,t)+\frac{1}{2}d_g\right)^2 + 2d^2 - 3d\left(w(L,t)+\frac{1}{2}d_g\right)w'(L,t)\right)}{4\pi \sqrt{w'(L,t)^2+1}\left(\left(w(L,t)+\frac{1}{2}d_g\right)^2+d^2\right)^{5/2}} \quad (14)$$

The material and geometric parameters in simulation and experiment are the same and are listed in Table 1. It should be noted that the stability of the neutral equilibrium position depends on the wind speed, i.e., it is unstable in the static state and will become stable in high-speed wind. When the wind blows through the DFEH, the aero-dynamical force acting on the rectangular wing will create a disturbance, leading the beam tip to move and then to be attracted by the fixed magnets. At this stable equilibrium position, owing to the deflection of the beam, the effect of aero-dynamical force will change and act as a restoring force, making the beam return to the neutral position. The repeat of this process could produce a large-amplitude vibration and thus generate a large output. To account for the aero-dynamical influence on potential energy, we define an equivalent aero-dynamic stiffness (EAS). The EAS is the force required to make the beam tip keep a unit displacement for a certain wind speed. As is known, the higher the wind speed is, the more force is needed to lift the beam's tip, i.e., EAS will increase with the wind speed. Figure 3 shows the influence of EAS on the potential energy, where SS is the structure stiffness. We can see that with the increase in EAS, the DFEH's potential well number turns to three from two, i.e., it goes to tri-stability from bi-stability.

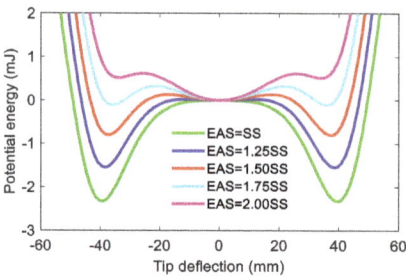

Figure 3. Variations of the potential energy function with tip deflection for different equivalent aero-dynamic stiffness (EAS) values.

Table 1. Model parameters used for simulation and experiment.

Symbol	Description	Value
Substrate		
L_s	Length	500 mm
b_s	Width	10 mm
h_s	Thickness	1 mm
ρ_s	Density	7800 kg/m^3
E_s	Young's modulus	205 Gpa
Rectangular wing		
L_{Fp}	Length	200 mm
b_{Fp}	Width	300 mm
h_{Fp}	Thickness	5 mm
ρ_{Fp}	Density	18 kg/m^3
Piezoelectric material (PZT-5H, Baoding Hongsheng Ltd.)		
L_p	Length	130 mm
b_p	Width	130 mm
h_p	Thickness	0.15 mm
ρ_p	Density	1785 kg/m^3
E_p (c_{xx}^E)	Young's modulus	2 Gpa
e_{zx}	Piezoelectric coupling coefficient	23×10^{-10} C/N
ϵ_{zz}^S	Permittivity constant	1.06×10^{-10} F/m
Magnets		
M_A, M_B	Magnetization intensity	1.25×10^5 Am^{-1}
V_A, V_B	Volume	5×10^{-7} m^3
μ_0	Permeability constant	$4\pi \times 10^{-7}$ NA^{-2}

From the potential energy analysis, it can be seen that the DFEH is a bi-stable system when it is subjected to a relatively low-speed incoming flow; then, when the flow speed increases and exceeds a threshold value, the DFEH will exhibit a tri-stable characteristic, whose three potential wells will make the snap-through motion easily occur. Therefore, the dynamic stability enables the DFEH to execute snap-through motion in both low and high wind speeds. Thus, the system could realize snap-through motion, and even coherence resonance in the environment of variable-speed wind.

3. Experimental Verification

The corresponding experiment was carried out to verify the advantages of DFEH. Figure 4 shows the prototype of DFEH and the experimental setup; the DFEH is composed of a cantilever substrate and a rectangular wing. A piezoelectric patch and a strain sensor were bonded to the cantilever. A data acquisition device (DH5922D, DONG HUA) was employed to measure and record the dynamic strain signal and the dynamic piezoelectric voltage across a resistive load (R = 5 MΩ), which were selected as the representative quantities for harvesting performance. The air speed was measured with a digital anemometer (AS8336, XI MA). Figure 5 illustrates the stable equilibrium positions of DFEH. As is clear from Figure 5, the DFEH has two static stable positions, one is near magnet A, the other is near magnet B. If the two fixed magnets are removed, the system degenerates to a linear flutter energy harvester (LFEH), which was put in the same wind field for comparison. The stable equilibrium position of LFEH is shown in Figure 6, which is at the neutral position.

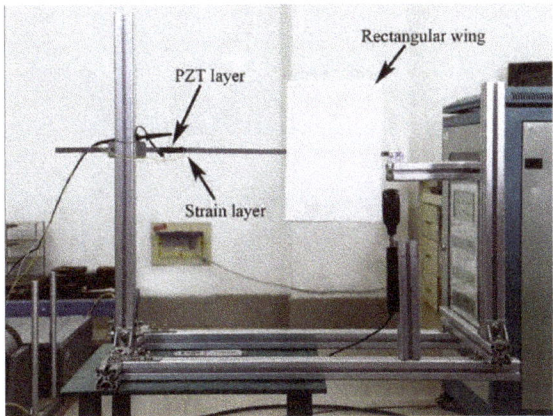

Figure 4. Experimental setup of the DFEH.

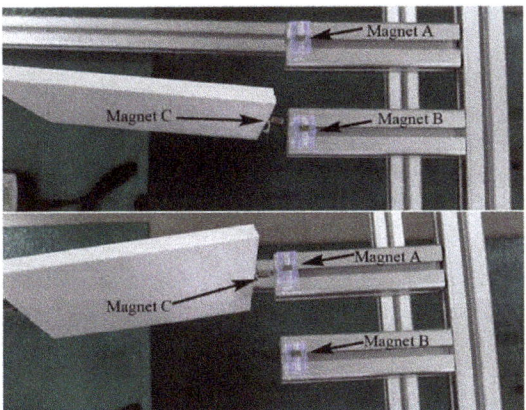

Figure 5. Two static stable positions of the DFEH.

Figure 6. One stable position of the linear flutter energy harvester (LFEH).

Figure 7 shows the variation in dynamic strain variance and root-mean-square (RMS) of voltage with the wind speed, which ranges from 1.5 to 7.5 m/s. It is apparent that the DFEH exhibits a much better performance than the LFEH. Moreover, in Figure 7a, there appears to be a sharp increase in the variance of strain for the DFEH, corresponding to the occurrence of snap-through motion. The snap-through motion could lead to an increase in output voltage. Moreover, the DFEH can maintain the snap-through motion over a wide range of wind speeds, thus keeping a large output. It should be noted that for the wind speeds larger than $v = 2.0$ m/s, the strain variance decreases monotonously with the wind speed, but the resulting voltage increases reversely and even reaches the maximum value at $v = 4.5$ m/s. This is due to the increase in EAS and the emergence of an additional neutral equilibrium position.

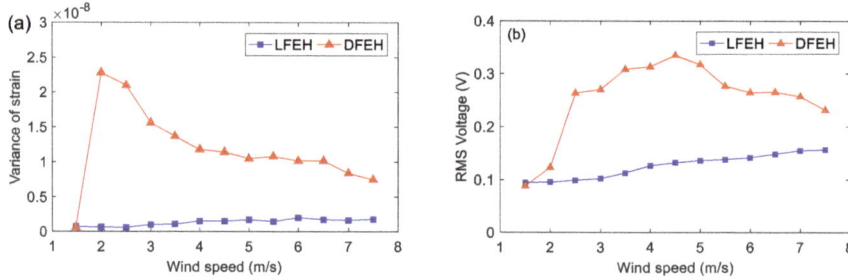

Figure 7. (**a**) Variance of dynamic strain and (**b**) root-mean-square (RMS) voltage versus wind speed (LFEH and DFEH).

In order to show the dynamic-stable characteristics of the DFEH for different wind speeds clearly, the dynamic strain and resulting voltage, in terms of time and spectrum domains, are shown in Figures 8–20, with the wind speed varying from 1.5 to 7.5 m/s.

First, at $v = 1.5$ m/s, i.e., a very low speed, the DFEH behaves like a linear one. Its strain and voltage are shown in Figure 8a,b. It can be seen that both DFEH and LFEH oscillate in one potential energy well now. The corresponding spectra of strain responses are shown in Figure 8c.

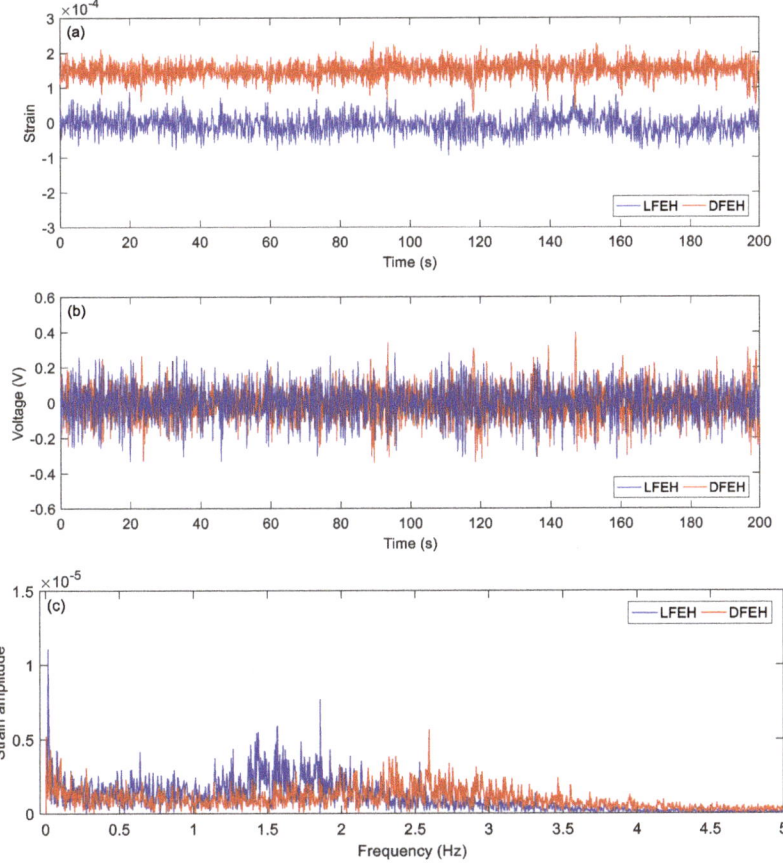

Figure 8. (**a**) Time histories of strain response, (**b**) time histories of voltage response, (**c**) spectrum of dynamic strain response ($v = 1.5$ m/s).

Then, increasing the wind speed to $v = 2.0$ m/s, the DFEH's dynamic strain shows that the DFEH begins to jump between two static equilibrium positions (Figure 9a). The snap-through motion results in a large amplitude and a high output voltage (as shown in Figure 9b). The spectra of strain show that there exists a large component in the low frequency region.

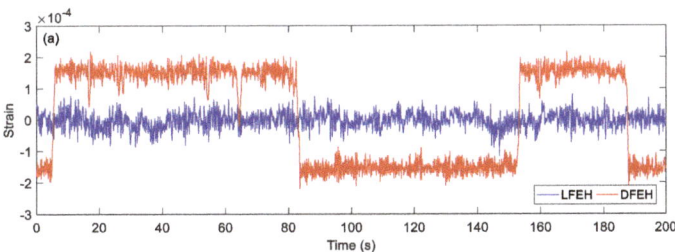

Figure 9. *Cont.*

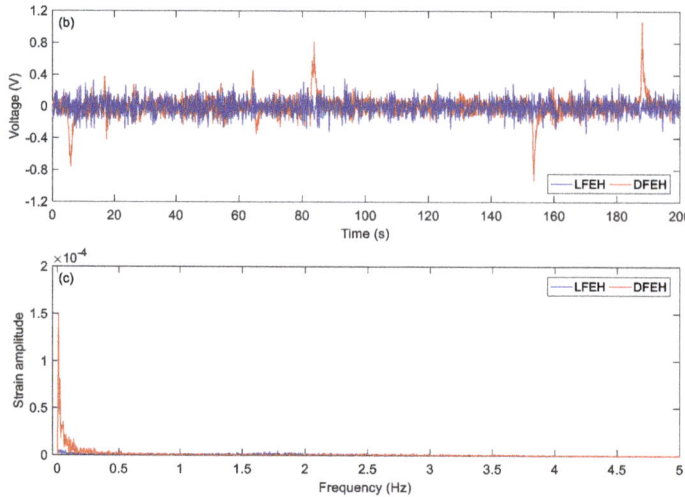

Figure 9. (**a**) Time histories of strain response, (**b**) time histories of voltage response, (**c**) spectrum of dynamic strain response ($v = 2.0$ m/s).

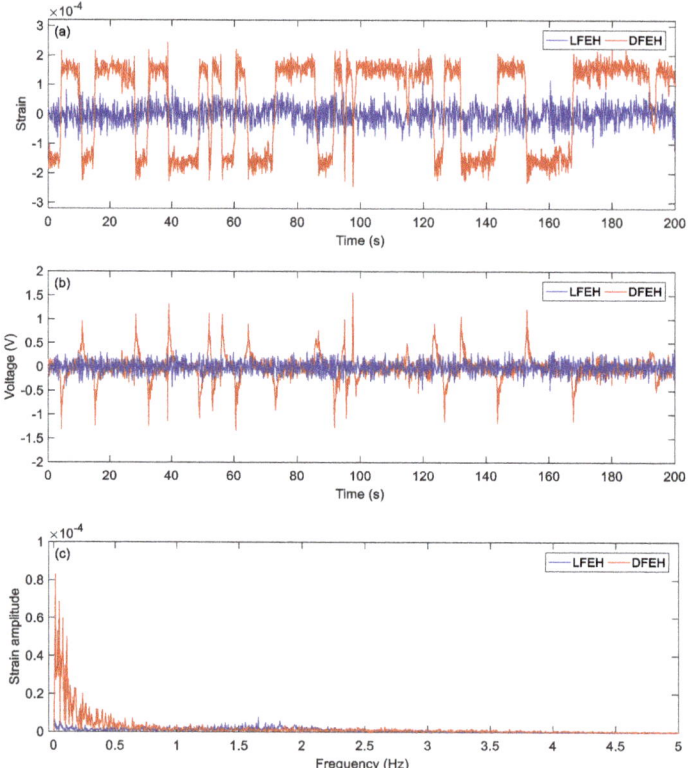

Figure 10. (**a**) Time histories of strain response, (**b**) time histories of voltage response (**c**) spectrum of strain ($v = 2.5$ m/s).

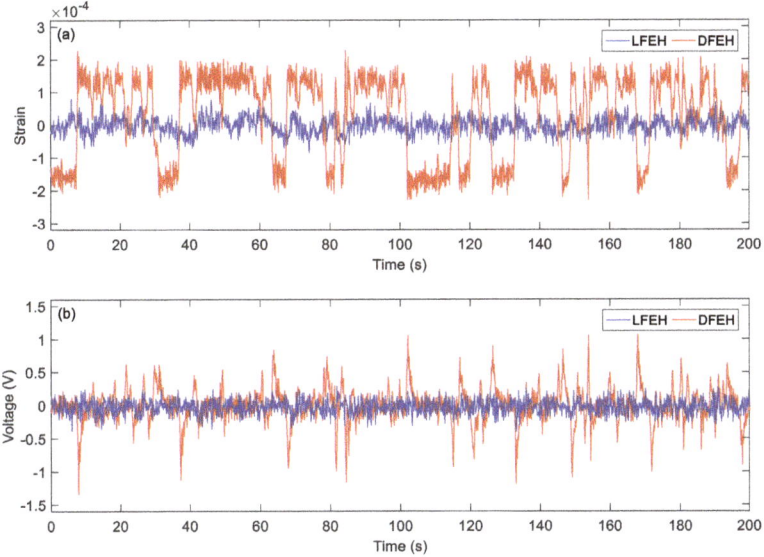

Figure 11. (a) Time histories of strain response, (b) time histories of voltage response ($v = 3.0$ m/s).

At $v = 2.5$ m/s, jumping happens more frequently. The jumping motion results in high pulses in the output voltage, as shown in Figure 10a,b.

When the wind speed reaches $v = 3.0$ m/s, it can be found that an additional equilibrium position emerges. In this case, the jumping motion could happen between any two of three equilibrium positions, as shown in Figure 11a. This phenomenon can be maintained till $v = 7.5$ m/s.

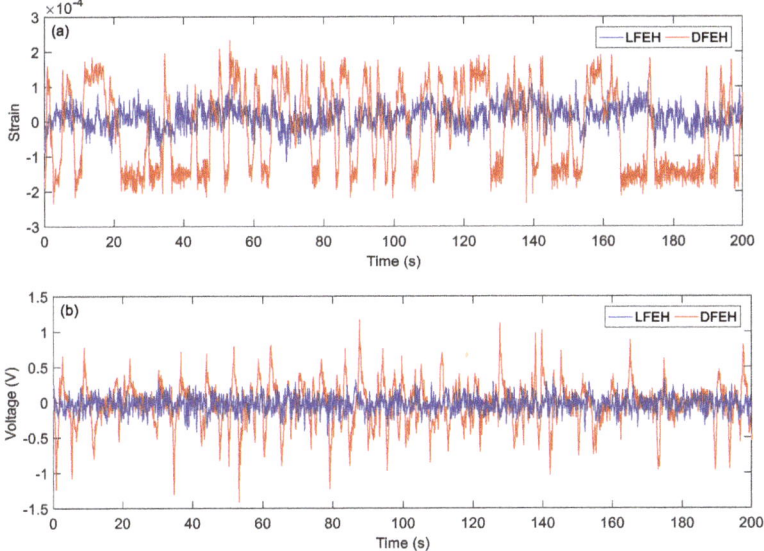

Figure 12. (a) Time histories of strain response, (b) time histories of voltage response ($v = 3.5$ m/s).

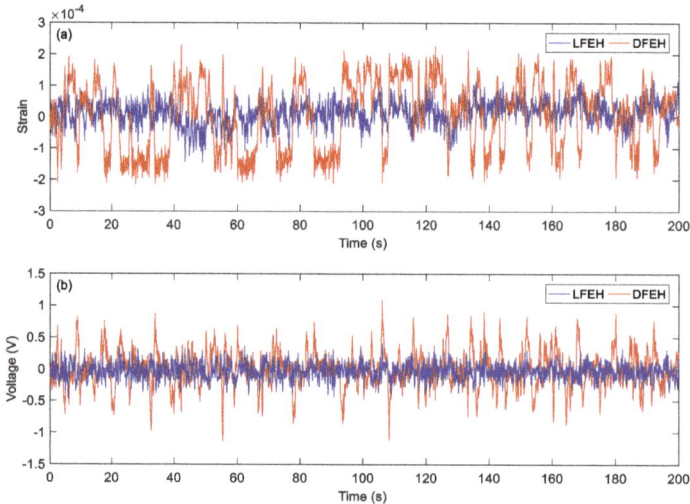

Figure 13. (**a**) Time histories of strain response, (**b**) time histories of voltage response (v = 4.0 m/s).

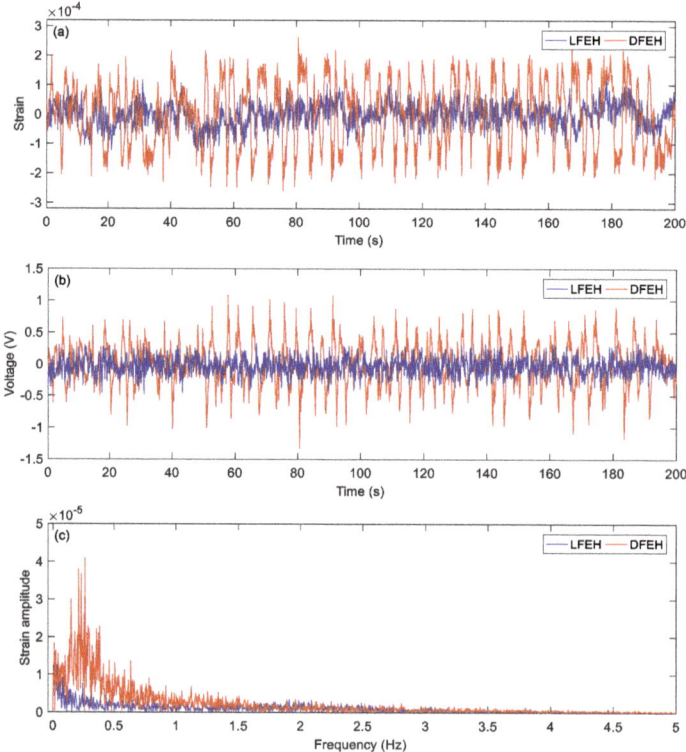

Figure 14. (**a**) Time histories of strain response, (**b**) time histories of voltage response, (**c**) spectrum of dynamic strain response (v = 4.5 m/s).

Then, at $v = 3.5$ m/s, the results show that two types of jumping could take place simultaneously. Although now jumping between the adjacent equilibrium positions produces relatively small pulse voltages, it can happen more frequently. Therefore, the output voltage is still quite large.

As the wind speed increases to $v = 4.0$ m/s, the jumping happens mostly between the static and the neutral equilibrium positions; since it happens densely, the output voltage increases accordingly.

When the wind speed reaches $v = 4.5$ m/s, it follows from Figure 14 that the jumping could happen densely between all three equilibrium positions. Accordingly, the corresponding output voltage increases greatly. From the spectrum, it can be observed that now there appears a peak in the low frequency region, indicating that the jumping exhibits a nearly periodic feature, i.e., the DFEH attains coherence resonance now.

At $v = 5.0$ m/s, as shown in Figure 15, the jumping between potential wells remains dense, thereby producing dense pulses in the output voltage. The corresponding spectra show that the DFEH has a dominant nearly periodic component, suggesting that the coherence resonance is happening.

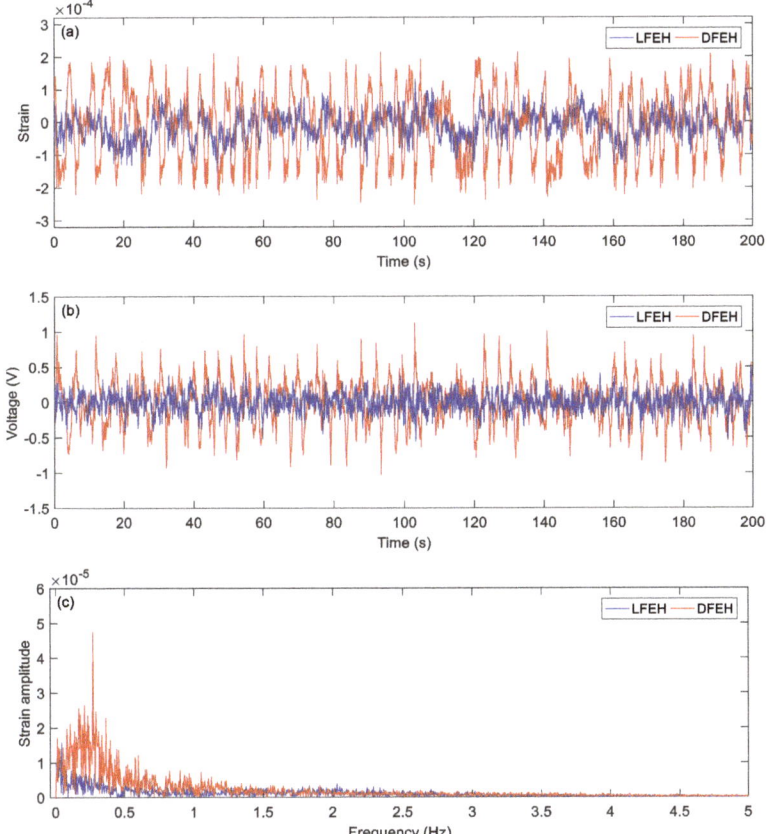

Figure 15. (a) Time histories of strain response, (b) time histories of voltage response, (c) spectrum of dynamic strain response ($v = 5.0$ m/s).

We increased the wind speed to $v = 7.5$ m/s. The corresponding results are shown in Figures 16–20. Since the wind speed is relatively high now, the emerged neutral equilibrium position becomes more stable, at which the response stays for more time. Thus, the jumping between the adjacent equilibrium positions becomes dominant. However, in this case, the output voltage is still quite large, since the adjacent jumping is easier to occur and the times of jumping increase.

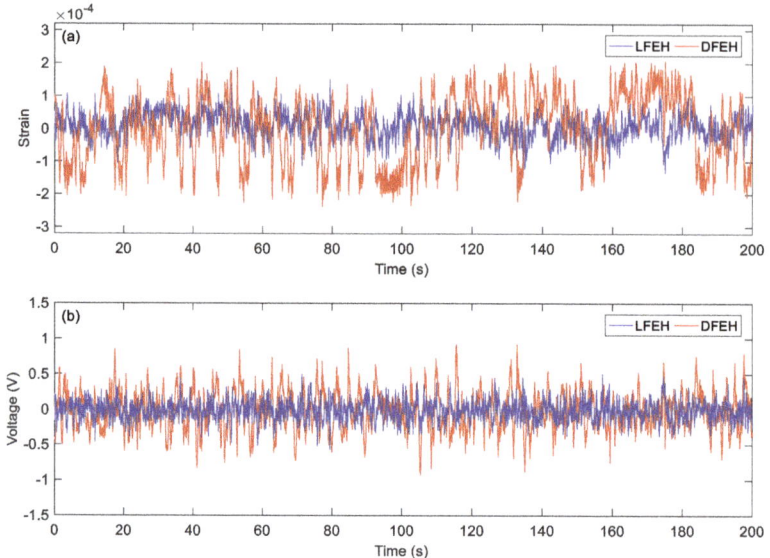

Figure 16. (**a**) Time histories of strain response, (**b**) time histories of voltage response (v=5.5 m/s).

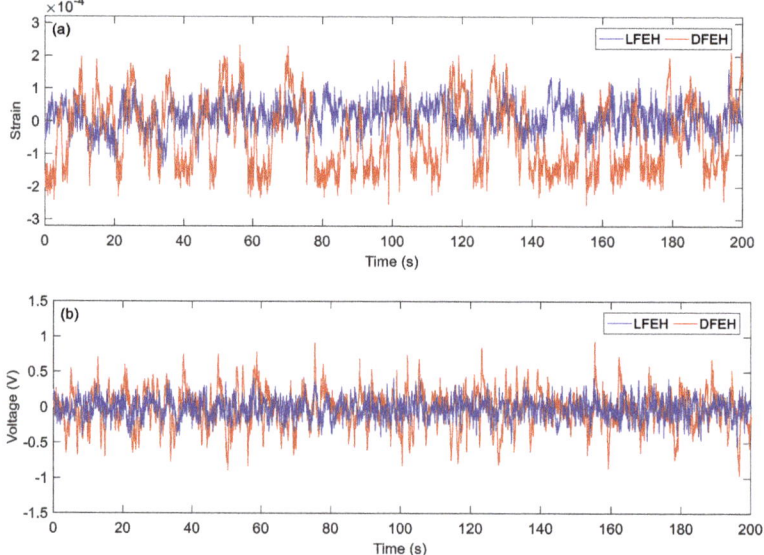

Figure 17. (**a**) Time histories of strain response, (**b**) time histories of voltage response (v = 6.0 m/s).

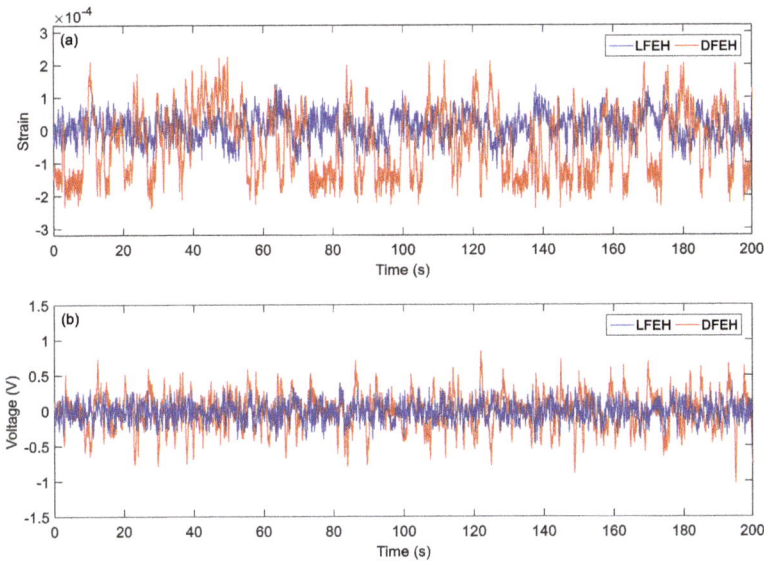

Figure 18. (**a**) Time histories of strain response, (**b**) time histories of voltage response (v = 6.5 m/s).

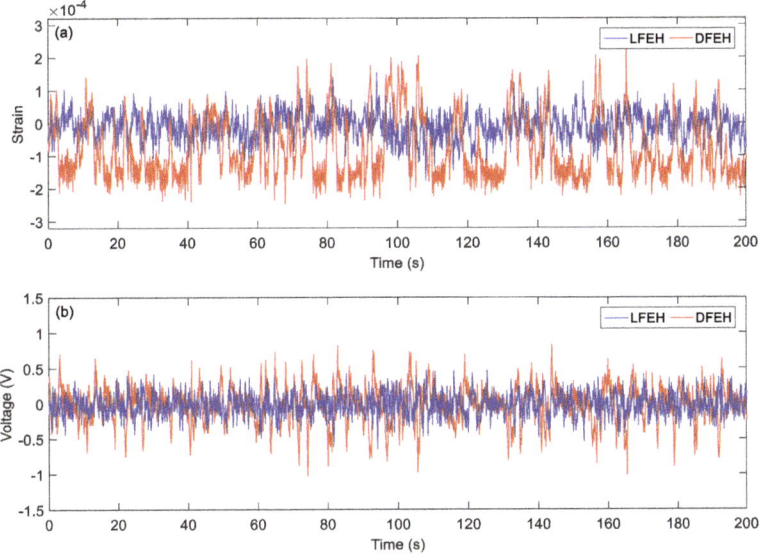

Figure 19. (**a**) Time histories of strain response, (**b**) time histories of voltage response (v = 7.0 m/s).

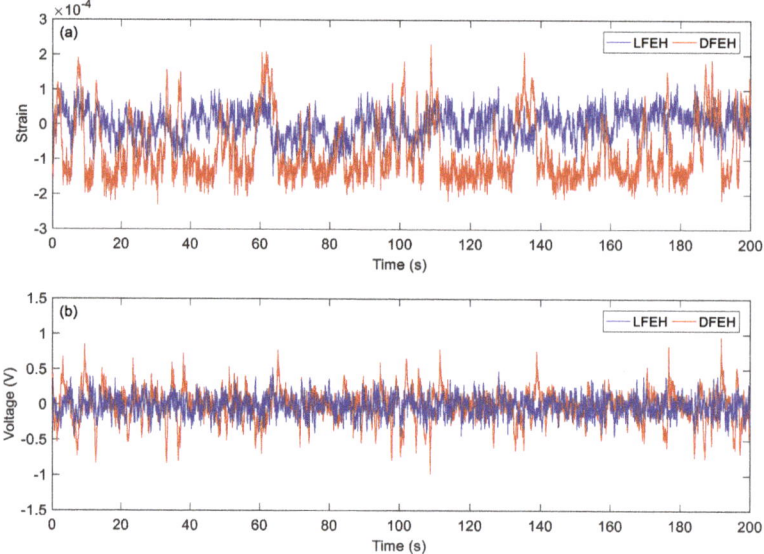

Figure 20. (a) Time histories of strain response, (b) time histories of voltage response ($v = 7.5$ m/s).

In order to show the jumping characteristics of the DFEH clearly, for two wind speeds, i.e., $v = 2.5$ m/s and $v = 4.5$ m/s, we magnify parts of their strain and voltage responses and plot them in Figures 21 and 22. As shown in Figure 21, at $v = 2.5$ m/s, the system exhibits the bi-stable characteristics, its response jumps between two equilibrium positions; in contrast, at $v = 4.5$ m/s, an additional equilibrium position emerges and the system becomes a tri-stable one, its response jumps between all three equilibrium positions. As for the output voltages, it is apparent that the jumping leads to a spike in the voltage response, thus the dense jumping could generate dense pulses in output voltage and promote the harvesting efficiency.

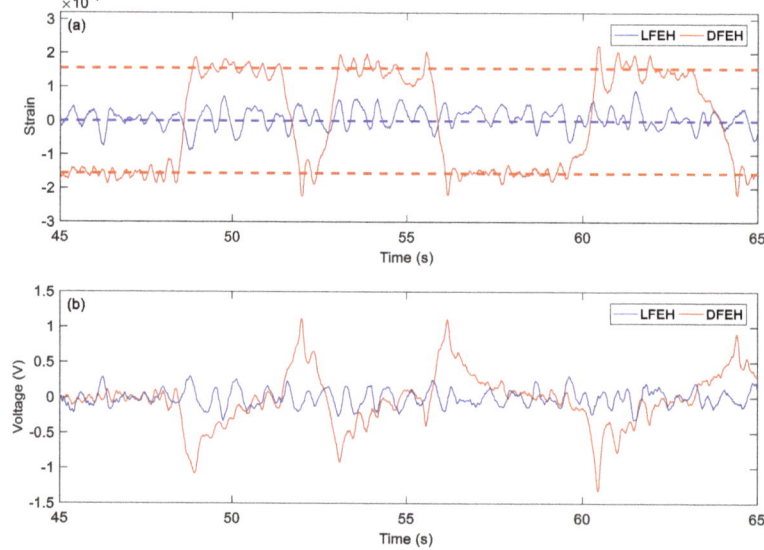

Figure 21. Bi-stable jumping (a) strain response and (b) voltage response ($v = 2.5$ m/s).

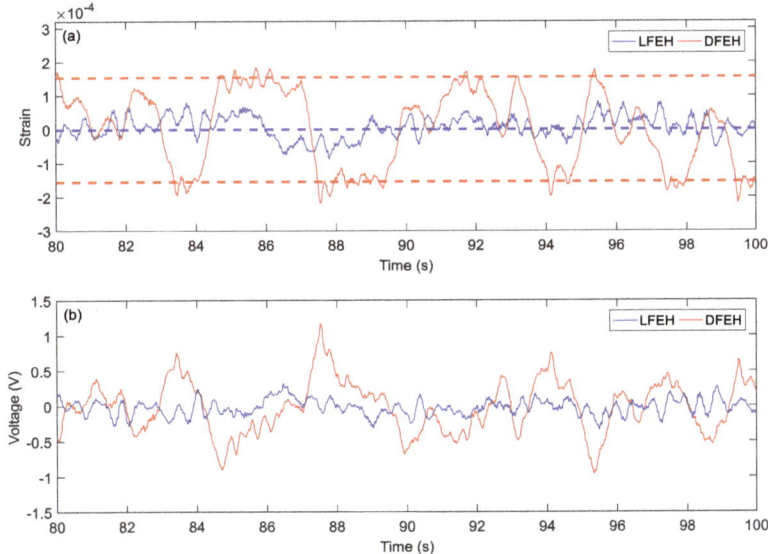

Figure 22. Tri-stable jumping (**a**) strain response and (**b**) voltage response ($v = 4.5$ m/s).

4. Conclusions

In this paper, a dynamic multi-stable flutter harvester is proposed to harvest the energy of variable-speed wind. The validation experiment was carried out. The experimental results prove that it could execute snap-through motion and generated high pulse voltages in the environment of variable-speed wind. At low wind speeds, the harvester jumps between two static equilibrium positions; whereas, at high wind speeds, with a neutral equilibrium position emerging, the harvester could jump between three equilibrium positions. Moreover, the harvester response can reach coherence resonance in a certain range of wind speeds. We believe that this harvester could be directly applied to the actual environment after improvement, due to its simple structure and cheap material. In practice, the size of the structure can be scaled down to adapt to the environment without affecting its performance. Of course, the storage circuit for electric energy needs further design and research, and the connections between the components should be improved.

Author Contributions: Conceptualization, data analysis and writing Y.W., Z.Z.; Experiments and interpretation Y.W., Z.Z., Q.L., P.Z.; Supervision, W.Q. All authors have read and agreed to the published version of the manuscript.

Funding: This research was supported under the National Natural Science Foundation of China (Grant No. 11672237).

Acknowledgments: We gratefully acknowledge the financial support from the National Natural Science Foundation of China (Grant No. 11672237).

Conflicts of Interest: The authors declare no conflict of interest.

References

1. Hu, G.; Tse, K.T.; Wei, M.; Naseer, R.; Abdelkefi, A.; Kwok, K.C.S. Experimental investigation on the efficiency of circular cylinder-based wind energy harvester with different rod-shaped attachments. *Appl. Energy* **2018**, *226*, 682–689. [CrossRef]
2. Abdelkefi, A.; Nayfeh, A.H.; Hajj, M.R. Enhancement of power harvesting from piezoaeroelastic systems. *Nonlinear Dyn.* **2012**, *68*, 531–541. [CrossRef]
3. Abdelkefi, A. Aeroelastic energy harvesting: A review. *Int. J. Eng. Sci.* **2016**, *100*, 112–135. [CrossRef]

4. Yang, K.; Wang, J.L.; Yurchenko, D. A double-beam piezo-magneto-elastic wind energy harvester for improving the galloping-based energy harvesting. *Appl. Phys. Lett.* **2019**, *115*, 193901. [CrossRef]
5. McCarthy, J.M.; Watkins, S.; Deivasigamani, A.; John, S.J. Fluttering energy harvesters in the wind: A review. *J. Sound Vib.* **2016**, *361*, 355–377. [CrossRef]
6. Allen, J.J.; Smits, A.J. Energy harvesting eel. *J. Fluids Struct.* **2001**, *15*, 629–640. [CrossRef]
7. Akaydin, H.D.; Elvin, N.; Andreopoulos, Y. Energy harvesting from highly unsteady fluid flows using piezoelectric materials. *J. Intell. Mater. Syst. Struct.* **2010**, *21*, 1263–1278. [CrossRef]
8. Akaydin, H.D.; Elvin, N.; Andreopoulos, Y. Wake of a cylinder: A paradigm for energy harvesting with piezoelectric materials. *Exp. Fluids* **2010**, *49*, 291–304. [CrossRef]
9. Goushcha, O.; Elvin, N.; Andreopoulos, Y. Interactions of vortices with a flexible beam with applications in fluidic energy harvesting. *Appl. Phys. Lett.* **2014**, *104*, 021919. [CrossRef]
10. Hobbs, W.B.; Hu, D.L. Tree-inspired piezoelectric energy harvesting. *J. Fluids Struct.* **2012**, *28*, 103–114. [CrossRef]
11. Lahooti, M.; Kim, D. Multi-body interaction effect on the energy harvesting performance of a flapping hydrofoil. *Renew. Energy* **2019**, *130*, 460–473. [CrossRef]
12. Tang, L.; Païdoussis, M.P.; Jiang, J. Cantilevered flexible plates in axial flow: Energy transfer and the concept of flutter-mill. *J. Sound Vib.* **2009**, *326*, 263–276. [CrossRef]
13. Dunnmon, J.A.; Stanton, S.C.; Mann, B.P.; Dowell, E.H. Power extraction from aeroelastic limit cycle oscillations. *J. Fluids Struct.* **2011**, *27*, 1182–1198. [CrossRef]
14. Akcabay, D.T.; Young, Y.L. Hydroelastic response and energy harvesting potential of flexible piezoelectric beams in viscous flow. *Phys. Fluids* **2012**, *24*, 054106. [CrossRef]
15. Michelin, S.; Doaré, O. Energy harvesting efficiency of piezoelectric flags in axial flows. *J. Fluid Mech.* **2013**, *714*, 489–504. [CrossRef]
16. Giacomello, A.; Porfiri, M. Underwater energy harvesting from a heavy flag hosting ionic polymer metal composites. *J. Appl. Phys.* **2011**, *109*, 084903. [CrossRef]
17. Li, S.; Yuan, J.; Lipson, H. Ambient wind energy harvesting using cross-flow fluttering. *J. Appl. Phys.* **2011**, *109*, 026104. [CrossRef]
18. Bryant, M.; Wolff, E.; Garcia, E. Aeroelastic flutter energy harvester design: The sensitivity of the driving instability to system parameters. *Smart Mater. Struct.* **2011**, *20*, 125017. [CrossRef]
19. Li, D.J.; Hong, S.; Gu, S.; Choi, Y.; Nakhmanson, S.; Heinonen, O.; Karpeev, D.; No, K. Polymer piezoelectric energy harvesters for low wind speed. *Appl. Phys. Lett.* **2014**, *104*, 012902.
20. Zhao, D.; Ega, E. Energy harvesting from self-sustained aeroelastic limit cycle oscillations of rectangular wings. *Appl. Phys. Lett.* **2014**, *105*, 103903. [CrossRef]
21. Aquino, A.I.; Calauti, J.K.; Hughes, B.R. A Study on the wind-induced flutter energy harvester (WIFEH) integration into buildings. *Energy Procedia* **2017**, *142*, 321–327. [CrossRef]
22. Shan, X.B.; Tian, H.G.; Chen, D.P.; Xie, T. A curved panel energy harvester for aeroelastic vibration. *Appl. Energy* **2019**, *249*, 58–66. [CrossRef]
23. Orrego, S.; Shoele, K.; Ruas, A.; Doran, K.; Caggiano, B.; Mittal, R.; Kang, S.H. Harvesting ambient wind energy with an inverted piezoelectric flag. *Appl. Energy* **2017**, *194*, 212–222. [CrossRef]
24. Jung, H.J.; Lee, S.W. The experimental validation of a new energy harvesting system based on the wake galloping phenomenon. *Smart Mater. Struct.* **2011**, *20*, 055022. [CrossRef]
25. Zhou, S.; Zuo, L. Nonlinear dynamic analysis of asymmetric tristable energy harvesters for enhanced energy harvesting. *Commun. Nonlinear Sci. Numer. Simul.* **2018**, *61*, 271–284. [CrossRef]
26. Alhadidi, A.H.; Daqaq, M.F. A broadband bi-stable flow energy harvester based on the wake-galloping phenomenon. *Appl. Phys. Lett.* **2016**, *109*, 033904. [CrossRef]
27. Zhang, L.B.; Abdelkefi, A.; Dai, H.L.; Naseer, R.; Wang, L. Design and experimental analysis of broadband energy harvesting from vortex-induced vibrations. *J. Sound Vib.* **2017**, *408*, 210–219. [CrossRef]
28. Huynh, B.H.; Tjahjowidodo, T. Experimental chaotic quantification in bistable vortex induced vibration systems. *Mech. Syst. Signal Process.* **2017**, *85*, 1005–1019. [CrossRef]
29. Naseer, R.; Dai, H.; Abdelkefi, A.; Wang, L. Comparative study of piezoelectric vortex-induced vibration-based energy harvesters with multi-stability characteristics. *Energies* **2020**, *13*, 71. [CrossRef]
30. Valipour, P.; Ghasemi, S.E.; Khosravani, M.R.; Ganji, D.D. Theoretical analysis on nonlinear vibration of fluid flow in single-walled carbon nanotube. *J. Theor. Appl. Phys.* **2016**, *10*, 211–218. [CrossRef]

31. Zhou, Z.Y.; Qin, W.Y.; Zhu, P.; Du, W.F.; Deng, W.Z.; Pan, J.N. Scavenging wind energy by a dynamic-stable flutter energy harvester with rectangular wing. *Appl. Phys. Lett.* **2019**, *114*, 243902. [CrossRef]
32. Stanton, S.C.; McGehee, C.C.; Mann, B.P. Nonlinear dynamics for broadband energy harvesting: Investigation of a bistable piezoelectric inertial generator. *Physica D* **2010**, *239*, 640–653. [CrossRef]

© 2020 by the authors. Licensee MDPI, Basel, Switzerland. This article is an open access article distributed under the terms and conditions of the Creative Commons Attribution (CC BY) license (http://creativecommons.org/licenses/by/4.0/).

Article

The Radial Piezoelectric Response from Three-Dimensional Electrospun PVDF Micro Wall Structure

Guoxi Luo [1,2,3], Yunyun Luo [1], Qiankun Zhang [1], Shubei Wang [2], Lu Wang [1], Zhikang Li [1,2], Libo Zhao [1,*], Kwok Siong Teh [4,*] and Zhuangde Jiang [1]

1. State Key Laboratory for Manufacturing Systems Engineering, School of Mechanical Engineering, Xi'an Jiaotong University, Xi'an 710049, China; luoguoxi@mail.xjtu.edu.cn (G.L.); luoyunyun@stu.xjtu.edu.cn (Y.L.); zhangqiankun3394@stu.xjtu.edu.cn (Q.Z.); wang.lu@stu.xjtu.edu.cn (L.W.); zhikangli@mail.xjtu.edu.cn (Z.L.); zdjiang@mail.xjtu.edu.cn (Z.J.)
2. Xi'an Jiaotong University Suzhou Institute, Suzhou 215123, China; wangwsb@mail.xjtu.edu.cn
3. State Key Lab of Digital Manufacturing Equipment & Technology, Huazhong University of Science and Technology, Wuhan 430074, China
4. School of Engineering, San Francisco State University, San Francisco, CA 94132, USA
* Correspondence: libozhao@mail.edu.cn (L.Z.); ksteh@berkeley.edu (K.S.T.)

Received: 15 February 2020; Accepted: 16 March 2020; Published: 18 March 2020

Abstract: The ability of electrospun polyvinylidene fluoride (PVDF) fibers to produce piezoelectricity has been demonstrated for a while. Widespread applications of electrospun PVDF as an energy conversion material, however, have not materialized due to the random arrangement of fibers fabricated by traditional electrospinning. In this work, a developed 3D electrospinning technique is utilized to fabricate a PVDF micro wall made up of densely stacked fibers in a fiber-by-fiber manner. Results from X-ray diffraction (XRD) and Fourier transform infrared spectra (FTIR) demonstrate that the crystalline structure of this PVDF wall is predominant in the β phase, revealing the advanced integration capability of structural fabrication and piezoelectric poling with this 3D electrospinning. The piezoelectric response along the radial direction of these PVDF fibers is measured while the toppled micro wall, comprised of 60 fibers, is sandwich assembled with a pair of top/bottom electrodes. The measured electrical output is ca. 0.48 V and 2.7 nA. Moreover, after constant mechanical compression happening over 10,000 times, no obvious reduction in the piezoelectric response has been observed. The combined merits of high-precision 3D fabrication, in situ piezoelectric poling, and high mechanical robust make this novel structure an attractive candidate for applications in piezoelectric energy harvesting and sensing.

Keywords: 3D electrospinning; PVDF fibers; piezoelectricity; energy harvesting; piezoelectric sensing

---

1. Introduction

Piezoelectricity has been demonstrated to be useful in energy harvesting devices and high-precision sensors due to its unique ability of mechanical-to-electrical conversion. Traditionally, piezoelectric materials can be divided into two classes, namely, ceramic and polymeric piezoelectric materials. Ceramic piezoelectric materials typically exhibit superior piezoelectric performance than that of polymeric materials due to their higher piezoelectric stress constants and coupling coefficients [1]. Piezoelectric polymers, on the other hand, have the advantages of being lightweight, low cost, biocompatible, and highly flexible, and hence find uses in specific applications that require large mechanical deformations, such as human-centered lifestyle applications, in which the rigid and brittle properties of inorganic piezoelectric ceramics cannot be easily implemented. This is the reason why the piezoelectric polymer-based microelectromechanical systems (MEMS) and microfluidics have attracted considerable attention in recent years [2]. Among the piezoelectric polymers, PVDF is undoubtedly the most application-ready material due to its high piezoelectric response [3,4]. Nevertheless, as-deposited PVDF materials usually exist in the nonpolar α crystalline phase and must be post-treated with electrical poling and mechanical stretching to obtain the polar β phase and piezoelectricity [5].

Electrospinning is a powerful and straightforward technology that utilizes high electrostatic force to produce continuous polymeric micro/nano fibers [6,7]. It has been well demonstrated that the coexisting strong electric fields and stretching force in the process of electrospinning can align the dipole moment in the PVDF fibers' crystal, endowing the unique piezoelectricity no matter how traditional far-field electrospinning or near-field electrospinning are utilized [8,9]. Therefore, many piezoelectric energy harvesters [10,11], actuators [12], and sensors [13,14] have been developed based on electrospun PVDF fibers. Regarding far-field electrospinning, Fang et al. demonstrated the radial piezoelectricity from randomly oriented electrospun PVDF nanofibers for the first time, and developed an electrical generator based on this nanofibrous membrane [8]. Gui et al. adopted parallel electrodes and a rotating-drum collector to align Poly(vinylidene fluoride-trifluoroethylene) (PVDF-TrFE) fibers, and the axial piezoelectric response along the far-field electrospun PVDF-TrFE fibers was successfully detected [15]. Even though the piezoelectric output performance from far-field electrospun PVDF fibers (in the order of several volts and micro amperes) satisfy most modern self-powered microsystems, the low controllability over deposition of the fibers and the macroscopic scale of the functionalized film (usually above 1 cm^2 area) limit the applications of this substitute to integrate with true micro devices. On the other hand, near-field electrospun PVDF fibers could be a good piezoelectric component in micro and nano systems concerning the high controllability and patterning capability. Chang et al. developed a PVDF nanogenerator based on one single PVDF electrospun nanofiber, which was accurately deposited on a pair of parallel electrodes using near-field electrospinning to harvest energy from mechanical vibrations along the axial direction of the fiber [9]. To amplify the electrical output performance, a near-field electrospun PVDF fibers array, with fiber spacing ca. 20 μm, was fabricated and arranged in series and in parallel to detect their axial piezoelectricity [16]. Although such a fiber array strategy could boost electrical outputs, the dispersed arrangement of fibers on the two-dimensional plane would inevitably cost more spaces and thus decrease the output power density. In the end, a method to realize high-density fiber structure, like controllable 3D arrangement, is critical to enhance piezoelectric performance for practical application.

Recently, a technique for achieving 3D near-field electrospinning on paper substrate for precise fabrication of arbitrarily shaped 3D electrospun structures has been successfully developed [17] by the authors. Using a printing paper as the fiber collector, the residual solvents from near-field electrospun fibers can infiltrate the paper substrate and a charge transfer path between the deposited fibers and ground plate can be established. Such a charge transfer grounds the locally deposited fibers and provides a low-potential site for the deposition of subsequent fibers, enabling a self-aligned fabrication process. In this work, utilizing this 3D electrospinning technology, we fabricated a free-standing, out-of-plane micro wall by densely and uniformly stacking PVDF fibers that were electrospun in a fiber-by-fiber manner on paper substrate. Then, this micro wall is toppled and laid flat on the paper, both of which are subsequently sandwiched between a pair of aluminum electrodes. The piezoelectric response along the radial direction of the fibers (normal to the wall surface) is measured. This, to the best of the authors' knowledge, is the first time that the radial piezoelectricity from near-field electrospun PVDF fibers is demonstrated. Furthermore, compared to the existing piezoelectric devices utilizing traditional electrospun PVDF fibers, this technique promises high precision over 3D fabrication, compact integration of functionalized fibers, enabling their enhanced piezoelectric performance for direct applications into other micro devices or systems.

2. Materials and Methods

2.1. Materials

The PVDF powder (Mw = 534, 000), dimethyl sulfoxide (DMSO), acetone, and anionic surfactant Capstone FS-66 were purchased from Sigma-Aldrich (Berkeley, CA, USA) for preparation of 3D electrospinning solution. The chemicals were directly used without any further treatment. Electrically conductive aluminum tapes (3302) were purchased from 3M corporation (Berkeley, CA, USA) and used as electrodes for fabrication of devices.

2.2. Electrospinning Solution Preparation

Firstly, 2.1 g PVDF powder was dispersed in 3 g acetone for 30 min using a magnetic stirrer, and then 7 g DMSO and 0.5 g anionic surfactant Capstone FS-66 were added into the PVDF-acetone suspension; finally, the mixture was stirred for more than 2 h to reach a good homogeneity. The solutions were prepared and stored at room temperature and 1 atm pressure. The container with PVDF solution was sealed with Parafilm (BEMIS 01852-AB, Berkeley, CA, USA) to minimize evaporation.

2.3. Three-Dimensional Electrospinning for Fabrication of Micro Wall

This 3D electrospinning technique actually was built on the near field electrospinning, the spinneret-to-collector distance was set at 1 mm, allowing deposition of fibers to obtain high positional accuracy. The applied voltage was 1.2 kV. A 30 g (inner diameter 150 μm) needle was utilized as spinneret for electrospinning. A programmable x-y translational motion stage with setting speed of 100 mm/s was used to repetitively deposit fibers for out-of-plane construction, and a z-axis manual linear stage (speed ca. 7 μm/min) was set to guarantee the spinneret-to-fiber distance at a constant. With pre-designing the trajectory of this x-y translational motion stage, a 3D wall construction in the fiber-by-fiber manner could be achieved with repetitive depositions, and the height can be adjusted by altering the repetition times.

2.4. Characterization and Measurements

The morphology and structure of the as-prepared micro wall were observed with a field scanning electron microscope (FESEM; Gemini SEM 500, Zeiss, Heidenheim, Germany). The crystal structure was characterized by X-ray diffraction (XRD; D8 Advanced, Bruker, Karlsruhe, Germany) using Cu Kα radiation over a 2θ range of 5°–40° through reflection mode. Fourier transform infrared spectroscopy (FTIR; Nicolet iS10, Thermo Fisher Scientific, Waltham, MA, USA) was used to obtain the absorption spectra and characterize the crystallographic structure. An electrochemical workstation (Reference 600+, Gamry Instruments, Berkeley, CA, USA) was utilized for detection of output voltages and currents. The mechanically dynamic compression and decompression on the as-prepared device was provided by a mechanical shaker (MS100, YMC Piezotronics Inc., Yangzhou, China).

3. Results and Discussion

3.1. Fabricaition and Characterization

Experimentally, as illustrated in Figure 1a, PVDF fibers were electrospun into an out-of-plane micro wall structure made up of 60 densely stacked PVDF fibers, in a fiber-by-fiber manner, on paper substrate. Figure 1d shows a SEM image of the fabricated micro wall of ca. 5 μm in width, an orderly out-of-plane 3D arrangement for fibers deposition can be clearly seen. In the next step, as illustrated in Figure 1b, the micro wall structure was trimmed by dipping acetone to dissolve and eliminate some redundant parts. After the wall was toppled onto paper substrate, a consistent fiber-by-fiber manner could be observed, and the height was over 200 μm, as shown in Figure 1e. Finally, to investigate the radial piezoelectric performance, the device was assembled with aluminum tapes as a pair of top and bottom electrodes, and the electrospun PVDF wall on paper substrate as the functional layer is sandwiched in between as illustrated in Figure 1c. The paper substrate was not removed as this insulating layer can prevent an electrical short between the top and bottom electrodes. Figure 1f shows the optical image of the as-fabricated device made up of a four-layer sandwich structure of top electrode/electrospun PVDF wall/paper substrate/bottom electrode.

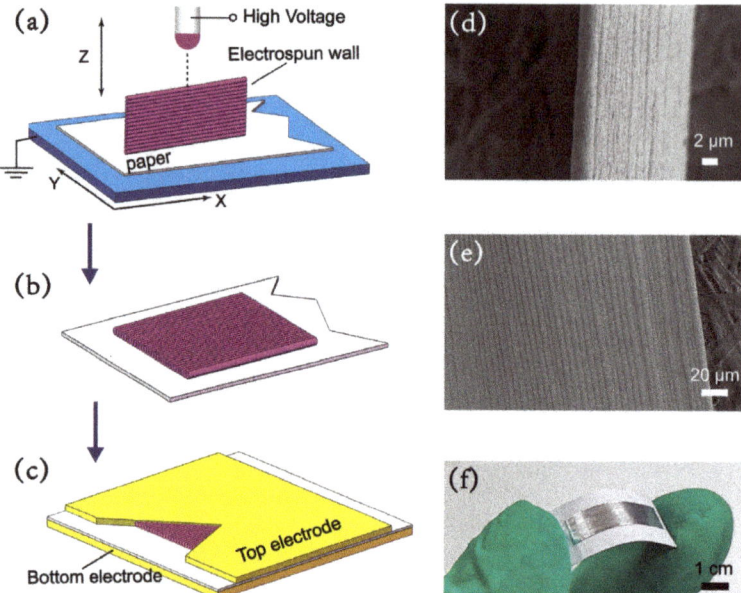

Figure 1. (**a**) Schematic of the fabrication process of wall structure by 3D electrospinning. (**b**) The wall is manually toppled and laid flat onto to paper substrate. (**c**) The toppled wall and paper are sandwiched in between a pair of top and bottom aluminum electrodes. SEM images of (**d**) straight wall and (**e**) toppled wall on paper substrate. (**f**) optical image of the assemble device.

To demonstrate the in situ piezoelectric poling of this 3D electrospinning process, XRD and FTIR are utilized for analysis of the crystal structure, and a spin-coated PVDF is used as reference. Figure 2a shows the XRD curves; it can be seen that the spin-coated PVDF film exhibits a prominent peak at $2\theta = 18.6°$, indicative of the (020) plane from the non-polar α phase [18]. In contrast, the electrospun wall structure shows a relatively weak and almost negligible peak at α phase and a prominent peak at $2\theta = 20.6°$, which is attributed to the sum of the diffraction at (110) and (200) planes unique to β phase [19]. The β phase is further confirmed by FTIR test with clear characteristic peak at 840 cm^{-1}, 1076 cm^{-1}, 1280 cm^{-1}, and 1431cm^{-1} as illustrated in Figure 2b. However, the spin-coated PVDF film only reflects weak peaks at 764 cm^{-1}, 796 cm^{-1}, 975 cm^{-1}, 1209 cm^{-1}, and 1385 cm^{-1}, corresponding to the characteristic absorption bands of *trans-gauche* linkages, configuration of the α phase [20,21]. Both the XRD and FTIR tests demonstrate that this 3D electrospinning process can transform the nonpolar α phase into polar β phase.

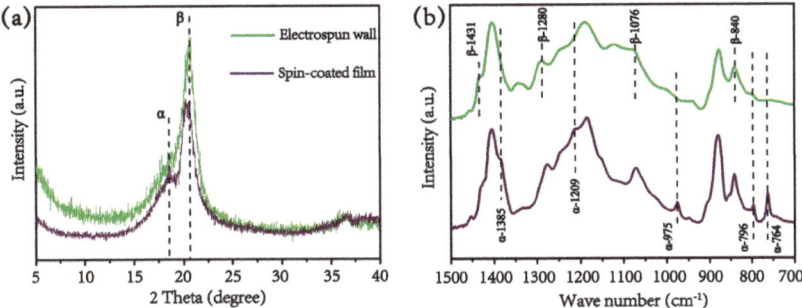

Figure 2. (a) X-ray diffraction XRD patterns and (b) Fourier transform infrared (FTIR) spectra for comparison between electrospun PVDF wall structure and spin-coated PVDF film.

3.2. Piezoelectric Performance

For testing the dynamic piezoelectric response, a mechanical shaker, operating at a specific frequency, was utilized to repetitively compress and decompress this device. The measurement platform is illustrated in Figure 3a, which includes a signal generator and a power amplifier to drive the shaker to work under a setting mode, and an electrochemical workstation is utilized to record the electrical outputs respectively. Figure 3b shows the optical image of this shaker. The mechanism for the piezoelectric output, normal to the electrospun wall surface, is illustrated in Figure 3c–e. Without the mechanical perturbation, no output current or voltage can be observed. As the electric dipoles randomly oscillate, the total spontaneous polarization from the electric dipoles is constant, as seen in Figure 3c. When mechanical impact is applied onto the device, each electrospun PVDF fiber is compressed, the spontaneous polarization of each fiber along the radial direction decreases significantly, a flow of electrons from the bottom electrode to the top electrode is produced due to the decrease in the induced charges, illustrated in Figure 3d. Conversely, when the impact is released, an opposite signal can be observed because the polarization is recovered, shown in Figure 3e. As such, under external mechanical perturbation, the electrospun device acts as a "charges pump" to generate electricity.

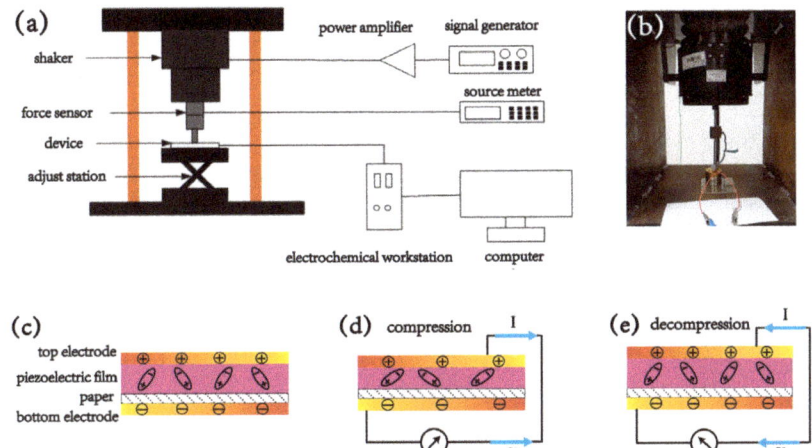

Figure 3. (a) Schematic of measurement platform for generation of dynamic compression and detection of electrical outputs. (b) Optical image of the mechanical shaker. (c) Schematic of the electric dipoles formed in the electrospun PVDF micro wall film. Principles of piezoelectric output under the (d) compression and (e) decompression mode.

Through setting the signal generator and power amplifier, a sinusoidal mechanical compression and decompression at a frequency of 0.5 Hz and a peak force of ca. 1 N is generated to drive the device. Upon this mechanical impact, Figure 4a,b record the dynamic output voltage and current, respectively. The typical peak electrical outputs for this device are ca. 0.48 V and ca. 2.7 nA. To verify that the measured signal is really produced from piezoelectric responses instead of artificial noises, voltage and current from "switching polarity" [22] tests are measured, as shown in Figure 4c,d. When the polarity of the fabricated device connected to the measurement system is reversed, the shapes of electrical output curves are flipped, confirming the piezoelectric response. However, the values of voltage and current in the reversed connection are almost reduced by a half compared with those in the forward connection; the result is similar to the previous reports based on micro or nano piezoelectric structures [16,20,23], possibly caused by a bias current in the measurement system.

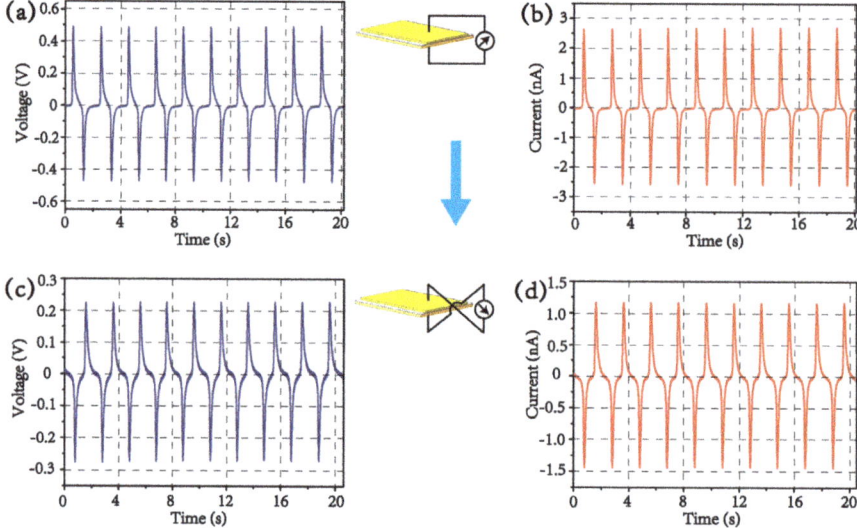

Figure 4. The electrical output of (**a**) voltage and (**b**) current under cyclic compression at 0.5 Hz. The switching polarity tests for output of (**c**) voltage and (**d**) current in the reversed connection. The inset is the schematic for the "switching polarity" tests.

Furthermore, different amplitudes of driving force can be easily produced by adjusting the magnification factor of power amplifier, and the force sensor in the measurement system can detect the real-time value of force imposed on device. As such, the relationship between driving force and generated outputs can be investigated. When the driving force is varied from 200 to 1000 mN, the peak values of output voltage/current are ca. 0.37 V/2.1 nA, 0.41 V/2.3 nA, 0.44 V/2.45 nA, 0.46 V/2.6 nA, 0.48 V/2.7 nA, respectively, as shown in Figure 5. It can be clearly seen that the generated output is almost linear to the driving force, revealing the high potential of this 3D electrospun PVDF microstructure as piezoelectric component for force/pressure sensing.

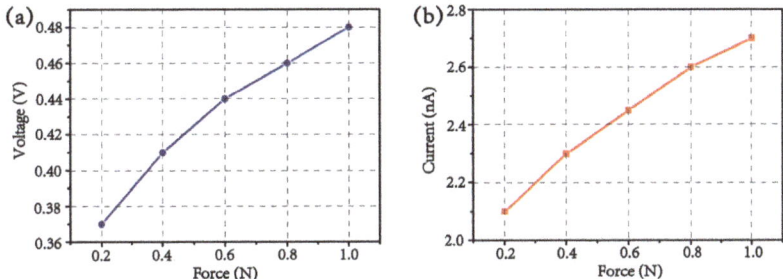

Figure 5. The curves of output (**a**) voltage and (**b**) current under different driving forces.

The durability of the output performance has also been examined by constant mechanical compression and decompression upon the as-prepared device for over 3 h, which means over 10,000 mechanical cycles when the impact frequency is set as 1 Hz. The output current is recorded as shown in Figure 6, which shows that no obvious reduction in the signal is observed, revealing the high stability and robust for our 3D electrospun structure. It is worth mentioning that the current exhibits a slight increase from ca. 2.9 nA to ca. 3.1 nA after 3 h compression as shown in the bottom two panels of Figure 6, this can be attributed to the repeated mechanical compression which leads to the decrease in contact resistance between the electrospun wall and top electrode. The excellent durability can enable some specific applications, such as heel energy harvesting and tire pressure detection, in which high and uninterrupted mechanical impact exists.

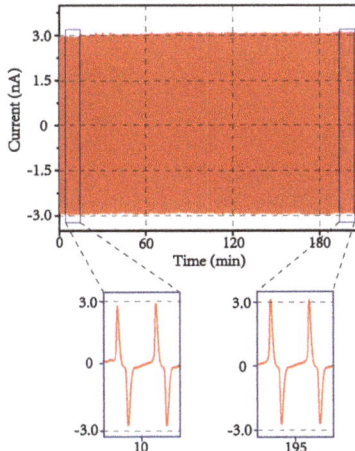

Figure 6. The current output curve of the as-prepared device operated for over 3 h under 1 Hz mechanical compression, the bottom two panels show the detailed shapes of current outputs.

The electrical response from spin-coated PVDF film on paper substrate is also measured with the sandwich-like electrodes configured under the same measurement conditions as recorded in Figure 7a,b. The output voltage and current from the spin-coated PVDF film are ca. 0.03 V and ca. 0.15 nA, respectively, both of which are more than one order of magnitude smaller than those from the 3D electrospun PVDF wall. The reason being that, without post poling, the spin-coated PVDF exists in the nonpolar α phase and no high enough piezoelectricity can be exhibited. This result further demonstrates the advantage of in situ poling of PVDF fibers through this 3D electrospinning technology, and eliminates the possibility that the measured signals are the result of the triboelectric effect [24] between aluminum electrodes and electrospun polymeric PVDF wall in this design.

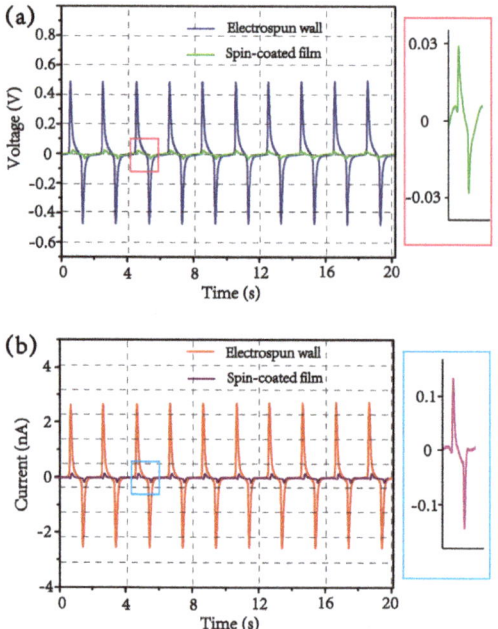

Figure 7. The outputs of (**a**) voltage and (**b**) current comparison between the electrospun PVDF wall and spin-coated PVDF film.

3.3. Enhancement of Electrical Outputs

In order to increase the total electrical outputs, either series or parallel connections of the devices can be utilized to result in multiplied voltage or current outputs, respectively. The output voltages are increased to ca. 1.35 and ca. 2.1 V when three and five devices are connected in series, respectively, as illustrated in Figure 8a. Similarly, Figure 8b shows the enhanced outputs of current by making parallel connections of devices when three and five devices are connected in parallel, the measured output currents constructively add and rise to ca. 7.1 and ca. 10.9 nA, respectively. As such, the net electrical outputs are almost linearly correlated to the number of piezoelectric components through "linear superposition" [22], the other criterion, which verifies that the electrical outputs are indeed produced by the piezoelectric effect.

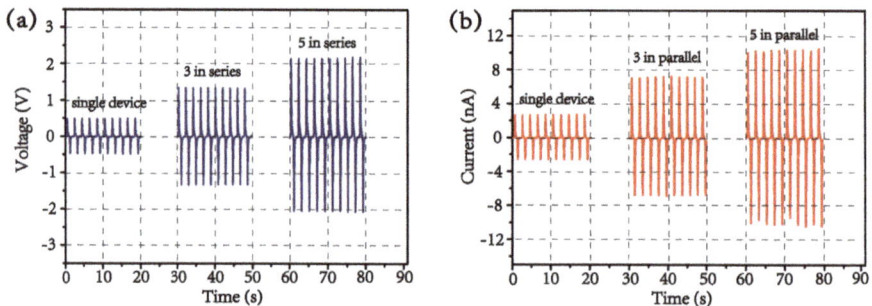

Figure 8. (**a**) Output voltages for one, three, and five devices that are electrically connected in series, and (**b**) output currents for one, three, and five devices that are electrically connected in parallel.

4. Conclusions

In summary, an advanced process combining high-precision 3D fabrication and in situ piezoelectric poling in a single step is demonstrated using the 3D electrospinning technology. The out-of-plane electrospun PVDF fibers are consistently stacked in a fiber-by-fiber manner on paper substrate, while the polar β phase is produced due to the electrical poling and the mechanical stretching during electrospinning. Piezoelectric responses along radial direction of the fibers are subsequently measured. Both "switching-polarity" and "linear superposition" criteria are tested to demonstrate that the measured signals are really generated due to piezoelectric response rather than from the noises in the measurement system. The combined merits of high controllability over fabrication, in situ poling of piezoelectricity, high integration of functionalized fibers, and excellent durability of output performance make this 3D electrospun structure promise great potential to integrate with other processes and structures as energy harvesting or high-precision sensing component.

Author Contributions: G.L., L.Z. and K.S.T. conceived the paper and designed the experiments; Y.L., Q.Z., L.W. and Z.L. developed the experiments; S.W., L.Z. and Z.J. performed the investigation and analysis; G.L. and K.S.T wrote the original draft while all the authors contributed to the final manuscript and approved the final paper. All authors have read and agreed to the published version of the manuscript.

Funding: This work is supported by the National Natural Science Foundation of China (grant numbers 51705409, 91748207, 51890884), the Natural Science Foundation of JiangSu Province (grant number BK20170415), the Open Fund of State Key Lab of Digital Manufacturing Equipment & Technology (grant number DMETKF2017009), the China Postdoctoral Science Foundation (grant number 2017M613120), the Fundamental Research Funds for the Central Universities (grant number xjj2017094), and the Natural Science Basic Research Plan in Shaanxi Province of China (grant number 2018JQ5050).

Acknowledgments: The authors thank Instrument Analysis Center of Xi'an Jiaotong University for the SEM, XRD, and FTIR facilities.

Conflicts of Interest: The authors declare no conflict of interest.

References

1. Brown, L.F. Design considerations for piezoelectric polymer ultrasound transducers. *IEEE Trans. Ultrason. Ferroelectr. Freq. Contr.* **2000**, *47*, 1377–1396. [CrossRef] [PubMed]
2. Ramadan, K.S.; Sameoto, D.; Evoy, S. A review of piezoelectric polymers as functional materials for electromechanical transducers. *Smart Mater. Struct.* **2014**, *23*, 033001. [CrossRef]
3. Bowen, C.R.; Kim, H.A.; Weaver, P.M.; Dunn, S. Piezoelectric and ferroelectric materials and structures for energy harvesting applications. *Energy Environ. Sci.* **2013**, *7*, 25–44. [CrossRef]
4. Vicente, J.; Costa, P.; L-Mendez, S.; Abete, J.M.; Iturrospe, A. Electromechanical properties of PVDF-based polymers reinforced with nanocarbonaceous fillers for pressure sensing applications. *Materials* **2019**, *12*, 3545. [CrossRef] [PubMed]
5. Cui, Z.; Drioli, E.; Lee, Y.M. Recent progress in fluoropolymers for membranes. *Prog. Polym. Sci.* **2014**, *39*, 164–198. [CrossRef]
6. Esfahani, H.; Jose, R.; Ramakrishna, S. Electrospun ceramic nanofiber mats today: Synthesis, properties, and applications. *Materials* **2017**, *10*, 1238. [CrossRef]
7. Xue, J.; Wu, T.; Dai, Y.; Xia, Y. Electrospinning and Electrospun nanofibers: Methods, materials, and applications. *Chem. Rev.* **2019**, *119*, 5298–5415. [CrossRef]
8. Fang, J.; Zhu, Y.; Lin, T. Electrical power generator from randomly oriented electrospun poly (vinylidene fluoride) nanofibre membranes. *J. Mater. Chem.* **2011**, *21*, 11088–11091. [CrossRef]
9. Chang, C.; Tran, V.H.; Wang, J.; Fuh, Y.K.; Lin, L. Direct-write piezoelectric polymeric nanogenerator with high energy conversion efficiency. *Nano Lett.* **2010**, *10*, 726–731. [CrossRef]
10. Mandal, D.; Yoon, S.; Kim, K.J. Origin of piezoelectricity in an electrospun poly (vinylidene fluoride-trifluoroethylene) nanofiber web-based nanogenerator and nano-pressure sensor. *Macromol. Rapid Commun.* **2011**, *32*, 831–837. [CrossRef]
11. Yu, H.; Huang, T.; Lu, M.; Mao, M.; Zhang, Q.; Wang, H. Enhanced power output of an electrospun PVDF/MWCNTs-based nanogenerator by tuning its conductivity. *Nanotechnology* **2013**, *24*, 405401. [CrossRef] [PubMed]

12. Pu, J.; Yan, Y.; Jiang, Y.; Chang, C.; Lin, L. Piezoelectric actuaton of direct-write electrospun fibers. *Sens. Actuators A* **2010**, *164*, 131–136. [CrossRef]
13. Ren, G.; Cai, F.; Li, B.; Zheng, J.; Xu, C. Flexible pressure sensor based on a poly(VDF-TrFE) nanofiber web. *Macromol. Mater. Eng.* **2013**, *298*, 541–546. [CrossRef]
14. Zhu, J.; Niu, X.; Hou, X.; He, J.; Chou, X.; Xue, X.; Zhang, W. Highly reliable real-time self-powered vibration sensor based on enhanced piezoelectric nanogenerator. *Energy Technol.* **2017**, *6*, 781–789. [CrossRef]
15. Gui, J.; Zhu, Y.; Zhang, L.; Shu, X.; Liu, W.; Guo, S.; Zhao, X. Enhanced output-performance of piezoelectric poly(vinylidene fluoride trifluoroethylene) fibers-based nanogenerator with interdigital electrodes and well-ordered cylindrical cavities. *Appl. Phys. Lett.* **2018**, *112*, 072902. [CrossRef]
16. Fuh, Y.K.; Chen, S.Y.; Ye, J.C. Massively parallel aligned microfibers-based harvester deposited via in situ, oriented poled near-field electrospinning. *Appl. Phys. Lett.* **2013**, *103*, 033114. [CrossRef]
17. Luo, G.; Teh, K.S.; Liu, Y.; Zang, X.; Wen, Z.; Lin, L. Direct-write, self-aligned electrospinning on paper for controllable fabrication of three-dimensional structures. *ACS Appl. Mater. Interfaces* **2015**, *7*, 27765–27770. [CrossRef]
18. Gregorio, R.; Ueno, E.M. Effect of crystalline phase, orientation and temperature on the dielectric properties of poly(vinylidene fluoride) (PVDF). *J. Mater. Sci.* **1999**, *34*, 4489–4500. [CrossRef]
19. Fang, J.; Niu, H.; Wang, H.; Wang, X.; Lin, T. Enhanced mechanical energy harvesting using needleless electrospun poly(vinylidene fluoride) nanofibre webs. *Energy Environ. Sci.* **2013**, *6*, 2196–2202. [CrossRef]
20. Fuh, Y.K.; Wang, B.S. Near field sequentially electrospun three-dimensional piezoelectric fibers arrays for self-powered sensors of human gesture recognition. *Nano Energy* **2016**, *30*, 677–683. [CrossRef]
21. Cai, X.; Lei, T.; Sun, D.; Lin, L. A critical analysis of the α, β, and γ phases in poly(vinylidene fluoride) using FTIR. *RSC Adv.* **2017**, *7*, 15382–15389. [CrossRef]
22. Yang, R.; Qin, Y.; Li, C.; Dai, L.; Zhang, Z.L. Characteristics of output voltage and current of integrated nanogenerators. *Appl. Phys. Lett.* **2009**, *94*, 022905. [CrossRef]
23. Yang, R.; Qin, Y.; Dai, L.; Wang, Z.L. Power generation with laterally packaged piezoelectric fine wires. *Nat. Nanotechnol.* **2009**, *4*, 34–39. [CrossRef] [PubMed]
24. Fan, F.R.; Tian, Z.Q.; Wang, Z.L. Flexible triboelectric generator. *Nano Energy* **2012**, *1*, 328–334. [CrossRef]

© 2020 by the authors. Licensee MDPI, Basel, Switzerland. This article is an open access article distributed under the terms and conditions of the Creative Commons Attribution (CC BY) license (http://creativecommons.org/licenses/by/4.0/).

Article

Triboelectric Energy Harvesting Response of Different Polymer-Based Materials

Tiago Rodrigues-Marinho [1], Nelson Castro [2], Vitor Correia [1,3], Pedro Costa [1,4,*] and Senentxu Lanceros-Méndez [2,5]

1. Center of Physics, Campus Gualtar, University of Minho, 4710-057 Braga, Portugal; tiagomarinho.fis@gmail.com (T.R.-M.); eng.v.correia@gmail.com (V.C.)
2. BCMaterials, Basque Center for Materials, Applications and Nanostructures, UPV/EHU Science Park, 48940 Leioa, Spain; nelsonjfcastro@gmail.com (N.C.); senentxu.lanceros@bcmaterials.net (S.L.-M.)
3. Algoritmi Research Center, University of Minho, 4800-058 Guimarães, Portugal
4. Institute for Polymers and Composites IPC/i3N, University of Minho, 4800-058 Guimarães, Portugal
5. IKERBASQUE, Basque Foundation for Science, 48009 Bilbao, Spain
* Correspondence: pcosta@fisica.uminho.pt

Received: 23 September 2020; Accepted: 31 October 2020; Published: 5 November 2020

Abstract: Energy harvesting systems for low-power devices are increasingly being a requirement within the context of the Internet of Things and, in particular, for self-powered sensors in remote or inaccessible locations. Triboelectric nanogenerators are a suitable approach for harvesting environmental mechanical energy otherwise wasted in nature. This work reports on the evaluation of the output power of different polymer and polymer composites, by using the triboelectric contact-separation systems (10 N of force followed by 5 cm of separation per cycle). Different materials were used as positive (Mica, polyamide (PA66) and styrene/ethylene-butadiene/styrene (SEBS)) and negative (polyvinylidene fluoride (PVDF), polyurethane (PU), polypropylene (PP) and Kapton) charge materials. The obtained output power ranges from 0.2 to 5.9 mW, depending on the pair of materials, for an active area of 46.4 cm^2. The highest response was obtained for Mica with PVDF composites with 30 wt.% of barium titanate (BT) and PA66 with PU pairs. A simple application has been developed based on vertical contact-separation mode, able to power up light emission diodes (LEDs) with around 30 cycles to charge a capacitor. Further, the capacitor can be charged in one triboelectric cycle if an area of 0.14 m^2 is used.

Keywords: triboelectric effect; polymer and composites; energy harvesting; low-power devices

1. Introduction

The world is experiencing a rapid revolution in the mode in which energy is being produced and consumed in daily life and industry [1,2]. Conventional ways to produce energy need to adapt to the environmental needs and concerns related to sustainability and, on the other hand, the energy consumption paradigm has also been strongly changing in the last decade based on increased mobility [2]. Thus, cell phones, tablets or related gadgets are common and ubiquitous nowadays. Hydroelectric energy generation remains the pillar of renewable energies [3,4], wind and solar energy generation are becoming increasingly important in the energy generation share [3].

In the last decade, with the fast development of the Internet of Things (IoT) [5] and portable electronics [5], the demand for a sustainable and environmentally friendly portable power supply is becoming very significant. In this context, energy generation systems are an interesting option for portable technologies, though still show low power output and, therefore, a low range of applications [3]. Piezoelectric, pyroelectric or thermoelectric energy generation are among the most studied technologies, though efficiency and power output [6] is limited compared to electrostatic or triboelectric energy harvesting devices [6].

Piezoelectric [3,7], electrostatic [8], electromagnetic [9] and triboelectric [10–12] energy harvesting technologies rely on the use of wasted mechanical energy (wind, wave, vibrations and even body movements) to produce electricity. These mechanical harvesters can be based on polymer and polymer composite materials, offering unique properties, such as lightweight and flexibility, combined with easy integration and environmentally friendly processability [3].

The power density per area of triboelectric devices is the largest among the aforementioned systems, reaching powers as high as 500 W·m^{-2} and an energy conversion efficiency of 70% has been demonstrated [9,13], further they are lightweight and cost-effective [13]. Compared with piezoelectric devices, triboelectricity can be more suited for environmentally friendly energy production for portable devices [3], low-power devices in remote or inaccessible places [14] or for needed IoT network of sensors [14].

Another advantage of the triboelectric phenomenon is the wide range of materials that can be used in the distinct triboelectric mechanical modes: contact-separation, lateral sliding, single electrode and free-standing triboelectric layer mode [5,14]. Being a process that can be carried out entirely with polymers and the corresponding composites, the overall properties of the materials can be tailored for each specific application, including dimensions, geometry and optical transparency. The triboelectric power output can also be strongly improved by tailoring the intrinsic properties of the polymers by synthesis and functionalization [15] or by reinforcing with high-dielectric or other functional fillers [9]. Further, geometrical dimensions (mainly the thickness) and roughness of the materials can also be designed to maximize the generated energy.

Literature reports different materials and order within the triboelectric series [9,16,17] in terms of relative triboelectric charge providing/receiving characteristics. The most interesting triboelectric materials are those with easy lose and gain electrons when in contact, leading to higher charge density between two different materials [18]. The most common materials in the literature are several polymers, but also some metals and crystalline materials [17–20]. Positive (losing electrons) materials include Mica (silicate) and glass, polyamide 6-6 (PA66) and polyamide 11 (PA11), polyethylene (PE) silk, aluminum, paper and polyvinylidene fluoride (PVDF) [9,12,16,17,20]. Negative materials (gaining electrons) include polytetrafluoroethylene (PTFE or Teflon), polyvinyl chloride (PVC), polyimide (Kapton), polystyrene, rubber-like or polyurethane (PU), among others [12,17–20]. The surface charge density of the triboelectric mode depends on the pair of the selected materials and range change from some nC.m^{-2} to mC.m^{-2} [18], being higher for polymeric pairs compared to metallic ones.

Triboelectric energy generation optimization includes, in addition to materials selection, tailoring the functional properties of the material reinforcing with dielectric fillers, selection of the triboelectric mode among contact-separation, sliding, single electrode or freestanding modes and, finally, the electronic circuit to harvest the electrical energy [14,18–21]. Ceramic or low amounts of conductive nanofillers [19,20] can be used to improve the dielectric properties and, correspondingly, the triboelectric performance of the materials. Triboelectric modes and electronic circuits allow the combination of two distinct modes or even, combine the piezoelectric and thermoelectric effects [21,22].

In this work, the triboelectric properties of the different materials, mostly polymers and polymer composites are evaluated. Pairs of materials in different places in the triboelectric series are evaluated, together with materials prepared by different technologies (solvent based, hot pressing and commercial materials). Finally, high-dielectric ceramic nanomaterial (barium titanate- BT) have been used to improve the dielectric and, consequently, the triboelectric power output of one of the materials. In this way, a complete set of materials and processing conditions are considered, allowing triboelectric output understanding, material selection and tailoring for specific applications.

2. Experimental

2.1. Materials

Thermoplastic elastomer styrene-ethylene/butylene-styrene (SEBS) copolymer (Calprene CH-6120, Madrid, Spain) with an ethylene-butylene/styrene ratio of 68/32 was supplied by Dynasol Gestión, S.I. (Madrid, Spain). Commercial polyvinylidene fluoride (PVDF) (Solef 6010), with density of 1.75 g/cm^3 was supplied by Solvay (Paris, France). The solvent used to process SEBS was cyclopentyl methyl ether (CPME) supplied by Carlo Erba Reagents (Val de Reuil, France) (density of 0.86 g/cm^3 at 20 °C; boiling point of 106 °C) and for PVDF, it was N,N-dimethylformamide (DMF, 99.5%) from Merck (Darmstadt, Germany).

A commercial sheet of PVDF (PVDF-c) with 1 mm of thickness was obtained from Swami Plast Industries (Gujarat, India) and Mica, Kapton and polyurethane (PU) were obtained from Agar Scientific (Essex, UK), Dupont (Faro, Portugal) and SWM-Engineered (Genk, Belgium), respectively. Polyamide 66 and polypropylene (PP) pellets were purchased from Merck (Sigma-Aldrich, St. Louis, MO, USA).

For the preparation of the polymer composite, barium titanate (BT), particles with an average size of 100 nm and a dielectric constant of 150 were obtained from Merck (Sigma-Aldrich, St. Louis, MO, USA).

2.2. Sample Preparation

Three types of different processed materials were investigated: solvent cast films for pristine polymers and polymer composite with BT, hot pressing and commercial polymers in sheet form.

The solvent casting method was similar for SEBS, PVDF and composites with solvent/polymer ratio of 80/20 v/v using about 1 g of polymer for 5 mL of solvent. For the SEBS dissolution, CMPE was used as the solvent, while DMF was used for the PVDF. Once the corresponding amount of polymer and solvent were added, the mixture was magnetically stirred for 3 h at 30 °C until complete polymer dissolution. For the PVDF composite, the corresponding amount of BT nanoparticles (30 weight percentage (wt.%) to maximize dielectric response were maintaining mechanically flexible films [23]) were homogeneously dispersed in DMF in an ultrasonic bath at 25–35 °C for 2 h, then the PVDF was added and the mixture was magnetically stirred for 3 h at 30 °C.

Thin films were obtained by spreading the mixtures on a clean glass substrate using the doctor blade technique with a 200 μm blade thickness. SEBS samples were dried at 30 °C for 12 h, whereas PVDF samples were melted in an oven at 210 °C for 20 min and recrystallized by cooling down to room temperature, promoting the crystallization of the PVDF in the α-phase and achieving complete solvent evaporation [24–26]. The different processed samples are represented in Table 1. The thicknesses of the films after complete evaporation of the solvent ranged from 40 to 60 μm.

The use of solvents was avoided in PA66 and PP, positive and negative triboelectric materials, respectively (Table 1). Both polymers were produced by the hot pressing method where 20 g of polymer pellets were placed in a hot-pressing machine (from Metalgrado LDA, Porto, Portugal) for 15 min at a temperature of 220 °C between two 40 × 40 cm sheets of Teflon. After removing, the film thickness was about 1 ± 0.1 mm.

Commercial films of different materials, including Mica, PVDF-c, PU and Kapton, with a thickness of about 1 mm, were also used. Samples produced by solvent casting and hot pressing and commercially available materials were also evaluated and compared. In this way, some of the most interesting triboelectric polymers have been comparatively evaluated to understand and optimize triboelectric output.

In order to collect the charge provided by the triboelectric effect, conductive silver ink (Electronic 131 paste DT1201, hunan LEED electronics Ink, Zhuzhou, China) was deposited on the outer surface of each material within a home-made screen-printing set-up using a squeegee over the screen placed at 1 mm distance from the substrate. After the printing step, the material and silver ink were dried at 60 °C for 60 min in an oven (Binder E, model 28, Binder, Tuttlingen, Germany). The printed electrode area was 8.0 × 5.8 cm, as shown in Figure 1. The final step of sample preparation was to place the active materials with electrodes on substrate support fabricated by 3D printing (Sigma R19 BCN3D with 20% PLA filling, BCN3D, Barcelona, Spain) with dimensions 10 × 6 × 1 cm (slightly larger than the active materials) to perform the triboelectric measures. To assure good adhesion of the active material to the substrate, double-sided adhesive tape (Tesa 4970, Tesa, Lisboa, Portugal) was used.

Table 1. (**A**) Materials within the triboelectric series used in the present work [27,28] and (**B**) pair of materials used in the contact-separation mode triboelectric experiments.

(A)	
Positive	**Negative**
Mica polyamide 6-6 (PA66) styrene-ethylene/butylene-styrene (SEBS)	polyvinylidene fluoride (PVDF) polyurethane (PU) polypropylene (PP) polyimide (Kapton)
(B)	
Mica	PVDF polyvinylidene fluoride commercial (PVDF-c) 30 wt.% barium titanate/polyvinylidene fluoride (30BT/PVDF) Kapton PU SEBS
PA66	SEBS PP

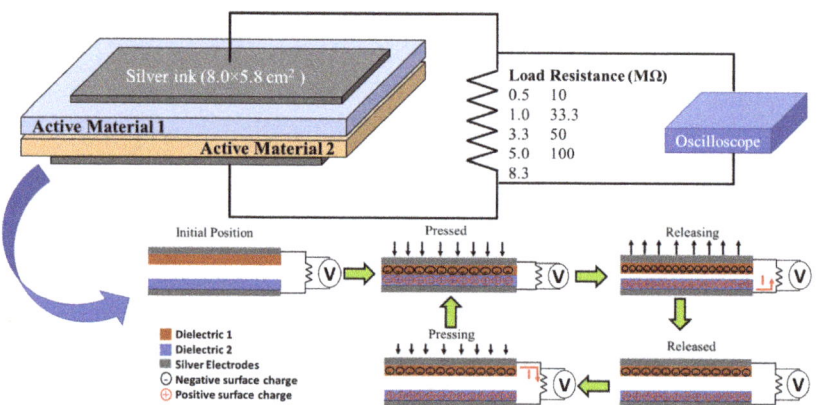

Figure 1. Illustration of the geometry of the samples with the active material, screen-printed silver ink electrodes and their connection to the Picoscope 2205A using load resistances of 0.5, 1, 3.3, 5, 8.3, 10, 33.3, 50 and 100 MΩ, above. Experimental method used for the triboelectric evaluation of the different pair of materials in a contact-separation mode, below. The force applied in each step is about 10 N.

Table 1 summarizes the materials used in this work considering the triboelectric series, including the composite with 30 wt.% BT/PVDF composite (30BT/PVDF).

2.3. Triboelectric Measurements

The triboelectric mode used to determine the output voltage, current and corresponding output power of the different pairs of materials (Table 1) was the contact-separation mode (Figure 1): a force of 10 N was applied in the contact mode, followed by a separation of 5 cm between the samples in each cycle.

The electrical response was obtained though load resistances of 0.5, 1, 3.3, 5, 8.3, 10, 33.3, 50 and 100 MΩ using a Picoscope 2205A (Picotech, Tyler, TX, USA) with resolution of 8 bit at 200 MS/s.

3. Results and Discussion

3.1. Triboelectric Output

The triboelectric voltage and current output of the different pairs of materials represented in Table 1 are shown in Figure 2 for a load resistance of 5 MΩ. The voltage is determined in open-circuit (Figure 2A,C,E) and the current in short-circuit (Figure 2B,D,F) for representative pair of materials, measured with a load resistance (R_L) of 5 MΩ under a constant force (10 N) in the contact-separation mode.

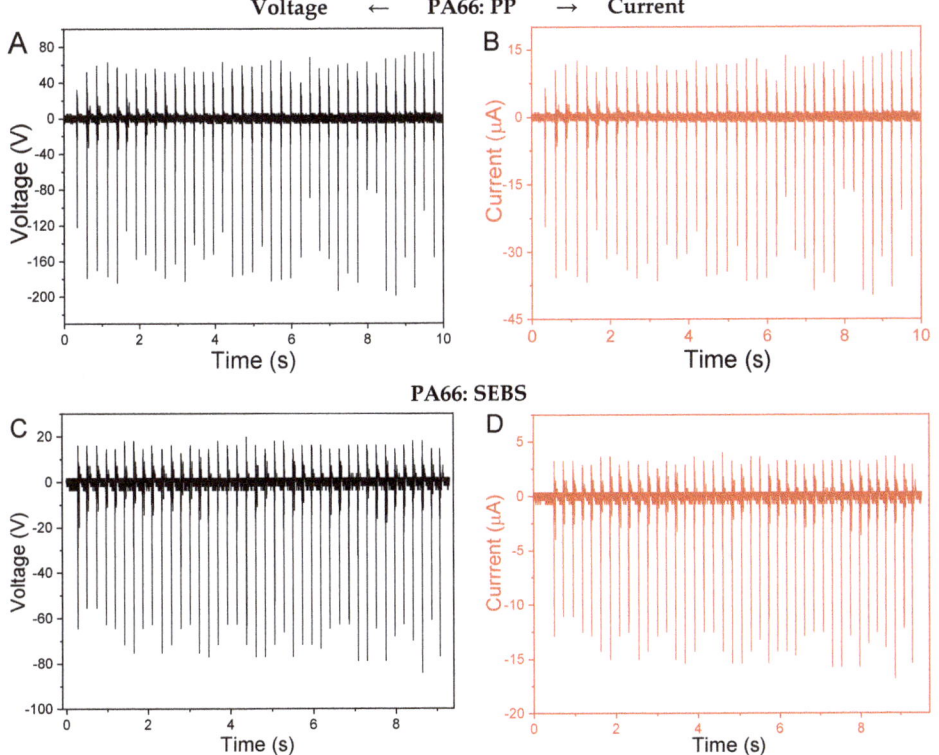

Figure 2. *Cont.*

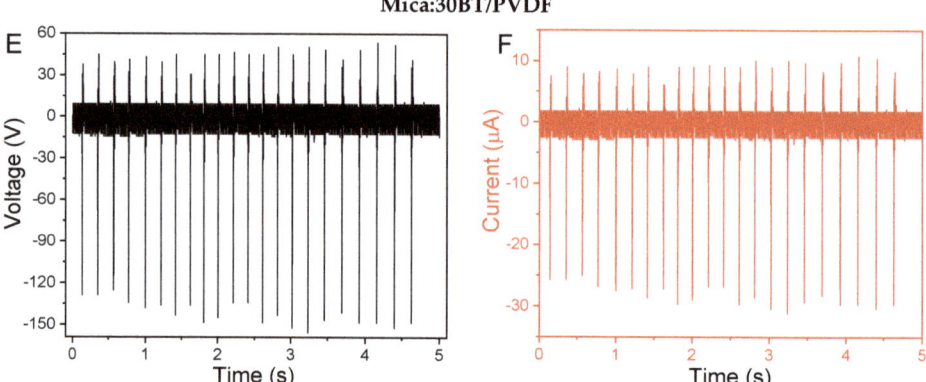

Figure 2. Open-circuit voltage (left) and the short-circuit current (right) measurements for different pairs of polymers. The pairs PA66:PP (**A,B**), PA66:SEBS (**C,D**) and Mica:30BT/PVDF (**E,F**), measured under with a R_L = 5 MΩ under constant force (10 N) in contact-separation mode.

Among the different pairs of materials measured (Table 1B), Figure 2 illustrates, as representative examples, the triboelectric performance of PA66:PP, PA66:SEBS and Mica:30BT/PVDF. The output voltages per cycle of these pair of materials are between approximately 60 and 150 V (in average for 40 cycles) and the current generated per cycle ranges between 12 to 30 µA, for the same experimental conditions. Further, it is to notice that the 30BT/PVDF sample shows piezoelectric properties (due to the piezoelectric ceramic material) that contribute to a piezoelectric voltage generation in each cycle (mechano-electrical conversion), together with the triboelectric energy generation, as can be observed in Figure 3E,F. It is to notice that the piezoelectric voltage generation is small in comparison to the triboelectric contribution.

Literature reports a wide amplitude of the voltages and currents generated in triboelectric systems, from some volts to thousands of volts [29–33], and current typically ranging up to hundreds of µA [31–34]. The voltage and current output values obtained in the present work are competitive with the literature, considering the use of pristine materials without any kind of surface treatment.

Thus, the above materials allow us to generate up to 150 V per cycle through the triboelectric effect in contact-separation mode, suitable for low-power devices [22], as it will be demonstrated later.

Based on the representative experimental results shown in Figure 2, the output voltage and current triboelectric output of all materials pairs are shown in Figure 3 and Table 2 as a function of the external load resistance (R_L) in the range from 0.5 to 100 MΩ. All systems show a similar electrical output response with a maximum output power for R_L = 3 to 10 MΩ. Increasing R_L leads to an output voltage increase and a decrease of the current, leading to maximum output power (Power (P) = voltage × current) at the interception of these [16].

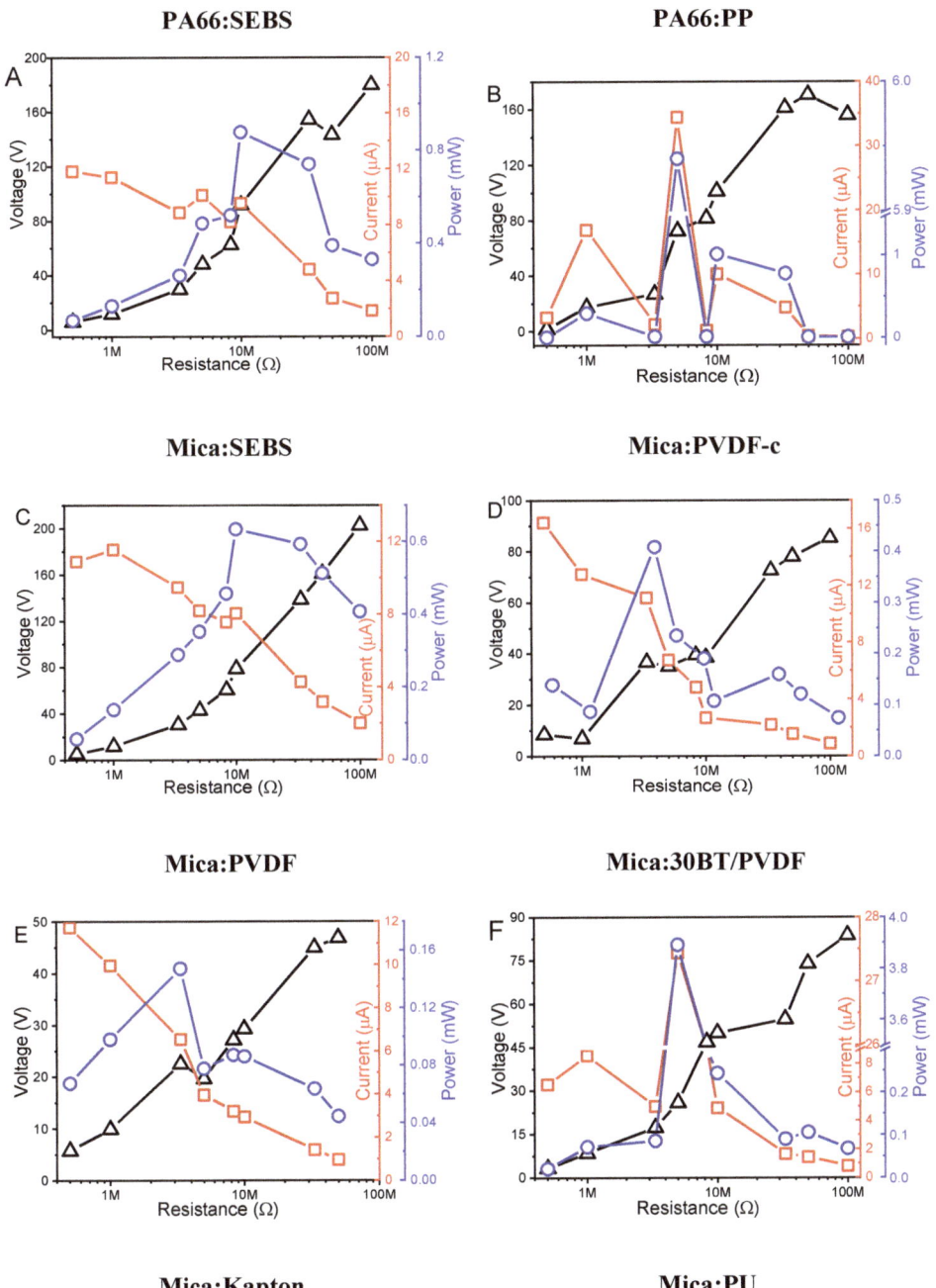

Figure 3. Cont.

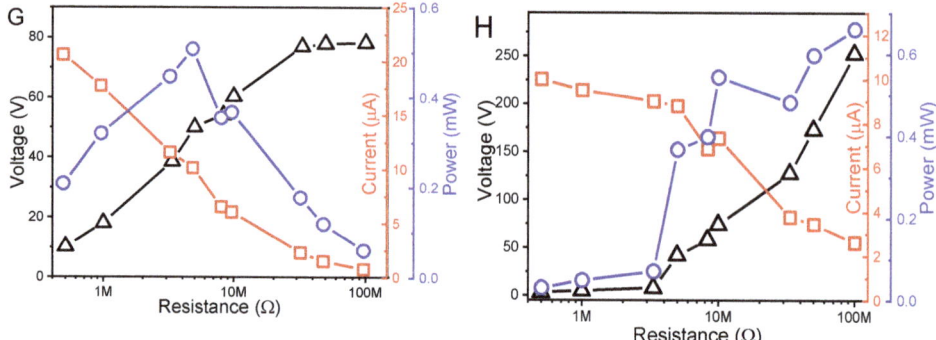

Figure 3. Triboelectric output voltage, current and power as a function of the load resistance for different material pairs: (**A**) PA66:SEBS; (**B**) PA66:PP; (**C**) Mica:SEBS; (**D**) Mica:PVDF-c; (**E**) Mica:PVDF; (**F**) Mica:30BT/PVDF; (**G**) Mica:Kapton and (**H**) Mica:PU.

Table 2. Summary of the output power, voltage and current for the different systems under evaluation, the larger output powers being for the PA66:PP and the Mica:30BT/PVDF pairs.

Materials	R_L (MΩ)	Voltage (V)	Current (μA)	Power (mW)
PA66:PP	5	172.5	34.4	5.9
PA66:SEBS	5	68.1	13.2	0.9
Mica:SEBS	10	78.6	8.1	0.6
Mica:PVDF-c	3.3	36.7	11.1	0.4
Mica:PVDF	3.3	22.5	6.6	0.2
Mica:30BT/PVDF	5	141.8	27.4	3.9
Mica:Kapton	5	50.1	10.2	0.5
Mica:PU	>100	252.4	2.6	0.7

The power output performance depends on several factors, such as triboelectric charge providing/receiving [11,34] and physical properties of the materials [35,36]. It is shown in the literature that materials further apart in providing/receiving electrons lead to a larger triboelectric output than materials close to each other, which may exchange small amounts of charge [34]. Roughness is also a key factor for triboelectric energy generation. It has been experimentally demonstrated for different polymer-based materials that increasing roughness leads to an increase in the output power of the triboelectric materials [35–37].

The triboelectric performance of the PA66:SEBS pair is represented in Figure 3A, showing a $P = 0.90$ mW at a $R_L = 5$ MΩ. It is to notice that these materials are close in the triboelectric series, but one prevalent factor, surface roughness, also plays a relevant role, as mentioned before. Solvent cast samples present higher roughness than commercial ones, leading to a higher surface area, resulting in improved triboelectric performance. Also, the samples prepared by hot-pressing, such as PA66, show a larger surface roughness and, as a consequence, the PA66:PP pair shows a $P = 5.94$ mW for $R_L = 5$ MΩ, as is shown in Table 2.

Thus, polyamide was used combined with SEBS and PP with a maximum power of 0.9 and 5.8 mW, respectively. Mica was also used to compare the triboelectric performance of the samples PVDF or PVDF-c, considering the processing method. Both have a maximum power around $R_L = 3.3$ MΩ with the commercial and solvent cast samples reaching the same order of magnitude for the output power, about 0.4 mW instead of 0.2 mW for PVDF-c and PVDF (Figure 3D,E), respectively. A similar performance indicates that the intrinsic properties of the PVDF materials, in particular the large dielectric constant ($\varepsilon \approx 6$) [38], overcomes the effects related to the manufacturing process and surface roughness variations.

The important role of the dielectric properties in the triboelectric output of the samples is demonstrated by the PVDF composites reinforced with barium titanate. The high dielectric constant of the ceramic nanoparticles (150) embedded into the PVDF matrix (dielectric constant around 6) leads to an increase of dielectric constant with an increase of filler content [39,40] up to $\varepsilon \approx 15$ for the composite PVDF, increasing the performance of the triboelectric system, as predicted by theoretical models [9,41,42] and experimental measurements [38]. Thus, a maximum power output of 0.2 and 3.9 mW was obtained for neat PVDF and 30BT/PVDF, respectively, as shown in Figure 3F, demonstrating that increasing the dielectric constant of a specific material leads to an increase of its triboelectric output. In conclusion, the roughness and dielectric permittivity of polymers influence the triboelectric performance, being that the dielectric properties are more preponderant in the charges transferred between opposite surfaces.

Commercial pair of materials Mica:Kapton presents a P = 0.5 mW for R_L = 5 MΩ, despite being one of the most opposite pairs within the triboelectric series. Contrary to PVDF, surface treatments in the surface (smooth surfaces) of the commercial materials decreases their triboelectric performance. A similar effect can be observed in Mica:PU with P = 0.7 mW. In this case, the output power continues to increase with increasing R_L, leading to an output voltage that increases up to 250 V. Mica:SEBS, on the other hand, reaches P = 0.6 mW at R_L = 10 MΩ (Figure 3C). SEBS with Mica, being similar to Mica:PA66, despite the proximity of these materials in the triboelectric series.

3.2. Energy Harvesting Application

A simple application was developed by harvesting the triboelectric energy into a capacitor and later powering a LED (Figure 4A). The two material pairs with the largest output powers, PA66:PP and Mica:30BT/PVDF, were used. The triboelectric pairs were connected to an electrical circuit containing 4 diodes in order to transform the AC to DC voltage and charge a capacitor of 15 µF, as illustrated in Figure 4B.

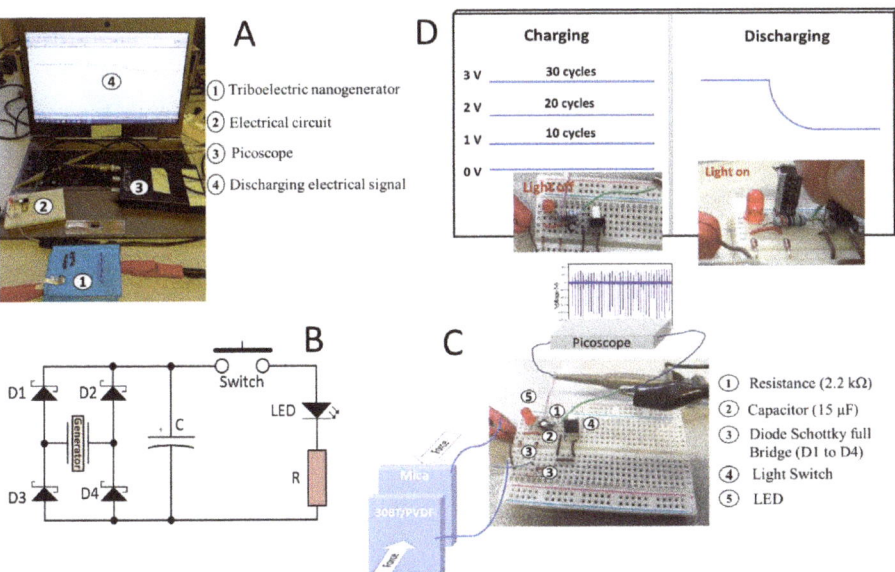

Figure 4. Illustration of the complete setup (**A**) with the pair of materials, detailed electronic circuit scheme (**B**) and Picoscope connected to a laptop. (**C**) Electric circuit for powering the LED and (**D**) charge-discharge cycles using triboelectric materials (PA66:PP or Mica:30BT/PVDF pairs) as nanogenerators.

The circuit follows a traditional DO-35 Schottky (D1 to D4) rectifier bridge topology with an output electrolytic capacitor for energy storage, powering a load composed by a manual switch button, the LED and the resistor (Figure 4C). This setup enables the energy to be stored and manually discharge over the load when the voltage level is suitable.

When the capacitor is charged, using a light switch, the LED was lighted on and the respective voltage drop at the capacitor ends was observed (Figure 4C).

By using Mica:30BT/PVDF or PA66:PP pairs it is possible to charge the 15 µF (capacitor with 25 to 30 cycles, the capacitor being able to turn on the LED for a few seconds) (Figure 4D). It is to notice that this is achieved with a small active area of 46.4×10^{-4} m^2 in each material. Thus, by increasing the active area of the materials to 0.14 m^2, the capacitor could be charged in just one cycle. Thus, implemented in an example, a human walking can generate in a few steps enough energy to power the LEDs, taking into account the weight and area of the shoe.

4. Conclusions

The triboelectric effect using a polymers as active materials can be used to harvest energy for low-power devices. This work compares pairs of materials in different places within the triboelectric series, showing that not just the place within the triboelectric series, but also the surface roughness and the dielectric constant play a critical role in determining, to some extent, triboelectric power output. Thus, PVDF composite with higher dielectric constant (2.5 × higher) than pristine polymer generates 10 × and 15 × times larger power (3.89 mW) when compared to commercial or solvent-based PVDF.

Pairs of polymers PA66:PP (P = 5.94 mW) and Mica:30BT/PVDF (P = 3.89 mW) show the larger triboelectric output power among the evaluated materials. Rubber-like material, such as SEBS, present good triboelectric performance with both negative and positive materials (PA66 and Mica).

Finally, it was shown that the generated triboelectric energy can be stored in a 15 µF capacitor and, after 30 cycles, allows the powering of a LED or other low-power application.

Author Contributions: Conceptualization, P.C. and S.L.-M.; methodology, T.R.-M. and N.C.; validation, T.R.-M., P.C. and S.L.-M.; formal analysis, T.R.-M. and N.C.; investigation, T.R.-M., N.C., V.C. and P.C.; resources, S.L.-M.; writing—original draft preparation, P.C. and V.C.; writing—review and editing, P.C. and S.L.-M.; visualization, S.L.-M.; supervision, P.C. and S.L.-M. All authors have read and agreed to the published version of the manuscript.

Funding: This research was funded by Fundação para a Ciência e Tecnologia under framework of the Strategic Funding UID/FIS/04650/2020 and projects UIDB/05549/2020 and TSSiPRO-NORTE-01-0145-FEDER-000015. The authors also thank the F.C.T. for financial support under grants SFRH/BD/140242/2018 (T.R.M.), SFRH/BPD/98109/2013 (P.C.). Finally, the authors acknowledge funding by Spanish State Research Agency (A.E.I.) and the European Regional Development Fund (ERFD) through the project PID2019-106099RB-C43/AEI/10.13039/501100011033 and from the Basque Government Industry and Education Department under the ELKARTEK, HAZITEK and PIBA (PIBA-2018-06) programs, respectively, are also acknowledged.

Acknowledgments: The authors thank the Fundação para a Ciência e Tecnologia for financial support under framework of the Strategic Funding UID/FIS/04650/2020 and projects UIDB/05549/2020 and TSSiPRO-NORTE-01-0145-FEDER-000015. The authors also thank the F.C.T. for financial support under grants SFRH/BD/140242/2018 (T.R.M.), SFRH/BPD/98109/2013 (P.C.). Finally, the authors acknowledge funding by Spanish State Research Agency (A.E.I.) and the European Regional Development Fund (E.R.F.D.) through the project PID2019-106099RB-C43/AEI/10.13039/501100011033 and from the Basque Government Industry and Education Department under the ELKARTEK, HAZITEK and PIBA (PIBA-2018-06) programs, respectively, are also acknowledged.

Conflicts of Interest: The authors declare no conflict of interest.

References

1. Li, J.; Zhang, X.; Ali, S.; Khan, Z. Eco-innovation and energy productivity: New determinants of renewable energy consumption. *J. Environ. Manag.* **2020**, *271*, 111028. [CrossRef] [PubMed]
2. Chen, C.; Pinar, M.; Stengos, T. Renewable energy consumption and economic growth nexus: Evidence from a threshold model. *Energy Policy* **2020**, *139*, 111295. [CrossRef]

3. Chandrasekaran, S.; Bowen, C.; Roscow, J.; Zhang, Y.; Dang, D.K.; Kim, E.J.; Misra, R.D.K.; Deng, L.; Chung, J.S.; Hur, S.H. Micro-scale to nano-scale generators for energy harvesting: Self powered piezoelectric, triboelectric and hybrid devices. *Phys. Rep.* **2019**, *792*, 1–33. [CrossRef]
4. Jiang, D.; Xu, M.; Dong, M.; Guo, F.; Liu, X.; Chen, G.; Wang, Z.L. Water-solid triboelectric nanogenerators: An alternative means for harvesting hydropower. *Renew. Sustain. Energy Rev.* **2019**, *115*, 109366. [CrossRef]
5. Luo, J.; Wang, Z.L. Recent advances in triboelectric nanogenerator based self-charging power systems. *Energy Storage Mater.* **2019**, *23*, 617–628. [CrossRef]
6. Nozariasbmarz, A.; Collins, H.; Dsouza, K.; Polash, M.H.; Hosseini, M.; Hyland, M.; Liu, J.; Malhotra, A.; Ortiz, F.M.; Mohaddes, F.; et al. Review of wearable thermoelectric energy harvesting: From body temperature to electronic systems. *Appl. Energy* **2020**, *258*, 114069. [CrossRef]
7. Costa, P.; Nunes-Pereira, J.; Pereira, N.; Castro, N.; Gonçalves, S.; Lanceros-Mendez, S. Recent Progress on Piezoelectric, Pyroelectric, and Magnetoelectric Polymer-Based Energy-Harvesting Devices. *Energy Technol.* **2019**, *7*, 1800852. [CrossRef]
8. Lu, Y.; O'Riordan, E.; Cottone, F.; Boisseau, S.; Galayko, D.; Blokhina, E.; Marty, F.; Basset, P. A batch-fabricated electret-biased wideband MEMS vibration energy harvester with frequency-up conversion behavior powering a UHF wireless sensor node. *J. Micromech. Microeng.* **2016**, *26*, 124004. [CrossRef]
9. Chen, J.; Wang, Z.L. Reviving Vibration Energy Harvesting and Self-Powered Sensing by a Triboelectric Nanogenerator. *Joule* **2017**, *1*, 480–521. [CrossRef]
10. Wang, S.; Lin, L.; Wang, Z.L. Triboelectric nanogenerators as self-powered active sensors. *Nano Energy* **2015**, *11*, 436–462. [CrossRef]
11. Wang, Z.L. Triboelectric Nanogenerators as New Energy Technology for Self-Powered Systems and as Active Mechanical and Chemical Sensors. *ACS Nano* **2013**, *7*, 9533–9557. [CrossRef]
12. Kim, D.W.; Lee, J.H.; Kim, J.K.; Jeong, U. Material aspects of triboelectric energy generation and sensors. *NPG Asia Mater.* **2020**, *12*, 6. [CrossRef]
13. Wang, Z.L. On Maxwell's displacement current for energy and sensors: The origin of nanogenerators. *Mater. Today* **2017**, *20*, 74–82. [CrossRef]
14. Dharmasena, G.R.D.I.; Silva, S.R.P. Towards optimized triboelectric nanogenerators. *Nano Energy* **2019**, *62*, 530–549. [CrossRef]
15. Saxon, D.J.; Luke, A.M.; Sajjad, H.; Tolman, W.B.; Reineke, T.M. Next-generation polymers: Isosorbide as a renewable alternative. *Prog. Polym. Sci.* **2020**, *101*, 101196. [CrossRef]
16. Yoon, J.H.; Ryu, H.; Kim, S.-W. Sustainable powering triboelectric nanogenerators: Approaches and the path towards efficient use. *Nano Energy* **2018**, *51*, 270–285. [CrossRef]
17. Wang, Z.L. *Triboelectrification, in Triboelectric Nanogenerators*; Springer International Publishing: Cham, Switzerland, 2016; pp. 1–19.
18. Pan, S.; Zhang, Z. Fundamental theories and basic principles of triboelectric effect: A review. *Friction* **2019**, *7*, 2–17. [CrossRef]
19. McCarty, S.L.; Whitesides, G.M. Electrostatic Charging Due to Separation of Ions at Interfaces: Contact Electrification of Ionic Electrets. *Angew. Chem. Int. Ed.* **2008**, *47*, 2188–2207. [CrossRef] [PubMed]
20. Friedrich, K. Polymer composites for tribological applications. *Adv. Ind. Eng. Polym. Res.* **2018**, *1*, 3–39. [CrossRef]
21. Lee, J.H.; Kim, J.; Kim, T.Y.; Al Hossain, M.S.; Kim, S.W.; Kim, J.H. All-in-one energy harvesting and storage devices. *J. Mater. Chem. A* **2016**, *4*, 7983–7999. [CrossRef]
22. Wang, Z.L.; Lin, L.; Chen, J.; Niu, S.; Zi, Y. Triboelectric Nanogenerator: Vertical Contact-Separation Mode. In *Triboelectric Nanogenerators*; Wang, Z., Lin, L., Chen, J., Niu, S., Zi, Y., Eds.; Springer International Publishing: Cham, Switzerland, 2016; pp. 23–47.
23. Marinho, T.; Costa, P.; Lizundia, E.; Costa, C.M.; Corona-Galván, S.; Lanceros-Méndez, S. Ceramic nanoparticles and carbon nanotubes reinforced thermoplastic materials for piezocapacitive sensing applications. *Compos. Sci. Technol.* **2019**, *183*, 107804. [CrossRef]
24. Martins, P.; Lopes, A.C.; Lanceros-Mendez, S. Electroactive phases of poly(vinylidene fluoride): Determination, processing and applications. *Prog. Polym. Sci.* **2014**, *39*, 683–706. [CrossRef]
25. Ribeiro, C.; Costa, C.M.; Correia, D.M.; Nunes-Pereira, J.; Oliveira, J.; Martins, P.; Goncalves, R.; Cardoso, V.F.; Lanceros-Mendez, S. Electroactive poly(vinylidene fluoride)-based structures for advanced applications. *Nat. Protoc.* **2018**, *13*, 681–704. [CrossRef] [PubMed]

26. Costa, P.; Silva, J.; Mendez, S.L. Strong increase of the dielectric response of carbon nanotube/poly(vinylidene fluoride) composites induced by carbon nanotube type and pre-treatment. *Compos. Part B Eng.* **2016**, *93*, 310–316. [CrossRef]
27. Wang, Z.L. *Triboelectric Nanogenerators*, 1st ed.; Springer: New York, NY, USA, 2016; pp. 1–517.
28. Diaz, F.A.; Felix-Navarro, R.M. A semi-quantitative tribo-electric series for polymeric materials: The influence of chemical structure and properties. *J. Electrost.* **2004**, *62*, 277–290. [CrossRef]
29. Lapčinskis, L.; Mālnieks, K.; Linarts, A.; Blūms, J.; Šmits, K.; Järvekülg, M.; Knite, M.; Šutka, A. Hybrid Tribo-Piezo-Electric Nanogenerator with Unprecedented Performance Based on Ferroelectric Composite Contacting Layers. *ACS Appl. Energy Mater.* **2019**, *2*, 4027–4032. [CrossRef]
30. Zhang, J.H.; Li, Y.; Du, J.; Hao, X.; Huang, H. A high-power wearable triboelectric nanogenerator prepared from self-assembled electrospun poly(vinylidene fluoride) fibers with a heart-like structure. *J. Mater. Chem. A* **2019**, *7*, 11724–11733. [CrossRef]
31. Singh, H.H.; Khare, N. Flexible ZnO-PVDF/PTFE based piezo-tribo hybrid nanogenerator. *Nano Energy* **2018**, *51*, 216–222. [CrossRef]
32. Cheng, X.; Tang, W.; Song, Y.; Chen, H.; Zhang, H.; Wang, Z.L. Power management and effective energy storage of pulsed output from triboelectric nanogenerator. *Nano Energy* **2019**, *61*, 517–532. [CrossRef]
33. Zhang, X.S.; Han, M.; Kim, B.; Bao, J.F.; Brugger, J.; Zhang, H. All-in-one self-powered flexible microsystems based on triboelectric nanogenerators. *Nano Energy* **2018**, *47*, 410–426. [CrossRef]
34. Zou, H.; Zhang, Y.; Guo, L.; Wang, P.; He, X.; Dai, G.; Zheng, H.; Chen, C.; Wang, A.C.; Xu, C.; et al. Quantifying the triboelectric series. *Nat. Commun.* **2019**, *10*, 1427. [CrossRef] [PubMed]
35. Neagoe, M.B.; Prawatya, Y.E.; Zeghloul, T.; Dascalescu, L. Influence of surface roughness on the tribo-electric process for a sliding contact between polymeric plate materials. In *IOP Conference Series: Materials Science and Engineering, Proceedings of the 13th International Conference on Tribology, Galati, Romania, 22–24 September 2016*; IOP Publishing: Bristol, UK, 2016; Volume 174, p. 012003.
36. Cheng, G.G.; Jiang, S.Y.; Li, K.; Zhang, Z.Q.; Wang, Y.; Yuan, N.Y.; Ding, J.N.; Zhang, W. Effect of argon plasma treatment on the output performance of triboelectric nanogenerator. *Appl. Surf. Sci.* **2017**, *412*, 350–356. [CrossRef]
37. Helseth, L.E. The Influence of Microscale Surface Roughness on Water-Droplet Contact Electrification. *Langmuir* **2019**, *35*, 8268–8275. [CrossRef]
38. Shao, Y.; Feng, C.P.; Deng, B.W.; Yin, B.; Yang, M.B. Facile method to enhance output performance of bacterial cellulose nanofiber based triboelectric nanogenerator by controlling micro-nano structure and dielectric constant. *Nano Energy* **2019**, *62*, 620–627. [CrossRef]
39. Araujo, C.M.; Costa, C.M.; Lanceros-Mendez, S. Evaluation of dielectric models for ceramic/polymer composites: Effect of filler size and concentration. *J. Non-Cryst. Solids* **2014**, *387*, 6–15. [CrossRef]
40. Mendes, S.F.; Costa, C.M.; Caparrós, C.; Sencadas, V.; Lanceros-Méndez, S. Effect of filler size and concentration on the structure and properties of poly(vinylidene fluoride)/BaTiO$_3$ nanocomposites. *J. Mater. Sci.* **2012**, *47*, 1378–1388. [CrossRef]
41. Lee, L.H. Dual Mechanism for Metal-Polymer Contact Electrification. *J. Electrost.* **1994**, *32*, 1–29. [CrossRef]
42. Niu, S.; Wang, S.; Lin, L.; Liu, Y.; Zhou, Y.S.; Hu, Y.; Wang, Z.L. Theoretical study of contact-mode triboelectric nanogenerators as an effective power source. *Energy Environ. Sci.* **2013**, *6*, 3576–3583. [CrossRef]

Publisher's Note: MDPI stays neutral with regard to jurisdictional claims in published maps and institutional affiliations.

© 2020 by the authors. Licensee MDPI, Basel, Switzerland. This article is an open access article distributed under the terms and conditions of the Creative Commons Attribution (CC BY) license (http://creativecommons.org/licenses/by/4.0/).

Article

Evaluating Energy Generation Capacity of PVDF Sensors: Effects of Sensor Geometry and Loading

Mohammad Uddin *, Shane Alford and Syed Mahfuzul Aziz

UniSA STEM, University of South Australia, Mawson Lakes, SA 5095, Australia; alfsc001@mymail.unisa.edu.au (S.A.); Mahfuz.Aziz@unisa.edu.au (S.M.A.)
* Correspondence: Mohammad.Uddin@unisa.edu.au; Tel.: +61-08-8302-3097

Abstract: This paper focuses on the energy generating capacity of polyvinylidene difluoride (PVDF) piezoelectric material through a number of prototype sensors with different geometric and loading characteristics. The effect of sensor configuration, surface area, dielectric thickness, aspect ratio, loading frequency and strain on electrical power output was investigated systematically. Results showed that parallel bimorph sensor was found to be the best energy harvester, with measured capacitance being reasonably acceptable. Power output increased with the increase of sensor's surface area, loading frequency, and mechanical strain, but decreased with the increase of the sensor thickness. For all scenarios, sensors under flicking loading exhibited higher power output than that under bending. A widely used energy harvesting circuit had been utilized successfully to convert the AC signal to DC, but at the sacrifice of some losses in power output. This study provided a useful insight and experimental validation into the optimization process for an energy harvester based on human movement for future development.

Keywords: PVDF; piezoelectric material; energy harvesting; human body movements; power generation

Citation: Uddin, M.; Alford, S.; Aziz, S.M. Evaluating Energy Generation Capacity of PVDF Sensors: Effects of Sensor Geometry and Loading. *Materials* **2021**, *14*, 1895. https://doi.org/10.3390/ma14081895

Academic Editor: Daniele Davino

Received: 24 March 2021
Accepted: 1 April 2021
Published: 10 April 2021

Publisher's Note: MDPI stays neutral with regard to jurisdictional claims in published maps and institutional affiliations.

Copyright: © 2021 by the authors. Licensee MDPI, Basel, Switzerland. This article is an open access article distributed under the terms and conditions of the Creative Commons Attribution (CC BY) license (https://creativecommons.org/licenses/by/4.0/).

1. Introduction

The use of portable electronics has been growing rapidly. As the emerging portable electronic devices are embedding more functionalities, they are becoming complex in nature, requiring higher amount of electrical energy for their operation. With higher levels of energy consumption, the batteries used to supply power to the portable devices deplete their energy quite quickly. Hence, there is increasing demand for alternative energy, and this has facilitated growth in the area of renewable energy sources, which can be harnessed to power portable devices 'on the go'. Current postulated application areas for portable energy harvesting include wireless sensor networks, biomedical (e.g., pacemakers) and military use [1,2]. In particular, energy harvesting via human body movements (e.g., running, jogging) is seen as a promising means to power portable devices carried by people. In this regard, different types of piezoelectric materials such as piezoelectric (PZT), polyvinylidene difluoride (PVDF), have been studied to understand and evaluate the efficacy of electrical power generation from the mechanical energy harvested from human movements. For example, piezoelectric materials tapped on shoe sole and wearable fabrics have been found to generate power [3]. As the amount of power generated through such movements is relatively small, the selection of appropriate piezoelectric material along with design of the energy harvesting circuits is crucial to generate usable power. In the past, with the aid of experimental and simulation tools, tremendous efforts have been made to study the design, implementation and evaluation of piezo sensors with the aim of increasing the amount of power generation [4]. Furthermore, PVDF is flexible, resistant to mechanical shock and highly compatible with the environment, which make it suitable for full PZT beam without needing any substrate layer. As such, researchers focused on gross energy scavenging capacity of PVDF based sensors in different applications, e.g., human walking, elbow bending. However, little research is reported on comprehensive parametric study of PVDF sensors in a holistic way. For example, studies focusing on the effects of different sensor parameters including geometry, mechanical loading, frequency and straining on electrical power generation capability of PVDF sensors have not been reported.

This paper presents the effect of the type and geometry of PVDF (polyvinyl difluoride) piezoelectric sensors on the amount of power generated, along with power conversion circuitry to be embedded in wearable-fabrics to harvest energy via human body movements. PVDF was chosen due to its flexibility, low cost, and good piezo-properties, such as high electric charge accumulation under small mechanical strain. Three piezoelectric sensor configurations, namely-parallel bimorph, series bimorph and unimorph sensors were investigated in this study. The effect of dielectric thickness, surface area and aspect ratio, loading type and frequency on the electrical power output of each sensor configuration were studied, with the aim of optimizing the senor parameters. A simple power conversion circuit without a voltage regulator was designed and implemented, and its efficiency in terms of power output and potential issues was discussed in this paper. The results are analyzed with reference to available literature on energy harvesting.

2. Related Works

The human body contains a wide variety of energy sources, for example body heat, breath, blood pressure, upper limb motion, walking and finger motion [5]. It is reported that upper arm motion could produce up to 60 W of power (i.e., arm lifts above head at 1.3 lifts/s by a man weighing 58 kg), of which up to 0.33 W is recoverable using piezoelectric fabric [6], while a man weighing 68 kg walking at 2 steps/s could produce 324 W of power, with up to 8.3 W recoverable using piezoelectric inserts. These estimations are under ideal conditions and any attempt at generating electricity using these methods is likely to produce far less power due to design inefficiencies and component energy loss. With the advancement of new materials, researchers have developed new PZT thin films to increase energy harvesting capacity. For example, Won et al. studied $Pb(Zr,Ti)O_3$ (PZT) thin film with a $LaNiO_3$ (LNO) buffer layer on an ultra-thin Ni-Cr-based austenitic steel metal foil substrate to generate electrical power from vibrational energy [7]. They reported that the maximum power and the corresponding peak voltage generated by the sensor were 5.6 µW and 690 mV, respectively. Yeo et al. [8] developed a flexible tactile sensor array consisting of aluminum nitride (AlN) thin film as PZT material deposited in Si wafer for energy generation when the sensor is subject to mechanical stimulus.

Despite the overwhelming success with PZT, PVDF is considered to be the most commonly used piezoelectric materials [9]. PZT is too stiff (Young's modulus of 63 GPa as opposed to 8.3 GPa for PVDF) to be of any use where flexibility is required, leaving PVDF as the best choice for use on wearable fabrics to harvest energy. Jones et al. [10] outlined the preparation of PVDF to increase the piezoelectric effect by poling and stressed that uniform crystal polarization must be achieved in order to maximise the potential power generation capabilities of the material. A PVDF coating was created by mixing, curing and poling PVDF solution and then coated onto conductive fabric to from a flexible piezo-electric membrane [2]. Paradiso and Starner [11] proposed a PVDF shell structure to increase power output, in which the PVDF was bonded to a pre-curved polyester film and electrodes were attached on either side. The maximum voltage generated was about 40 V. The sensitivity and power output of three different types of piezo-electric actuators were investigated [12]. The three actuator configurations were: (1) parallel triple layer (parallel bimorph), (2) series triple layer (series bimorph) and (3) unimorph. A detailed illustration of the configurations is presented in [12]. Most research in the field of piezoelectric energy harvesting had focused on high frequency loading. Low frequency loading may produce different output characteristics similar to static loading. Further, Ozeri and Shmilovitz [13] analyzed the time response properties of the piezoelectric harvesters.

Using 3D simulation, a unimorph PZT generator consisting of a 200 µm thick steel substrate and 10 µm to 400 µm dielectric was investigated [14]. It was observed that the generated current increased with increasing dielectric thickness until a maximum was reached after which it decreased with increasing thickness. The voltage followed a different trend, which increased continually with increasing dielectric thickness. Granstrom et al. [15] tested the power generating capability of a PVDF "shoulder strap" to be integrated into an energy generating backpack. The average power generated was 45.6 mW, and a PVDF sample of 28 μm thickness consistently produced a greater power output than that of 56 μm sample by an average of 0.9 mW or 62%.

Waqar et al. [16] performed a dual-field computational analysis on a simulated PVDF patch bonded to a flexible fabric and studied the effect of various sensor parameters on the electrical power output. They used a PVDF film of 0.2 mm thickness, where electrodes on either side connected to a resistor "that was matched to the piezo properties" reported in [2]. It was observed that a surface area of 400 mm^2 (with aspect ratio of 4 and 0.005 N load), aspect ratio of 8 (with surface area of 1600 mm^2 and 0.005 N load) and load of 0.05 N (with surface area of 1600 mm^2 and aspect ratio of 4) produced the highest power output. Only one variable was changed at a time, and no attempt was made to maximize the power output using the optimum value for all three parameters at the same time. Further, only cantilever load was utilized in this analysis, which may not reflect actual loading scenario a PVDF being subject to due to human body movements.

An adaptive energy harvesting circuit was presented to maximize the efficiency of piezoelectric energy harvesting [17]. Through testing with a Quickpack® actuator element attached to a shaker operating at 53.8 Hz, the system was shown to increase power transfer significantly. The major drawback of this system is that it uses a relatively large amount of energy to power the microcontroller required for adaptive control, thus negating the benefit of renewable energy generation.

It is well known that piezoelectric devices generate electrical signals only in response to a change in the applied force, because under static stress, the free carriers drift toward the dipoles, eventually discharging the device [18]. Piezoelectric devices can be modelled by the Butterworth Van Dyke model, where a series connection of resistors (R_s), inductors (L_s) and capacitors (C_s) models the mechanical resonance. The design of a proper energy harvesting circuit is thus essential to minimize the loss of input energy, thus maximizing the storage of harnessed electrical energy.

The above review analysis reveals that researchers have attempted on gross energy scavenging capacity of PVDF based sensors in different applications, e.g., human walking, elbow bending. However, studies focusing on the effects of different sensor parameters, including geometry, mechanical loading, frequency and straining on electrical power generation capability of PVDF sensors have been very limited. Thus, this paper aims to propose a comprehensive experimental study in investigating the effect of sensor geometry and loading parameters on electrical power generation of PVDF sensors.

3. Materials and Method

3.1. Piezoelectric Sensor Design and Construction

Three piezoelectric sensor configurations investigated in this study were unimorph (U), parallel bimorph (PB) (parallel triple layer), and series bimorph (SB) (series triple layer). The corresponding circuit models are shown in Figure 1. A simple unimorph sensor is essentially a capacitor C_p and resistor R_p connected in series with a voltage source V_s and a total output voltage of V_{TU} (Figure 1a). As the name suggests, a parallel bimorph consists of two dielectrics connected in parallel, so the equivalent capacitive circuit model is two capacitors connected in parallel, effectively doubling the capacitance (Figure 1b). On the other hand, the series bimorph is two dielectrics connected in series, so the equivalent capacitive circuit model is two capacitors connected in series (Figure 1c).

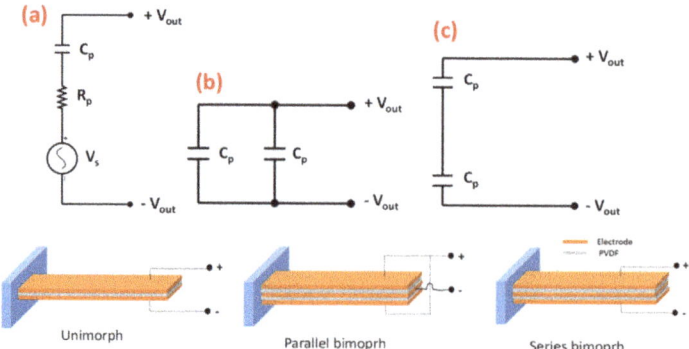

Figure 1. Voltage generator piezoelectric circuit model (**a**) unimorph (**b**) parallel bimorph (**c**) series bimorph [12].

The sensor prototypes were constructed using sandwiched layers of PVDF and copper shim (3M™ Red Dot™, 3M Ltd, Sydney, Australia) joined by a 2-part epoxy adhesive with the whole structure laminated (Figure 2a). PVDF used in this study is in β phase with polarization characteristics. Copper shims were used as electrodes, while the purpose of epoxy was to keep the dielectric in contact with the conductive copper. Ideally, the epoxy would be spread as thinly and evenly as possible to minimize the contact resistance and maintain uniformity across the whole sensor. The PVDF and copper were first cut to the required size (Figure 2b) and joined using epoxy. Two electrode-dielectric (copper shim-PVDF) pairs were prepared first by joining them with epoxy and placing under a uniform load of 20 kg for a period of two hours to make sure the epoxy was spread as thinly and evenly as possible. For the parallel bimorph, the two end pairs, as prepared above, were joined to a center electrode and placed under an identical load. For the series bimorph, the end pairs were joined directly together on the PVDF side. For the unimorph, a PVDF layer sandwiched between two copper shims (of 0.025 mm thick) were joined at once. Once the sensors were fabricated, the leads were soldered to the electrodes (Figure 2c). The final step was to laminate the sensor to create greater structural integrity and to provide isolation for the electrodes.

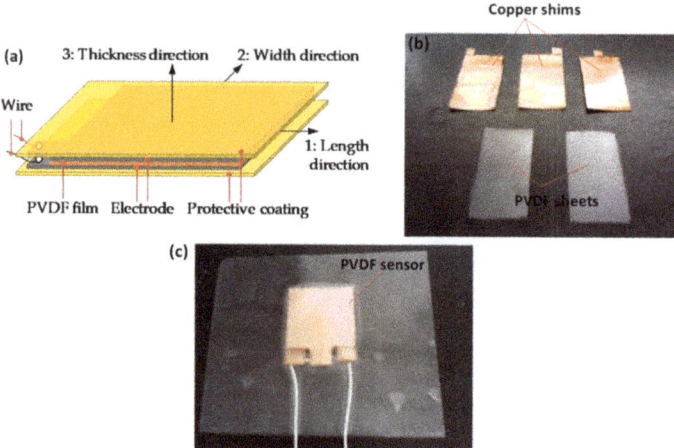

Figure 2. (**a**) Schematic of a typical sensor design (**b**) PVDF and copper sections before being joined and (**c**) laminated PVDF bimorph sensor.

3.2. Sensor Geometry

For a given sensor, the effect of thickness, aspect ratio and surface area on the power output was investigated. The range of values of each parameter and the corresponding sensor type considered are summarized in Table 1, where PB, SB and U stand for parallel bimorph, series bimorph and unimorph, respectively. The aspect ratio is defined as the ratio of the length, L to the width, W of the PVDF film used in the sensor. Note that, in this study, we first aimed to demonstrate and compare the performance of three configurations for the PVDF sensor to decide which configuration gives the best power output.

Table 1. Sensor types and geometries studied.

Sensor Type	Thickness, t (mm)	Aspect Ratio, L/W	Surface Area, A (mm^2)
PB	0.0254	1	100
PB	0.0254	1	400
PB	0.0254	1	625
PB	0.0254	25:12	300
PB	0.0254	4:3	300
PB	0.0254	3	300
PB	0.254	1	225
PB	0.508	1	225
PB	0.0254	1	225
U	0.0254	1	225
SB	0.0254	1	225

3.3. Applied Loading and Frequency

Cantilever tensile/compressive loading was used to evaluate the power generation capabilities of the sensors. Two different forms of cantilever loading were applied to simulate the types of motions likely to be caused by the movement of the human body. One is called 'Bending' and the other 'Flicking'.

'Bending' simulates low impact and low velocity loading. To apply this loading, the sensor was placed on a stepped surface with approximately half of the area of the sensor fastened to the top 'step' with a masking tape to allow very low freedom of movement (0.5 mm). The surface of the bottom step is approximately 1.5 mm below the sensor (Figure 3a). The fastening of the sensor to the top step was not completely rigid to decrease the likelihood of creasing in the copper shim and to decrease the likelihood of reduction in elasticity. A quasi-static distributed load via displacement was applied to the suspended mid-section of the sensor, which was enough to bend the sensor so that it touched the lower step, i.e., the sensor's free end has moved by 1.5 mm. It was then released at a certain velocity in a periodic manner at a certain frequency. The 'bending' loading followed approximately a typical sine wave. No actual reaction force due to applied displacement of the sensor was measured in the work.

'Flicking' simulates a high impact and high velocity loading. In this case, similar to the 'bending' loading, the same stepped surface was used. A strip of 3M's double-sided copper tape of the same thickness as that of the top step was placed between the top step and the sensor, thus increasing the distance between the bottom step and the sensor by another 1.5 mm (Figure 3b). The purpose of the flexible tape spacer is to allow the sensor to have a greater degree of movement, while reducing the likelihood of creeping. The larger gap between the bottom step and the sensor also allows for a greater displacement angle which assists with a 'flicking' motion. Similar to 'bending', a sudden load via displacement was applied to the suspended section of the sensor, and then released quickly by sliding off and allowing the sensor to spring back rapidly due to the elastic properties of the structure. The characteristic of 'flicking' loading appeared to be a decaying wave as shown in Figure 3c. No reaction force due to cyclic displacement of the sensor because of flicking motion was measured. Note that, in addition to the loading type, the loading frequency was varied from 1 to 5 Hz. Loading or excitation frequency was defined as the number of bending or flicking motion cycles being applied onto the sensor per second.

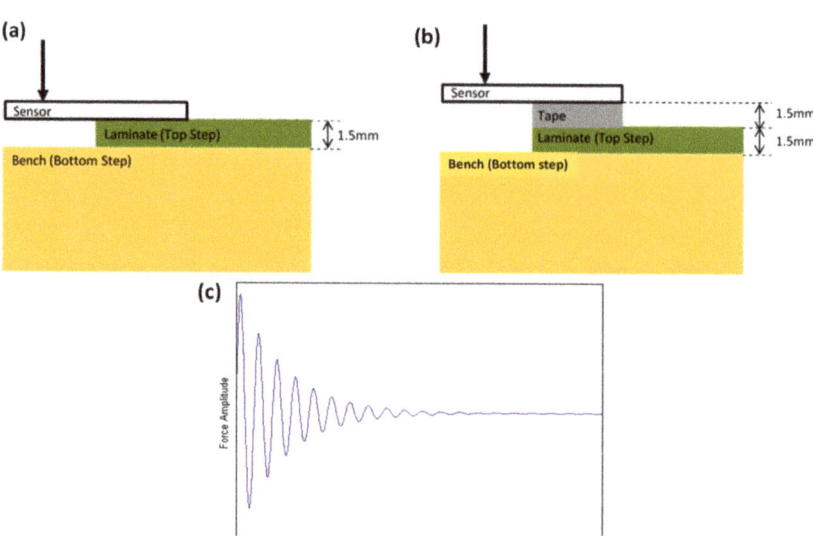

Figure 3. (a) Cross sectional view of "bending" loading setup and (b) cross sectional view of "flicking" loading setup (c) "Flicking" loading representation of force over time.

3.4. Experiments and Energy Harvesting Circuits

The experimental setup developed for testing the energy harvesting circuits is shown in Figure 4a. The information captured by the data acquisition (DAQ) card was processed using LabVIEW (National Instruments, Sydney, Australia). The maximum, minimum, peak-to-peak and rms currents (and hence voltage and supply power) were calculated and displayed. The captured data was recorded for post processing and for analyzing waveform characteristics.

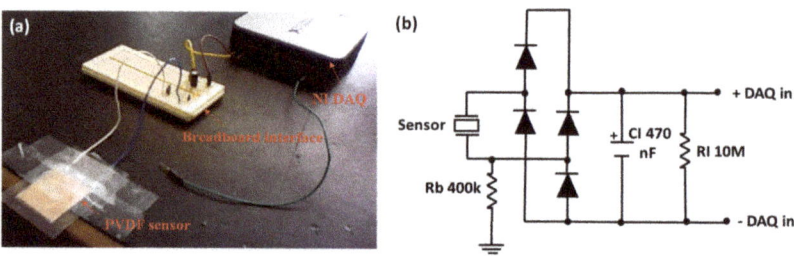

Figure 4. (a) Experimental setup for sensor testing and (b) energy harvesting circuit with a 470 nF load capacitor.

Effective energy harvesting circuit is crucial to harness the maximum power output from a piezo sensor at the expense of a small mechanical strain energy. Figure 4b shows the energy harvesting circuit used in this study. The sensor electrodes were attached to a 10 MΩ resistor, which was connected in shunt to a DAQ differential input. A large resistance was used to force as much current as possible into the DAQ to create a more accurate measurement. Hence, one electrode was connected to ground through a combined laminate and body resistance R_b of approximately 400 kΩ. For the AC-DC conversion, a full-wave diode bridge rectifier and with a 470 nF capacitor in parallel was connected between the sensor and the load resistor. 1N5711 Schottky diodes (Element14, Sydney,

Australia) were used due to its low forward voltage (0.41 V maximum) and satisfactory reverse repetitive peak voltage (70 V maximum). The full wave diode bridge rectifier circuit was used, because researchers have observed it to show high electrical power efficiency, especially in low voltage situations [19].

In this paper, poling was not carried out for the sensors studied. Other researchers have studied the effects of poling on energy harvesting PZT sensors [3]. As a preliminary work, we aimed to investigate the fundamental behaviour of PVDF in energy generation, without applying any external electrical field or heat. While poling might have improved the overall power generation, the trends for the effects of sensor geometry, loading, frequency, and strain on power generation will largely remain similar to those reported in this study and so will the conclusions of the paper. Study of the effects of poling on sensor performance remains as a scope of our future work. Note that all sample preparations and experiments to test the sensor performance were conducted at a room temperature of 23 °C at atmospheric pressure.

4. Results and Discussion

4.1. Capacitance

As a measure of charge storing capability, the capacitances of the sensors with different geometric configurations were measured using a digital multi-meter. For the purpose of comparison, capacitance values were also estimated using the capacitive circuit model of the sensors [20].

Table 2 summarizes the measured and calculated capacitances of the sensors. Clearly, for small PVDF thickness (t = 0.0254 mm), the calculated capacitance values were considerably larger than the measured capacitances. This is expected because the theoretical calculation does not take into account the impedance of the epoxy joiner. Sensors with a larger surface area showed a higher capacitance, which was to be expected, and sensors with the same surface area but different aspect ratios had very similar capacitances. For much larger PVDF thickness (t = 0.254 and 0.508 mm), the calculated capacitance value is very close to the measured capacitance. This is due to the fact that in the practical sensor the impedance of the epoxy joiner has lesser effect on the overall capacitance. The capacitance can give us an indication of charge decay time of the actuator and hence its energy harvesting capability. For instance, the largest measured capacitance in Table 2 was 0.34 nF for the PVDF sensor with A = 25 mm × 25 mm and t = 0.0254 mm. This capacitance was approximately 2% of the capacitance of the PVDF sample with A = 240 mm × 240 mm and t = 0.027 mm studied in [15]. However, it is noted that the surface area of our sensor is only 1.1% of that of [15], which resulted in much lower capacitance. Therefore, it can be argued that the capacitance generated by the proposed sensor is consistent and reasonable. Results on measured capacitance per unit area as shown in Table 2 follow the similar trend and are consistent to that of [15] as well.

Table 2. Capacitance results for different sensor geometries and circuits.

Sensor Type	Thickness t (mm)	Aspect Ratio L/W	Surface Area, A (mm^2)	Calculated Capacitance (nF)	Measured Capacitance (nF)	Measured Capacitance per Unit Area (pF/mm^2)
PB	0.0254	1:1	100	0.6	0.06	0.6
PB	0.0254	1:1	225	1.4	0.1	0.44
PB	0.0254	1:1	400	2.5	0.3	0.75
PB	0.0254	1:1	625	3.9	0.34	0.54
PB	0.0254	25:12	300	1.9	0.25	0.83
PB	0.0254	4:3	300	1.9	0.34	1.13
PB	0.0254	3:1	300	1.9	0.23	0.77
PB	0.254	1:1	225	0.14	0.12	0.53
PB	0.508	1:1	225	0.071	0.09	0.4
U	0.0254	1:1	225	0.71	0.16	0.71
SB	0.0254	1:1	225	0.36	0.1	0.44

4.2. Effect of Sensor Type

To measure the effect of sensor type on the electrical power generation capability, three different sensor types with the same dielectric thickness and surface area were tested under 'Flicking' and 'Bending' loading at a frequency of 3 Hz. Figure 5 shows the mean current (Irms) and power outputs (Pout) for PB, SB, and U sensors.

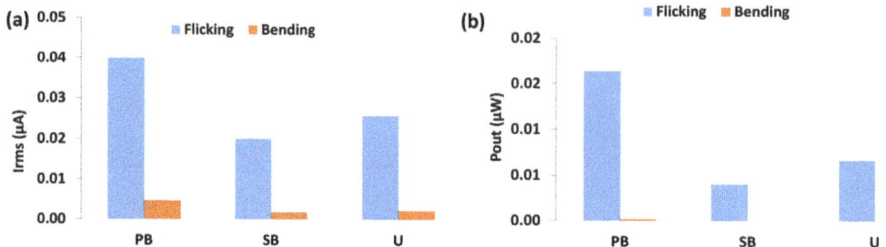

Figure 5. (a) Mean current and (b) power output for parallel bimorph (PB), series bimorph (SB) and unimorph (U) sensors.

For both loading types, clearly the electrical output for parallel bimorph sensor was the highest among all three sensor types. The unimorph sensor exhibited the second highest electrical output. The results are consistent the work of [12]. Since the parallel bimorph sensor showed the highest current and power output, we used this sensor type for the remainder of the experiments, i.e., for parametric studies, which are described in the following sections.

4.3. Effect of PVDF Thickness

To measure the effect of PVDF thickness, parallel bimorph sensors with three different thicknesses (0.0254 mm, 0.254 mm and 0.508 mm) were tested for the same surface area under 'bending' and 'flicking' loading at a frequency of 3 Hz. Figure 6 shows the mean current and power output with respect to the thickness. It is seen that the current (Figure 6a) and power output (Figure 6b) decreases with the increase in dielectric thickness under both types of loading.

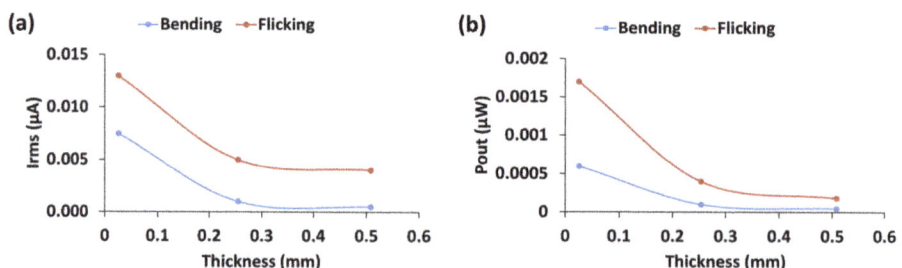

Figure 6. (a) Mean current and (b) power output for increasing PVDF thickness.

The results are comparable with the work of [14] who reported that at an optimal dielectric thickness of the sensor, after which, the stored electrical energy decreases with the increase of dielectric thickness. They found that the optimal piezoelectric thickness to substrate thickness ratio was 1.05 for a fixed (200 μm thick) steel substrate, which was independent of geometric dimensions of the sensor. It was postulated that the maximum power output was achieved before the rigidity of the piezoelectric material had a significant effect on the overall rigidity of the sensor. As can be seen from Figure 6, the power output decreases with the increase of dielectric thickness and the maximum output occurs at substrate thickness of 0.025 mm (first data point in Figure 6). PVDF thicknesses we studied were beyond the optimal ratio, which caused a decreasing trend of power output due to the potential dominant stiffness effect of the dielectric material on the overall stiffness of the sensor. For flexible substrate similar to copper shims used in our study, a sharp decreasing trend of power output with the increase of stiffness of polyvinylidene fluoride-trifluoroethylene (PVDF-TrFE) films was reported in [21]. Therefore, the overall trend of power generation with respect to the dielectric thickness seems reasonably accurate. Note that only three PVDF thickness values were studied in this paper and thus the results obtained are not exhaustive, and a detailed study encapsulating wider range of thickness including thinner PVDF films (<0.025 mm) would be of greater interest to conclusively determine an optimum characteristic of the sensors for future work.

4.4. Effect of Aspect Ratio

To measure the effect of aspect ratio (L/W) on electrical power generation capability, three sensors with different ratios (4:3, 25:12, and 3:1) at constant thickness, surface area and configuration were tested at a frequency of 3 Hz for both bending and flicking loads. Mean current and power output with respect to aspect ratio are shown in Figure 7. As can be seen from the figure, power output increases up to a maximum value and then decreases as the aspect ratio increases.

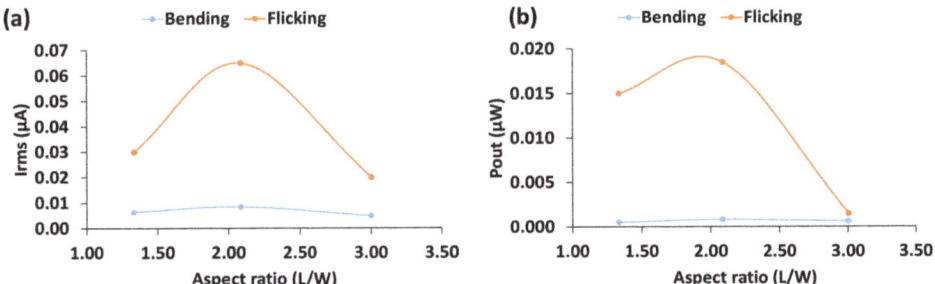

Figure 7. (**a**) Mean current and (**b**) power output with respect to aspect ratio (L/W).

There are two factors to consider when analyzing the power output characteristics: one is the moment, which is the product of the force vector multiplied by the beam length, and the other is the width of the sensor. A longer beam length results in a greater moment, which generates larger electrical energy. On the other hand, with a smaller width, the applied load is distributed across a smaller area at the edge of the bend line, which results in a smaller power output. This would explain the parabolic output characteristics of the sensors (Figure 7). The power increased and reached a peak point, followed by a decreasing trend, as L/W ratio increases. The ideal L/W ratio was found to be about 2:1. The findings are slightly in contrast with that of [16], where they reported a noticeable increasing trend of power output only when the L/W ratio was greater than 3.

4.5. Effect of Surface Area

To measure the effect of surface area on electrical power generation capability, sensors with four different surface areas of 100, 225, 400, and 625 mm^2 with the same thickness and configuration were tested at a frequency of 3 Hz. As can be seen in Figure 8, mean current and power increase with the increase of surface area.

Interestingly, the relationship between power output and surface area is slightly different to that found from the simulation in [16]. The authors reported that the power output was greatest for an area of 400 mm^2 and then decreased with increasing surface area. They further suggested that this was due to the increased flexural stiffness for larger areas, which would result in less strain for the same applied load. It must be noted that they applied a constant magnitude of loading, whereas our method of loading was aimed at maintaining constant strain. This fundamental difference in loading method could be responsible for the discrepancy in the trends of the results.

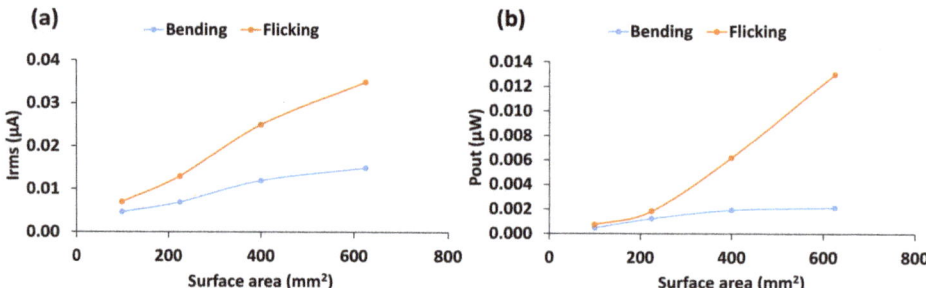

Figure 8. (a) Mean current and (b) power output with respect to surface area.

4.6. Effect of Loading Frequency

To measure the effect of loading frequency on electrical power generation capability, a sensor was tested at frequencies ranging from 1 to 5 Hz, which encompasses equivalent loading frequency on wearable fabrics due to human body movements. All other geometric parameters of the sensor were kept constant. Figure 9 shows the test results for different loading frequencies.

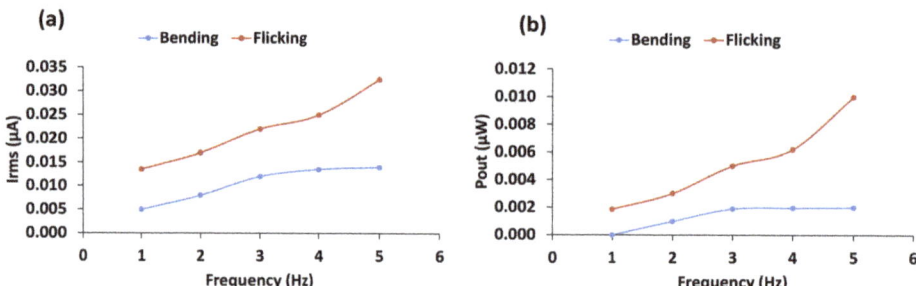

Figure 9. (a) Mean current and (b) power output with respect to loading frequency.

The effect of loading frequency had a similar effect to that of surface area for "bending", the output current increased modestly with frequency up to a point (4 Hz) and then dropped slightly at higher frequencies. The power output increased slightly with frequency up to point (4 Hz) before dropping marginally at higher frequencies. With "flicking", the current and power output increased almost linearly with frequency all the way up to 5 Hz. The increases were much more pronounced than seen in the case of "bending".

The trends can be explained by carefully observing the instantaneous current waveforms for 3, 4, and 5 Hz, as can be seen in Figure 10. When the sensor was stressed, the current spiked initially and then discharged. This is consistent with the observations of [13], although the force application is slightly different. The discharging of a sensor can be characterised by the time constant τ, which indicates how quickly the sensor is able to discharge. The time constants for both 'flicking' and 'bending' loading at 3–5 Hz were calculated by analyzing the data associated with the instantaneous waveforms shown in Figure 10. There was a distinct time gap between the positive and negative currents for the "up" and "down" motions with 'flicking', whereas the two regions were much closer together for 'bending'. In the case of "bending", when the loading was done at 3 Hz, the induced current had enough time between each 'down' and 'up' motion to discharge. Once the loading frequency reached 5 Hz, there was not enough time for the current to discharge naturally. Therefore, potential energy from that motion was not completely transferred into electrical energy. This led to the logarithmic increase in power output with the increase of frequency. Indeed, the power even dropped at 5 Hz due to this effect. As Figure 10 demonstrates, with the "Flicking" motion, the loading was much more sudden and the induced current had ample time to return to 0 before the next load is applied, so the 'pulse-cutting' was not seen.

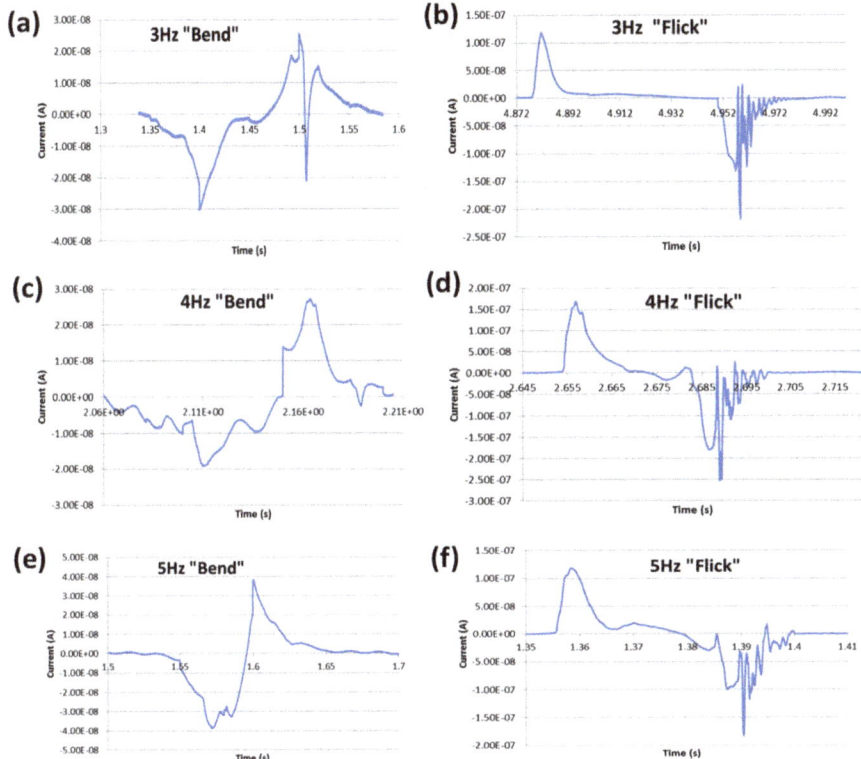

Figure 10. Current waveforms with respect to loading frequency for "Bending" (a,c,e) and "Flicking" (b,d,f).

As summarized in Table 3, the time constants for 3 Hz and 4 Hz of 'bending' were larger than that for 5 Hz. This reflects the fact that the 3 Hz and 4 Hz pulses are longer than the 5 Hz pulse, and are able to impart a greater amount of energy per pulse. For all frequencies, the time constants for 'bending' were consistently greater than that for 'flicking'. The results are in line with the findings of [13] which reported that an increase in static pressure or stress in PZT sensor results in a decrease in decay time constant, and thus suggest that 'flicking' imparts greater pressure than 'bending'. Moreover, the mechanical stability of the flexible electrodes can be compromised under both loading types when subjected to large number of cycles of operation. Eventually, this can cause permanent deformation of the sensor device, which can affect the overall power generation efficiency of the sensor [22]. This warrants further investigation and remains within the scope of our future work.

Table 3. Time constants for increasing load frequency.

Loading Frequency (Hz)	Time Constant (τ)	
	Bending	Flicking
3	0.0157866	0.00376225
4	0.01664428	0.0044029
5	0.00989307	0.00300784

4.7. Effect of Mechanical Strain

To measure the effect of strain on electrical power generation capability, two sensors with the same PVDF thickness (of 0.0254 mm) and two different surface areas (of 400 and 625 mm^2) were tested with different loading in terms of maximum displacements of PVDF sensor end at 1.5, 3.0, 5, and 6.5 mm, which corresponds to strains of 0.00602, 0.0122, 0.0209 and 0.0283 for the 400 mm^2 sample, and 0.00603, 0.0123, 0.0216, and 0.0303 for the 625 mm^2 sample. The values of strains are chosen based on human body movements at key flex points such as the inside of the elbow and behind the knee. The strain was calculated using a simple formula to find the angle of separation between the bent sensor and the bottom step when fully loaded, which is as follows.

$$\Delta l = 2\pi t_s \left(\frac{\theta}{360}\right) \quad (1)$$

where θ is the angle of separation and, Δl is the change in length, and t_s is the total sensor thickness. Figure 11 shows the mean current and power output with respect to mechanical strain.

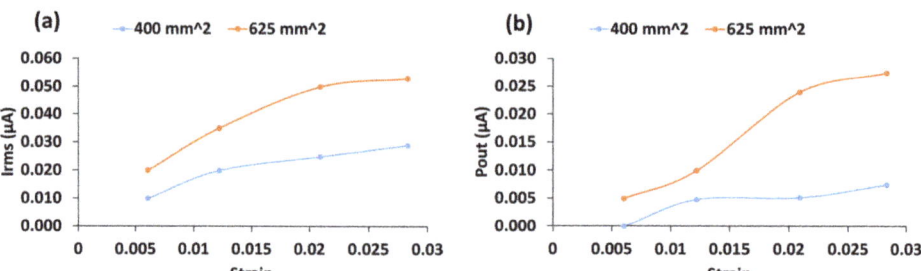

Figure 11. (a) Mean current output (b) mean power output with respect to strain.

It can be seen from the figure that for the sensor with surface area of 400 mm^2, clearly, mean voltage and current increased when the strain increases from 0.00602 to 0.0122 (first two data points on the graphs). They, however, did not increase as rapidly beyond past the strain of 0.0122. For the sensor with surface area of 625 mm^2, the voltage and current curves started to flatten after the strain of 0.0216. The limiting strain seems to be higher for sensor with larger surface area. Our results suggest that an upper limit exists on the amount of deformation a material may experience before reaching its optimum charge generating capability. Defects in the sensor construction may also limit the maximum strain for power generation. With a large deformation, the dielectric material may become detached from the conducting substrate, hence increasing the impedance of the gap between the two materials and decreasing the energy generating potential. Therefore, a more thorough analysis would be required to investigate the full effect of this.

4.8. Analysis of Energy Harvesting Circuit Efficiency

To test the efficiency of the energy harvesting circuit, the electrical output parameters were measured for different system configurations with a 625 mm^2 sensor with 3 Hz

'Bending' loading at strains of 0.0123 and 0.0216. Table 4 summarizes the test results on the performance of the energy harvesting circuit that was presented in Figure 4b. As the capacitors take a certain amount of time to charge as determined by the RC time constant, $\tau_{RC} = R_l C_l$, where $R_l = 10$ MΩ and $C_l = 470$ nF and 1 µF, the measurements could not be taken from the start of the loading, but rather taken from the time at which the capacitor became fully charged. The charging time was determined by collecting data from a "cold start" and determining the point at which the charge stopped increasing in post-analysis as shown in Figure 12 (for $\varepsilon = 0.0216$). As less strain generates less current, the charging time for the capacitors was higher for the lower strain measurement ($\varepsilon = 0.0123$). After the capacitor is fully charged (the time of which is dependent on the energy of the signal being fed into the capacitor) we would expect the I_{rms} output to hold steady under continuous loading of the same frequency and magnitude. This concept was reiterated in [12] for energy.

Table 4. Electrical output characteristics with and without energy harvesting circuitry.

Strain	Rectifier	Capacitor (nF)	* I_{pp} (nA)	I_{rms} or $^\$$ I_{dc} (nA)	P_{out} (nW)	Charging Time (s)
0.0123	No	-	135	34.0	11.0	-
0.0216	No	-	294	49.4	24.5	-
0.0123	FW	-	116	35.6	12.8	-
0.0216	FW	-	140	47.6	22.7	-
0.0123	FW	470	2.88	26.5	7.01	9.92
0.0216	FW	470	4.64	34.7	12.0	7.34
0.0123	FW	1000	1.42	20.3	4.12	13.91
0.0216	FW	1000	1.93	33.3	11.1	11.42

* AC current with no rectifier circuit, $^\$$ DC current with rectifier circuit.

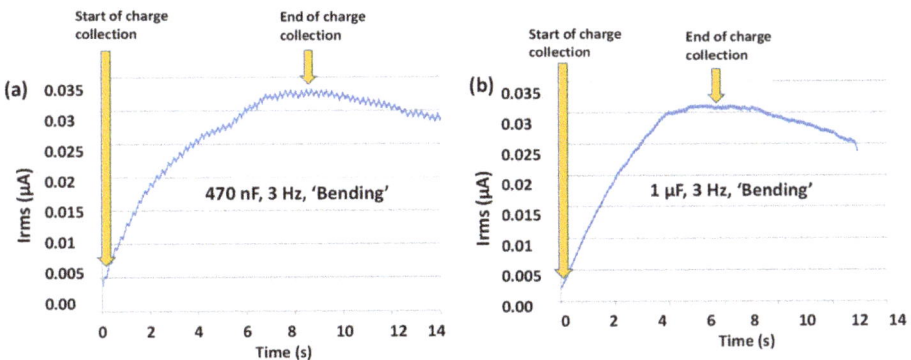

Figure 12. Current over time for constant force loading for (a) 470 nF and (b) 1 µF capacitors.

However, what actually occurs is that after the current stops increasing due to the capacitor's energy storage capacity, it starts to decrease, as shown in Figure 12. It may be due to some parasitic element of the capacitor itself, such as leakage resistance or current leakage through the rest of the system. It is clear that once the capacitor reaches its maximum energy storage capability, the current is being drained faster than it is being produced under constant loading.

As shown in Table 4, at $\varepsilon = 0.0123$, the power generated at the output of the rectifier (refer to Figure 4b) was slightly higher than the power generated with no rectifier. As the values were very close, we can put this down to small changes in the applied force. For $\varepsilon = 0.0216$, the power generated at the output of the rectifier is less than the power generated with no rectifier, which is what we would expect due to the voltage drop and associated power dissipation through the diodes.

As Table 4 shows, the greatest constant DC current generated in this study was 34.7 nA corresponding to a generated power of 12 nW when the load capacitor was 470 nF. If such sensor device was used to charge a typical 3.7 V, 1000 mAh phone battery from flat with a 100% efficient voltage step up circuit, it would take approximately 307 million hours to charge fully. Charging a 2.8 V, 1000 mAh pacemaker battery would take 233 million hours.

The largest AC current generated was 49.4 nA with a corresponding power output of 24.5 nW. However, the power output reduces to almost half for the same strain (0.0216) when a full-wave rectifier is used with a 470 nF capacitor. This power output capability is far lower than that predicted (1800 µW) in [16], and the theoretical recoverable power of 0.33 W from upper arm motion calculated by [6]. It is however to be noted that the simulated PVDF patch in [16] had a much larger surface area (1600 mm^2 as opposed to 625 mm^2 in our study), which should explain the reduced power generated in our case.

The main reason behind our generator's poor power generation capability is likely to be the absence of poling in the test procedure. Noise in the system was a big issue during testing. Table 5 displays the different earthing configurations with the associated noise levels. Figure 13a shows direct grounding of the sensor, while Figure 13b shows grounding through the body. The sensor used was a 20 × 20 mm parallel bimorph with 0.0254 mm thick PVDF. Clearly, the best configuration to minimize system noise is with the sensor connected to ground through the body. The most noise occurs when the sensor is connected to the body with no grounding at all (shown as "Direct" in Table 5). As the sensor is unlikely to remain completely out of contact from the human body, the best solution is to ground the body. For future applications, this may require a ground point integrated into the clothing or grounding directly from the electronic device, which is being charged.

Table 5. Noise measurements for different connections.

V_{peak}	V_{rms}	Grounding	Sensor
2.37 mV	2.16 mV	Through Body	Yes
75.8 mV	53.1 mV	Direct	Yes
3.66 mV	3.01 mV	Ungrounded	Yes
4.94 mV	4.12 mV	Through Body	No
88.7 mV	61.3 mV	Direct	No
10.1 mV	6.96 mV	Ungrounded	No

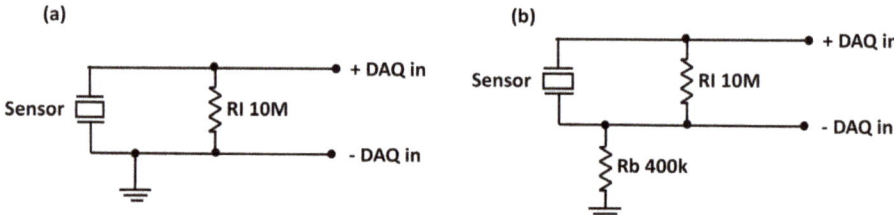

Figure 13. (a) Circuit grounding directly to the sensor and (b) circuit grounding through the body.

5. Conclusions and Outlook

In this paper, the effects of PVDF sensor type, geometry, loading frequency and strain on the electrical power generation capability were studied. Two types of movements were emulated, namely 'bending' and 'flicking'. The major findings from the study are summarized as follows.

- Among the three sensor configurations, parallel bimorph sensor was shown to be the best energy harvester in that it generated the highest amount of electrical power. Practically measured capacitance values for PVDF sensors with large thicknesses were found to be close to the theoretical values [23].

- Given the range of PVDF thickness studied, power output seems to decrease with increasing PVDF thickness. Such trend could be because the stiffness of the PVDF dielectric might have dominated power generation, and the sensor thickness passed the optimum value after which the power output decreased.
- Power output increased consistently with the increase of sensor surface area. Flicking movement showed greater power output than bending for all surface areas as well as far greater increase in power output with surface area. As for the effect of geometry, power output increased initially with increasing aspect ratio, reaching the maximum value at an aspect ratio of around 2:1, and then decreasing for higher aspect ratios.
- Mean current and power output showed a significant increase with the increase in loading frequency for flicking. However, for bending, modest increases were noticed initially (3 Hz and 4 Hz) with the output dropping for higher frequency (5 Hz).
- Power output also followed an increasing trend with the increase of mechanical strain up to a certain value of strain. In addition, an increase in loading frequency resulted in an increase in power generation, however, it did not increase linearly as "wave-cutting" decreased the potential for charge generation at higher frequencies.
- A simple energy harvesting circuit was found to successfully convert the AC signal generated by the sensor into DC signal, but at the cost of a significant loss in power output.

The above findings for PVDF sensors were largely in line with the trends observed using sensors made from other piezoelectric materials, for example, PZT. Although the PVDF sensors studied were not able to generate significant power output to be readily useful in practical applications, some credible relationships and insights on the performance of PVDF sensors were obtained. To realise the full potential of PVDF sensors, future work will focus on more precise preparation of the PVDF sensors with poling, increasing the efficiency of the energy harvesting circuit, optimizing the geometry of the parallel bimorph configuration, and deploying efficient noise reduction techniques.

In this study, the resistance load was kept constant for all parametric analysis of the sensors. However, the change in resistance load in the energy harvesting circuit has a significant influence on the current and power output. Using a coupled piezoelectric circuit-finite element method (CPC-FEM) on vibration-based PZT energy harvesting devices, Zhu et al. [24] reported that when the resistive load within the circuit increases, power output increases up to a peak (maximum) point, and after which, decreases. On the other hand, vibrational amplitude decreases and then picks ups. They found that maximum power occurs at a resistive load of 488 kΩ and vibrational amplitude of 100 µm. The findings imply that the electrical power generation is dependent on the resistive load. Frequency and phase response of the output current and power with respect to the changing resistive load should be analysed to better understand, explain and optimize the power generation capacity of the PVDF sensor [25]. In addition to noise reduction via grounding through the body, the variation of resistive load must be taken into account to elude the efficiency of the energy harvesting PZT sensors. Furthermore, our future work will aim to address theoretical background and simulation to reveal the efficacy of the sensors. Though, in this study, all experiments were conducted in ambient temperature, the efficiency of power generation can be impacted by the change in temperature. For instance, a recent study by Bernard et al. [26] reported the increase of dielectric constant of the PVDF with the increase of temperature from 30 °C to 60 °C and strain up to 2.5%, which caused a decrease in efficiency down to 60–75%, as opposed to ideal conversion efficiency of 90%. They suggested that the circuit output capacitance must be decreased in proportion to be adapted to operation conditions in order to restore higher power generation capacity.

Author Contributions: Conceptualization and methodology, M.U. and S.M.A.; validation and experiments, S.A.; formal analysis, S.A.; data curation, M.U. and S.A.; writing—original draft preparation, S.A.; writing—review and editing, M.U. and S.M.A.; visualization, S.A.; supervision, M.U. and S.M.A. All authors have read and agreed to the published version of the manuscript.

Funding: This research received no external funding.

Institutional Review Board Statement: Not applicable.

Informed Consent Statement: Not applicable.

Data Availability Statement: Data sharing is not applicable to this article as this is an on-going project.

Conflicts of Interest: The authors declare no conflict of interest.

References

1. Chiu, Y.-Y.; Lin, W.-Y.; Wang, H.-Y.; Huang, S.-B.; Wu, M.-H. Development of a piezoelectric polyvinylidene fluoride (PVDF) polymer-based sensor patch for simultaneous heartbeat and respiration monitoring. *Sens. Actuators A* **2013**, *189*, 328–334. [CrossRef]
2. Hackworth, R.; Moreira, J.R.; Maxwell, R.; Kotha, R.; Ayon, A. Piezoelectric charging for smart fabric applications. In Proceedings of the Symposium on Design, Test, Integration and Packaging of MEMS/MOEMS (DTIP), Aix-en-Provence, France, 11–13 May 2011.
3. Caliò, R.; Rongala, U.B.; Camboni, D.; Milazzo, M.; Stefanini, C.; de Petris, G.; Oddo, C.M. Piezoelectric energy harvesting solutions. *Sensors* **2014**, *14*, 4755–4790. [CrossRef] [PubMed]
4. Pallapa, M.; Aly Saad Aly, M.; Aly, S.; Chen, A.; Wong, L.; Wong, K.W.; Abdel-Rahman, E.; Tze, J.; Yeow, W. Modeling and simulation of a piezoelectric micro-power generator. In Proceedings of the COMSOL Conference, Boston, MA, USA, 7 October 2010.
5. Ottman, G.K.; Hofmann, H.F.; Bhatt, A.C.; Lesieutre, G.A. Adaptive piezoelectric energy harvesting circuit for wireless remote power supply. *IEEE Trans. Power Electron.* **2002**, *17*, 669–676. [CrossRef]
6. Starner, T. Human-powered wearable computing. *IBM Syst. J.* **1996**, *35*, 618–629. [CrossRef]
7. Won, S.S.; Seo, H.; Kawahara, M.; Glinsek, S.; Lee, J.; Kim, Y.; Jeong, C.K.; Kingon, A.I.; Kim, S.-H. Flexible vibrational energy harvesting devices using strain-engineered perovskite piezoelectric thin films. *Nano Energy* **2019**, *55*, 182–192. [CrossRef]
8. Yeo, H.G.; Jung, J.; Sim, M.; Jang, J.E.; Choi, H. Integrated piezoelectric AlN thin film with SU-8/PDMS supporting layer for flexible sensor array. *Sensors* **2020**, *20*, 315. [CrossRef] [PubMed]
9. Song, H.J.; Choi, Y.-T.; Wereley, N.M.; Purekar, A. Comparison of monolithic and composite piezoelectric material-based energy harvesting devices. *J. Intell. Mater. Syst. Struct.* **2014**, *25*, 1825–1837. [CrossRef]
10. Jones, G.; Assink, R.; Dargaville, T.; Chaplya, P.; Clough, R.; Elliott, J.; Martin, J.; Mowery, D.; Celina, M. *Characterization, Performance and Optimization of PVDF as a Piezoelectric Film for Advanced Space Mirror Concepts*; Technical Report No. 876343; Sandia National Laboratories: Livermore, CA, USA, 2005.
11. Paradiso, J.A.; Starner, T. Energy scavenging for mobile and wireless electronics. *IEEE Pervasive Comput.* **2005**, *4*, 18–27. [CrossRef]
12. Ng, T.; Liao, W.-H. Sensitivity analysis and energy harvesting for a self-powered piezoelectric sensor. *J. Intell. Mater. Syst. Struct.* **2005**, *16*, 785–797. [CrossRef]
13. Ozeri, S.; Shmilovitz, D. Static force measurement by piezoelectric sensors. In Proceedings of the IEEE International Symposium on Circuits and Systems, Island of Kos, Greece, 21–24 May 2006.
14. Guizzetti, M.; Ferrari, V.; Marioli, D.; Zawada, T. Thickness optimization of a piezoelectric converter for energy harvesting. In Proceedings of the COMSOL Conference, Boston, MA, USA, 8–10 October 2009.
15. Granstrom, J.; Feenstra, J.; Sodano, H.A.; Farinholt, K. Energy harvesting from a backpack instrumented with piezoelectric shoulder straps. *Smart Mater. Struct.* **2007**, *16*, 1810–1820. [CrossRef]
16. Waqar, S.; McCarthy, J.M.; Deivasigamani, A.; Wang, C.H.; Wang, L.; Coman, F.; John, S. Dual field finite element simulations of piezo-patches on fabrics: A parametric study. In Proceedings of the 4th International Conference on Smart Materials and Nanotechnology in Engineering, Gold Coast, Australia, 10–12 July 2013; p. 87931K.
17. Kasyap, V.S.A.; Lim, J.; Ngo, K.; Kurdila, A.; Nishida, T.; Sheplak, M.; Cattafesta, L. Energy Reclamation from a Vibrating Piezoceramic Composite Beam. Available online: https://www.img.ufl.edu/publication/1604/energy-reclamation-vibrating-piezoceramic-composite-beam (accessed on 26 May 2020).
18. Arnau, A.; Soares, D. Fundamentals of piezoelectricity. In *Piezoelectric Transducers and Applications*; Vives, A.A., Ed.; Springer: Berlin/Heidelberg, Germany, 2008; pp. 1–38. ISBN 978-3-540-77508-9.
19. Kashiwao, T.; Izadgoshasb, I.; Lim, Y.Y.; Deguchi, M. Optimization of rectifier circuits for a vibration energy harvesting system using a macro-fiber composite piezoelectric element. *Microelectron. J.* **2016**, *54*, 109–115. [CrossRef]
20. Songsukthawan, P.; Jettanasen, C. Generation and storage of electrical energy from piezoelectric materials. In Proceedings of the IEEE 3rd International Future Energy Electronics Conference and ECCE Asia (IFEEC 2017-ECCE Asia), Kaohsiung, Taiwan, 3–7 June 2017; pp. 2256–2259.
21. Chen, D.; Chen, K.; Brown, K.; Hang, A.; Zhang, J.X.J. Liquid-phase tuning of porous PVDF-TrFE film on flexible substrate for energy harvesting. *Appl. Phys. Lett.* **2017**, *110*, 153902. [CrossRef]
22. Elahi, H.; Israr, A.; Swati, R.; Khan, H.; Tamoor, A. Stability of piezoelectric material for suspension applications. In Proceedings of the 5th International Conference on Aerospace Science and Engineering, Islamabad, Pakistan, 14–16 November 2017.
23. Park, S.; Kim, Y.; Jung, H.; Park, J.-Y.; Lee, N.; Seo, Y. Energy harvesting efficiency of piezoelectric polymer film with graphene and metal electrodes. *Sci. Rep.* **2017**, *7*, 17290. [CrossRef] [PubMed]

24. Zhu, M.; Worthington, E.; Njuguna, J. Analyses of power output of piezoelectric energy-harvesting devices directly connected to a load resistor using a coupled piezoelectric-circuit finite element method. *IEEE Trans. Ultrason. Ferroelectr. Freq. Control* **2009**, *56*, 1309–1317. [CrossRef] [PubMed]
25. Erturk, A.; Inman, D.J. *Piezoelectric Energy Harvesting: Erturk/Piezoelectric Energy Harvesting*; John Wiley and Sons, Ltd.: Chichester, UK, 2011; ISBN 978-1-119-99115-1.
26. Bernard, F.; Gimeno, L.; Viala, B.; Cugat, O. The effect of temperature and strain on power conversion efficiency of PVDF-based thermal energy harvesters. *Proceedings* **2017**, *1*, 576. [CrossRef]

Article

An Approach toward the Realization of a Through-Thickness Glass Fiber/Epoxy Thermoelectric Generator

George Karalis, Christos K. Mytafides, Lazaros Tzounis, Alkiviadis S. Paipetis and Nektaria-Marianthi Barkoula *

Department of Materials Science and Engineering, University of Ioannina, 45110 Ioannina, Greece; gkaralis@uoi.gr (G.K.); cmytafides@uoi.gr (C.K.M.); latzounis@uoi.gr (L.T.); paipetis@uoi.gr (A.S.P.)
* Correspondence: nbarkoul@uoi.gr; Tel.: +30-265100-8003

Abstract: The present study demonstrates, for the first time, the ability of a 10-ply glass fiber-reinforced polymer composite laminate to operate as a structural through-thickness thermoelectric generator. For this purpose, inorganic tellurium nanowires were mixed with single-wall carbon nanotubes in a wet chemical approach, capable of resulting in a flexible p-type thermoelectric material with a power factor value of 58.88 µW/m·K^2. This material was used to prepare an aqueous thermoelectric ink, which was then deposited onto a glass fiber substrate via a simple dip-coating process. The coated glass fiber ply was laminated as top lamina with uncoated glass fiber plies underneath to manufacture a thermoelectric composite capable of generating 54.22 nW power output at a through-thickness temperature difference of 100 K. The mechanical properties of the proposed through-thickness thermoelectric laminate were tested and compared with those of the plain laminates. A minor reduction of approximately 11.5% was displayed in both the flexural modulus and strength after the integration of the thermoelectric ply. Spectroscopic and morphological analyses were also employed to characterize the obtained thermoelectric nanomaterials and the respective coated glass fiber ply.

Keywords: glass fiber-reinforced polymer composite; multifunctional structural laminate; thermal energy harvesting; through-thickness thermal gradient; thermoelectric generator (TEG)

Citation: Karalis, G.; Mytafides, C.K.; Tzounis, L.; Paipetis, A.S.; Barkoula, N.-M. An Approach toward the Realization of a Through-Thickness Glass Fiber/Epoxy Thermoelectric Generator. *Materials* **2021**, *14*, 2173. https://doi.org/10.3390/ma14092173

Academic Editor: Daniele Davino

Received: 22 March 2021
Accepted: 22 April 2021
Published: 23 April 2021

Publisher's Note: MDPI stays neutral with regard to jurisdictional claims in published maps and institutional affiliations.

Copyright: © 2021 by the authors. Licensee MDPI, Basel, Switzerland. This article is an open access article distributed under the terms and conditions of the Creative Commons Attribution (CC BY) license (https://creativecommons.org/licenses/by/4.0/).

1. Introduction

Nowadays, there is a continuously increasing rate of global energy consumption. Although efforts have been made toward the exploitation of renewable or alternative energy sources, their use is still limited [1]. Moreover, eco-friendly solutions are required not only in terms of the source of energy but also in the way the power is supplied. For example, low power-consuming electronics such as wireless sensor networks typically use batteries as a power source. The limited lifetime of batteries results in increased total costs associated with their replacement in remote areas [2,3]. Next to that, high amounts of energy losses, often in the form of heat, could be partially recovered as power by proper energy conversion methodologies [4]. A promising way toward eco-friendly and autonomous structures is, thus, to broadly embed self-powered energy harvesting solutions, such as solar cells, vibration-based or thermal energy harvesters in structural materials [5–9]. This could be extremely relevant for boilers and steam piping systems, especially in large industrial and power plants, which show high amounts of wasted heat [10]. As a consequence, a reduction of operating and control costs could be achieved, also contributing to the new global requirements for CO_2 emissions reduction [11].

The basic principle behind thermal energy harvesting is the thermoelectric effect (i.e., the Seebeck effect). It is well known that when a thermoelectric (TE) material is exposed to a temperature difference (ΔT), it spontaneously generates a potential difference (ΔV) due to the motion of free electrons (n-type semi-conductor material) or holes (p-type) toward specific directions. The magnitude of the Seebeck effect is expressed by the Seebeck coefficient (*S*), which is employed for the estimation of the power factor (*PF*) (see Equations

(1) and (2)). This quantity is used for the direct comparison of various materials' TE efficiency. The overall TE performance is classified by the dimensionless figure of merit (ZT) (see Equation (3)).

$$S = -\frac{\Delta V}{\Delta T} \quad (1)$$

where:
S = the Seebeck coefficient in μV/K
ΔV = the generated TE voltage in mV
ΔT = the externally applied temperature difference in K

$$PF = \sigma \times S^2 \quad (2)$$

where:
PF = the power factor in μW/mK^2
σ = the electrical conductivity in S/m

$$ZT = \frac{\sigma \times S^2 \times T}{\kappa} \quad (3)$$

where:
ZT = dimensionless figure of merit
T = the absolute temperature
κ = thermal conductivity in W/m·K

Depending on their intrinsic carrier mobility and concentration, efficient TE materials present high values of ZT, combining high electrical and low thermal conductivity (κ) [12,13]. Traditional TE materials typically consist of low bandgap semiconductors, e.g., Te, Bi$_2$Te$_3$, PbTe, etc. [14,15]. Recently, hybrid or organic nanostructured materials have been suggested as auspicious candidates for TE applications [16,17]. The scientific community is highly interested in blends of conductive polymers with inorganic thermoelectric crystals, bulk and 1-D superlattice nanostructures, etc., due to their ability to tune the carrier transport via, i.e., energy filtering mechanisms, inherent low thermal conductivity, tailored electrical conductivity, facile processing, relatively moderate large-scale production cost and superior flexibility properties [18–20].

Large-scale TE energy harvesting and conversion to sustainable electrical power is realized by TE generators (TEGs) devices. A common TEG device comprises single-type, or p-/n-type thermoelements interconnected electrically in series and thermally in parallel. Advancements in flexible and wearables TEGs have been recently reported, presenting desired power output values for practical applications [21–25]. Scientific works related to bulk or structural TEGs targeting different application areas have also been published [26–28]. Special interest has been concentrated on polymer nanocomposites [29–31], and fiber-reinforced polymer (FRP) composites [32,33] since such materials are widely used in aerospace, automotive, renewable energy, etc. applications. FRPs offer the potential for flexible design and novel manufacturing approaches with significantly improved specific properties, such as high strength to weight ratio [34,35]. Additionally, a variety of secondary functionalities can be introduced, transforming these materials into smart and multifunctional structures. This can be realized via the integration of dispersed nanomaterials in a polymer matrix or through hierarchical coatings deposited onto the reinforcing phases, such as glass and carbon fibers. Functionalities may include increased interfacial adhesion strength [36], increased interlaminar shear strength [37,38], non-destructive structural diagnostics [39,40], self-healing perspectives [41], energy storage capabilities [42], lightning-strike protection [43] and energy harvesting solutions [28].

More specifically, in the area of structural polymer composites, research has focused on the targeted enhancement of FRP's TE properties through the introduction of nanomaterials [44]. Previous studies on in-plane and through-thickness TE properties of FRP laminates mainly focused on the polymer matrix-interfaces modification with nano or micro-scale fillers [45]. For instance, Han et al. reported increased Seebeck coefficient values from 8 to 163 µV/K by brushing a mixture of tellurium and bismuth microparticles onto carbon fiber prepregs of a polymer matrix-based structural composite [46].

Based on the above, the existing literature findings are limited to the bulk through-thickness interface modifications, mainly with inorganic microparticles, to achieve enhanced TE response at the laminate level. Instead of modifying the matrix and/or the interface of a structural composite, the main goal of the current work is to develop, for the first time, an approach that involves the integration of a proper architecture acting as a through-thickness structural TEG device in a composite laminate. To achieve this goal, an inorganic-organic nanomaterial based on tellurium nanowires (TeNWs) with single-wall carbon nanotubes (SWCNTs) added during growth is deployed. The nanostructured TE material is produced following a surfactant-assisted chemical reduction reaction based on previously well-established synthetic routes [47–49] with a few variations. The rationale behind this selection is to obtain enhanced performance through the combination of the TE properties of the two nanomaterials and flexibility at a film level via the use of highly durable SWCNTs. By redispersing the synthesized nanomaterials, an aqueous TE ink is prepared and used to coat a glass fiber (GF) unidirectional (UD) fabric via a simple dip-coating and oven-drying process. Eventually, the coated ply is purposely laminated to manufacture a 10-ply GF reinforced polymer (GFRP) laminate. The structure of the obtained nanomaterials is characterized using Raman spectroscopy and X-ray diffraction (XRD) analyses. Successful GF coating is confirmed via scanning electron microscopy (SEM). The TE performance of the nanomaterials and TEG laminates is also assessed. Finally, the effect of the TE ply integration on the mechanical performance of the obtained laminates is investigated under flexural loading. The obtained results reveal that it is possible to modify a conventional thin laminate with inorganic-organic nanomaterials on a ply level to enable efficient through-thickness thermal energy harvesting capabilities without eliminating the structural integrity of the obtained structure. Thus, the current paper deals with the demonstration of the ability of FRPs to act, by design, as through-thickness TEGs, with the aim to harvest thermal energy during the operational lifetime in the presence of temperature gradients.

2. Materials and Methods

2.1. Materials

For the preparation of the TE nanomaterials, ascorbic acid (AA) with 99% purity, sodium dodecylbenzenesulfonate (SDBS, 348.48 g/mol), and sodium tellurite (Na_2TeO_3) ~100 mesh with >99% purity were purchased from Sigma Aldrich (Missouri, USA). SWCNTs dispersion (TUBALL, INK H_2O 0.2%) was acquired by OCSiAl (Novosibirsk, Russia). All chemicals were analytical grade and used as received without any further purification procedure. Distilled (DI) water was used throughout this research. PVDF membrane (pore size 45 µm) was purchased from Merck (Darmstadt, Germany) and used for the preparation of Te-based buckypapers.

For the manufacturing of the GFRP laminates, unidirectional (UD) glass fabric 320 gr/m^2 with a single-ply thickness of 0.26 mm from Fibermax (Volos, Greece) was used. The epoxy resin system Araldite LY 5052/Aradur 5052 was purchased from Huntsman (The Woodlands, TX, USA) and was used as the matrix of the composite. To facilitate the fabrication of the TEG device, silver paste (ORGACON™ Nanosilver Screen Printing Ink SI-P2000) was received from Agfa (Mortsel, Belgium), while silver foil tape (thickness of 0.055 mm) with conductive adhesive was acquired by 3M™ (Saint Paul, MN, USA).

2.2. Synthesis of TE Nanomaterial and Ink Preparation

The whole process, from the synthesis of the TE material to the formation of the TE ink is illustrated in Figure 1. Initially, TeNWs were synthesized according to the following procedure: 4.93 g AA was dissolved in 200 mL of DI water in a reaction flask followed by the addition of 0.10 g SDBS. SDBS has been introduced due to its high dispersion efficiency that results in the prevention of agglomeration phenomena [50,51]. After the homogenization of the solution, 0.28 g Na$_2$TeO$_3$ was added to the vigorously stirred mixture. For the synthesis of the inorganic-organic TE nanomaterial, 2.5 mL of the SWCNTs commercial ink was added to the previous mixture. Consequently, the mixture was raised up to 90 °C for 20 h and then left to cool down. The cleaning procedure included centrifugation at 8000 rpm for 30 min and removal of the sediment by dilution with DI water and pouring off repeatedly the SDBS rich supernatant side products and the residual reagents. The final precipitated material was redispersed and via a vacuum filtration process through a PVDF membrane filter (0.45 μm filter pore size) collected in the form of buckypaper while being kept finally for drying at 80 °C overnight. Finally, the buckypaper was redispersed in DI water (40 mg/mL) via bath-sonication for 30 min, resulting in a homogenous dispersion, hereafter denoted as TE ink. For comparison reasons, a buckypaper film was developed based on TeNWs before the in-situ growth in the presence of SWCNTs and used as reference material.

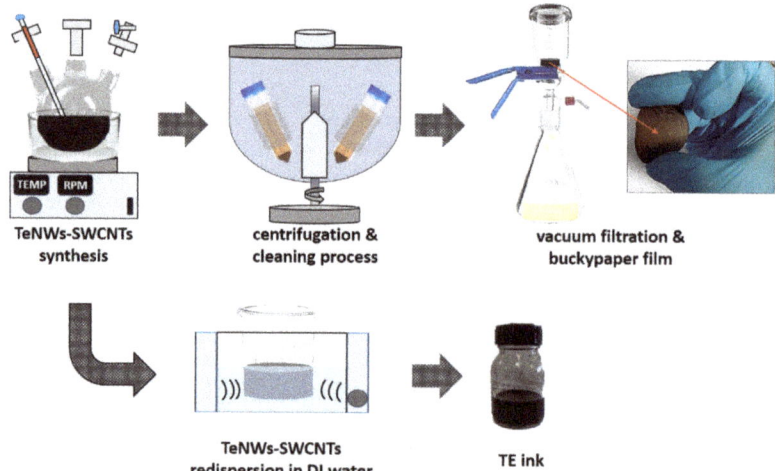

Figure 1. The steps followed for the synthesis of the inorganic-organic TE material and respective ink including: the solvothermal reaction step, the centrifugation and cleaning procedure with DI water, the vacuum filtration and buckypaper preparation procedure, the re-dispersion process, and the final TE ink.

2.3. Manufacturing of the GFRP Laminate with the Through-Thickness TEG Functionality

The fabrication of a single thermoelement TEG required firstly the integration of highly conductive electrode-like plies that will function as the interconnection between the internal TEG structure and the external electrodes. To do so, 2 GF plies were one-sided blade-coated with silver (Ag) paste, as shown schematically in Figure 2a. Then, the Ag-coated GF laminae were transferred in a ventilated oven and cured for 10 min at 150 °C. Afterward, Ag tape stripe was adapted to each Ag-coated GF lamina using a conductive adhesive to create the external electrodes. The next step was the incorporation of the TE functional ply into the laminate, as depicted in Figure 2b. This was achieved employing a facile dip-coating process, where the GF ply was immersed into the TE ink and subsequently dried overnight at 80 °C. For the manufacturing of the 10-ply GFRP TEG (50 × 50 mm^2), the TE-coated GF ply was sandwiched between the Ag-coated GF plies to create the top layer of the composite, while 7 unmodified GF plies were added below the internal Ag-coated GF ply, following a cross-ply lamination (see Figure 2c). A plain 10-ply GFRP laminate was also developed for comparison. Both the TEG GFRP laminate, as well as the reference plain GFRP laminate, were manufactured by hand lay-up epoxy resin impregnation and thermopressing. According to the specifications of the thermoset system, the resin to hardener weight ratio was set at 100:38 w/w. Curing was conducted for 24 h at room temperature (RT) under 3 MPa pressure using a hydraulic press, and the post-curing was performed at 100 °C for 4 h. Based on the technical datasheet of the manufacturer, the T_g of the resin system, after this curing cycle, is in the range of 120 to 134 °C, while dynamic mechanical analysis unpublished data on the reference GFRP laminate (before the incorporation of the electrodes and the functional ply) indicated a T_g of 131 °C. Attention was paid to avoid any direct contact between the 2 Ag-coated GF laminae. This was ensured through the strict alignment of the functional GF ply during the fabrication of the TEG device. The manufactured GFRP TEG is presented in Figure 2d.

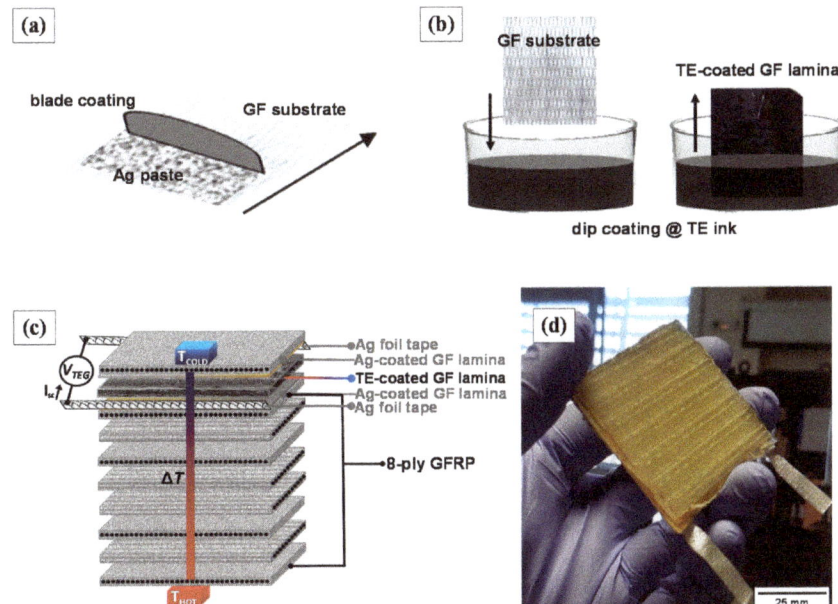

Figure 2. Schematic illustration of (**a**) blade coating for silver paste deposition onto the GF ply for the production of highly conductive electrode-like plies, (**b**) dip-coating of the GF ply within the TE ink to produce the coated functional ply, (**c**) the detailed lamination of the multifunctional GFRP and (**d**) photo of the manufactured through-thickness TEG GFRP laminate.

2.4. Characterization Methodologies

The structure of the synthesized inorganic-organic TE nanomaterial was characterized via Raman spectroscopy and XRD. Both Raman and XRD spectra were obtained from the TE buckypaper films. The spectroscopic measurements were carried out with a Labram HR (Horiba, Kyoto, Japan) scientific micro-Raman system. The 514.5 nm line of an Ar^+ ion laser operating at a power of 1.5 mW at the focal plane was employed for the Raman excitation. An optical microscope equipped with a 50× long working distance objective served both for delivering the excitation light and collecting the back-scattered Raman light. Raman spectra in the range of 90–3500 cm^{-1} were collected. XRD analysis was performed with a D8 ADVANCE system (Bruker, Billerica, MA, USA) in symmetric step-scan mode with $2\theta = 0.05°$ in transmission mode. The diffractometer operated at 40 kV and 30 mA with Kα radiation (λ = 1.5406 Å), diffraction angle (θ, 10° < 2θ < 80°), and a step size of 5° at room temperature. The morphology investigation of the coated GF ply was performed using JEOL JSM 6510 LV SEM/Oxford Instruments (JEOL, Tokyo, Japan) with an operating voltage of 3.5 kV.

The electrical resistivity values of the produced buckypaper films were obtained using a typical 4-probe sheet resistance commercial system (Ossila Ltd., Sheffield, UK).

The generated TE voltage (ΔV) of the produced buckypaper films (in-plane) and the TEG laminate (through-thickness) was measured with a 34401A multimeter (Agilent, Santa Clara, CA, USA). As illustrated in Figure 2c, the voltage was measured using the metallic connectors-electrodes (Ag-coated GF laminae) located in the 8th and 10h plies of the TEG laminate. Thus, the through-thickness TE voltage output measurements were defined in the transverse direction of the device based on a ~0.27 mm interelectrode distance (thickness of the TE ply). A custom-made setup consisting of two metal blocks was developed for the generation of a temperature gradient (see Figure 3). For all measurements, one block was kept at room temperature (~25 °C) via water circulation, while the other was heated at higher temperatures via calibrated temperature-controlled resistors, allowing the generation of a ΔT. Three different levels of the thermal gradient were applied (i.e., ΔT of 50, 75, and 100 K), which result in temperatures below or close to the T_g of the TEG laminates to avoid any substantial degradation in the structural integrity upon heating. The temperature of the two blocks was constantly measured with K-type thermocouples. Figure 3 demonstrates the generated short-circuit current at ΔT of 100 K. Note that optical inspection for the TEG GFRP laminates indicated the absence of obvious evidence for any kind of degradation after several testing hours of continuous operation at the enforced maximum temperature gradient of 100 K.

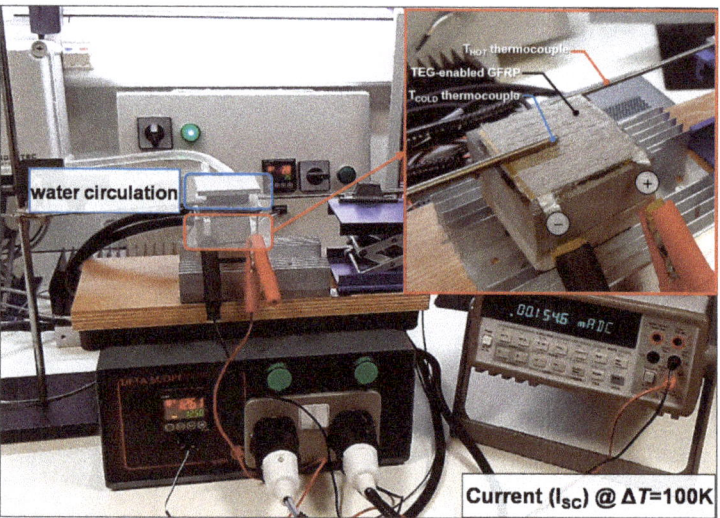

Figure 3. Demonstration of the through-thickness TEG GFRP TE performance during exposure to $\Delta T = 100$ K.

This enables the calculation of *S*, *PF*, and *ZT* according to Equations (1)–(3).

Consequently, the thermal to electrical energy conversion efficiency (Carnot efficiency—η) can be determined by Equation (4) [52]:

$$\eta = \left(\frac{T_H - T_C}{T_H}\right) \frac{\sqrt{1+\overline{ZT}}-1}{\sqrt{1+\overline{ZT}}+(T_C/T_H)} \tag{4}$$

where:
T_H = the temperature of the hot side in K
T_C = the temperature of the cold side in K
$\overline{ZT}$ = ZT calculated at the average temperature between the hot and the cold side

Based on the above measurements, it is also possible to calculate the maximum TE power output of the TEG GFRP laminate according to the following Equation (5) [28]:

$$P_{max} = \frac{V_{TEG}^2}{4xR_{TEG}} \quad (5)$$

where:

P_{max} = maximum power output in nW
V_{TEG} = the TE open-circuit voltage in mV
R_{TEG} = internal electrical resistance of the TEG in Ohm

Oriented to a TEG device characterization for practical applications, power output measurements as a function of the externally applied load resistances (R_{LOAD}) were carried out. Thus, the through-thickness TEG GFRP device power output characteristics have been evaluated using a custom-built fully automated electronic system based on a LabVIEW-PC interface with a range of applied external loads from 1 to 10,000 Ohm with discretion capability down to 1 Ohm. Apart from the experimental power output values, it is possible to obtain calculated ones using Equation (6) [52]:

$$P = I^2 x R_{LOAD} = \left(\frac{V_{TEG}}{R_{TEG} + R_{LOAD}}\right)^2 x R_{LOAD} \quad (6)$$

where:

P = the output power in nW
I = the output current that passes through the load in μA
R_{Load} = externally applied load resistances in Ohm

All tests were performed at ambient conditions (1 atm, T_C ~ 25 °C, relative humidity: 40 ± 5% RH).

Finally, the mechanical performance of the unmodified and TEG GFRP laminates was evaluated under flexural loading according to ASTM D 790-03 standard [53] using a 100 KN Universal Testing Machine (Jinan Testing Equipment IE Corporation, Jinan, China). Five rectangular specimens (50 × 10 × ~2.6 mm³) were tested for each type of laminate at a deformation rate of 1 mm/min. All specimens were conditioned in an oven at 40 °C overnight prior to testing.

3. Results and Discussion

3.1. Characterization of the Inorganic-Organic Nanomaterial and the Coated GF Fabric

Raman and XRD spectra were obtained from the developed buckypaper films to verify the presence of hybrid TeNWs-SWCNTs inorganic-organic thermoelectric material. As illustrated in Figure 4a, the existence of TeNWs is identified through the Raman peaks at 118.1 cm^{-1} and 138.2 cm^{-1}. Furthermore, the peak at ca. 120 cm^{-1} is attributed to the Te content and the A$_1$ vibrational mode response of TeNWs [54]. SWCNTs are also visible via the D (1337 cm^{-1}), G (1588 cm^{-1}), and 2D (2666 cm^{-1}) peaks. A characteristic peak at ca. 1588 cm^{-1} is related to the vibration of sp^2-bonded and in-plane stretching E$_{2g}$ mode of carbon atoms of the SWCNTs. Raman spectra of SWCNTs located this characteristic peat at ca. 1590 cm^{-1} [30]. As discussed previously, the slight shift of the peak position of the G-band can be attributed to the interaction between the TeNWs with SWCNTs, leading to a decreased conjugation [47]. A relatively lower peak at ca. 1337 cm^{-1} is correlated with sp^3 hybridization [55].

Moreover, as illustrated in Figure 4b, the XRD pattern of the inorganic-organic buckypaper film is in agreement with the Te reference spectrum (36–1452, black inset bars) [49]. Thus, the XRD spectra confirmed the crystal hexagonal structure of tellurium with three atoms per unit cell and cell constants equal to 4.46 Å for a and 5.92 Å for c [56].

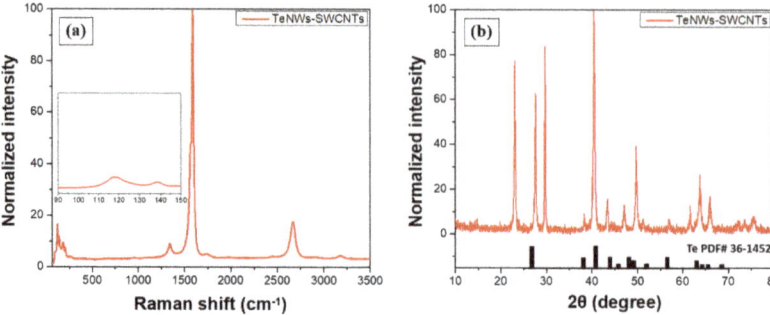

Figure 4. (a) Raman spectrum and (b) X-ray diffraction pattern of the TeNWs-SWCNTs buckypaper film.

For comparison purposes, the in-plane TE properties of the reference inorganic and the developed inorganic-organic buckypaper film are included in Table 1. The reported mean TE values are referred to five measurements of different buckypaper films for each case. The positive values of the Seebeck coefficient indicate the p-type semiconducting behavior for both materials. Furthermore, as observed in Table 1, the TeNWs-based film presented a relatively high Seebeck coefficient of approximately +302 µV/K combined with a low electrical conductivity of 8.4 S/m, which corresponds to a PF of 0.77 µW/m·K^2. At the same time, the inorganic-organic film presented a Seebeck coefficient of approximately +80 µV/K combined with an electrical conductivity of 9200 S/m, resulting in a two order of magnitude higher PF of 58.88 µW/m·K^2, compared to the inorganic TE film. Since κ values were not experimentally, the calculation of the ZT values was obtained using respective values from other studies [48] (see Table 1). We believe that this is a good approximation since similar synthetic routes were followed for the manufacturing of the TeNWs and TeNWs-SWCNTs as those presented in [48]. It should also be noted that an average value of ZT was obtained since the actual temperature was approximated with the externally applied ΔT. As observed, the energy conversion efficiency for the reference system is as low as 0.008%. On the contrary, the combination of the TeNWs with SWCNTs resulted in approximately 74 times higher efficiency of the obtained TE nanomaterials. Based on the above, it can be concluded that the addition of the highly conductive SWCNTs within the TeNWs was beneficial for the overall TE performance of the developed films. This can be attributed to bridging phenomena that create conductive paths between the two nanomaterials, resulting in a reduction of the contact resistance without eliminating the overall TE performance [57–59]. Moreover, the extremely brittle nature of TeNWs buckypaper films renders this material unsuitable for further processing and practical applications. Therefore, the inorganic-organic nanomaterial was employed for TE ink production.

Table 1. TE values of the inorganic and inorganic-organic buckypaper films at ΔT of 100 K. Note that κ values used to calculate ZT are taken from elsewhere [48].

TE Material	σ	S	PF	κ	ZT	η
	S/m	µV/K	µW/m·K^2	W/(m·K)	-	%
TeNWs	8.4 ± 0.6	+302 ± 8	0.77	0.28	0.001	0.008
TeNWs-SWCNTs	9200 ± 5	+80 ± 4	58.88	0.26	0.080	0.590

Images of the TE-coated GF fabric at different magnifications are presented in Figure 5. Based on these images, satisfying adhesion properties between the TE coating and the fibrous substrate is elucidated by the continuously distributed nanostructures onto the surface of the GFs. Consequently, the created through-thickness uniform coating introduces multiple interconnected TE paths due to GF fabric porosity and voids between the stitched GF tows, as observed in Figure 5a,b. Subsequently, the highest magnification image (Figure 5c) reveals a dense network of high aspect ratio typical 1D nanostructures with a diameter of a few nm, which correspond to the TeNWs-SWCNTs hybrid system. The absence of any aggregates implies the preparation of a high-quality TE ink and the application of an efficient coating process. This is expected to result in superior bulk TE properties of the coated GF fabric.

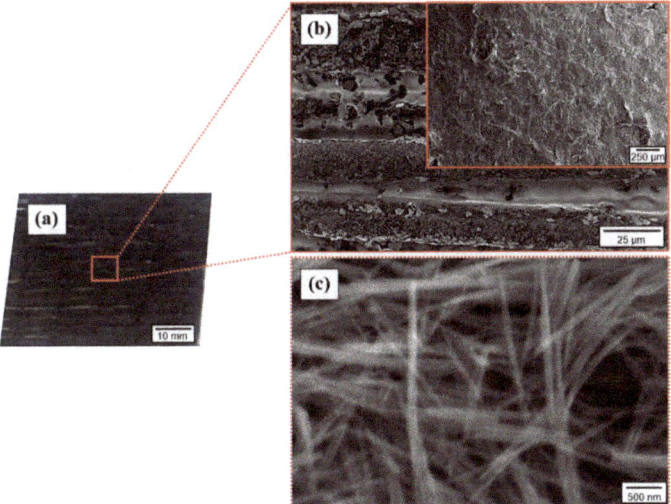

Figure 5. (a) Image of the TE-coated GF and (b,c) SEM images at different magnifications of the coated GF ply.

3.2. Characterization of the TEG GFRP Laminate

Table 2 presents the experimentally measured TE values and the calculated maximum TE power output of the TEG GFRP laminate at different thermal gradients. The average output values correspond to TE measurements, which arose from four manufactured TEG GFRP laminates. At this point, it is worth mentioning that the measured through-thickness internal electrical resistance (R_{TEG}) value of the laminate system prior to the impregnation of the epoxy resin was as low as ~3 Ohm at RT. Subsequently, after the hardening of the epoxy matrix, the composite laminates' R_{TEG} value was slightly increased at 8.3 Ohm. This increase could be mainly ascribed to possible interactions between the epoxy resin, the thin coating of the TE GF lamina, and the Ag-coated GF laminae. Thus, the internal TEG interconnection marginally affected the electrical characteristics of the manufactured laminates.

Table 2. TE measurements of the through-thickness TEG GFRP laminate at various ΔT.

ΔT (K)	R_{TEG} (Ohm)	V_{TEG} (mV)	I_{sc} (μA)	P_{max} (nW)
0	8.30 ± 0.10	-	-	-
50	9.15 ± 0.15	0.76 ± 0.13	82.67 ± 0.18	15.78
75	9.33 ± 0.24	1.10 ± 0.17	119.84 ± 0.24	32.42
100	9.56 ± 0.32	1.44 ± 0.22	154.60 ± 0.38	54.22

As observed in Table 2, the manufactured through-thickness TEG GFRP laminate can harvest thermal energy upon exposure to a temperature gradient and under enforced cooling to sustain the design desired ΔT. In the case where there is no enforced cooling, the power output of the TEG appears to be insufficient. For instance, a ΔT of 50 K generates a sufficient open-circuit voltage (Seebeck voltage) of 0.76 mV, which increases with a further increase of the ΔT. Inevitably, a short-circuit current of 82.67 µA and a power output of 15.78 nW could be achieved at a temperature difference of 50 K that can be easily attained between a car engine in operation and the outside air that surrounds the car during its movement [60,61]. The maximum power output of 54.22 nW of the through-thickness TEG GFRP laminate, derived at ΔT = 100 K, corresponds to a power density of 0.02 W/m^2. The power density value was calculated by dividing the maximum TE power output with the cross-sectional active area of the coated GF ply.

Figure 6a,b, shows the power output characteristics for the TEG GFRP laminate. In more detail, the multifunctional GFRP laminate exhibits an open-circuit voltage (V_{TEG}) of 1.44 mV and short-circuit current (I_{SC}) of 154.6 µA at ΔT of 100 K with an internal resistance (R_{TEG}) of 9.56 Ohm. Figure 6a depicts the measured TE performance in various ΔT. Specifically, output voltage-current (V-I), output power-current (P-I) curves. Figure 6b depicts the output voltage-external load (V-R_{LOAD}) and output power-external load (P-R_{LOAD}) curves with the application of different external load resistances. The continuous lines in all cases have been derived from calculations, while the dots correspond to the experimental values that were acquired through the specially designed custom-built measuring unit. A maximum power generation of 54.22 nW at a through-thickness ΔT of 100 K is dissipated at the applied external electrically in-series connected R_{LOAD}. As it was noticed, when the external load is compatible with the R_{TEG} experimental value of 9.53 Ohm, the maximum output power is matched with the external load resistance of ca. 9.5 Ohm, which is equal to the R_{TEG} value. As expected, the output voltage for the different R_{LOAD} was inversely proportional to the output current, presenting a typical parabolic behavior.

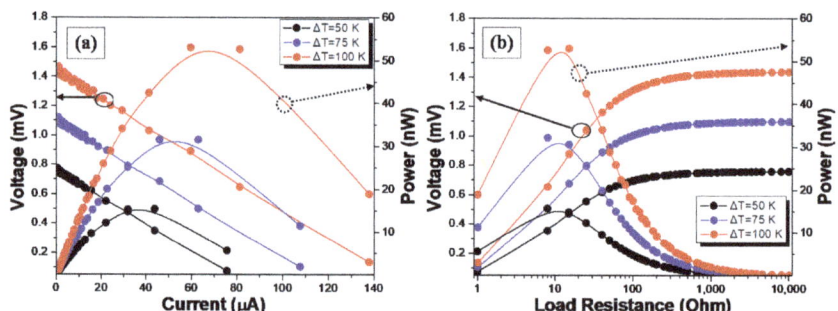

Figure 6. (a,b) Power output characteristics for ΔT of 50, 75, and 100 K.

Conventional bulk or micro-scale through-thickness TEG designs during application suffer from thermal gradient equilibrium during time evolution, especially in the case where the heat dissipation is spontaneous, without being sustained artificially [21,59]. For comparison purposes, it is important to mention that the in-plane TE output of the inorganic-organic buckypaper was 8 ± 0.12 mV, at 25 mm for an applied ΔT of 100 K. The respective through-thickness TE voltage of the TEG laminate was 1.44 mV, as stated in Table 2. Thus, the TEG laminate shows a lower through-thickness voltage by ~82% compared to the voltage obtained from the in-plane measurement of the buckypaper. This could be attributed mainly to the extremely compact interelectrode distance ~0.27 mm in the case of the TEG laminate, as has also been previously pointed out, e.g., [41]. Partially, the thermal insulating character of the GFRP laminate could also negatively affect the internal ΔT distribution, resulting in lower power output values in relation to the expected ones. Additionally, it is worth noting that the abovementioned values are the result of a single thermoelement. Indicative, Inayat et al. succeeded a TE power output of 112 nW for a ΔT of ~20 K resulting from a four-thermoelement inorganic nanomaterial-based through-thickness TEG prototype window glass [62]. Similarly, Lu et al. developed a fabric-based through-thickness TEG prototype consisting of 12 inorganic nanostructured thermoelements that were able to achieve a TE power output of ca. 15 nW at a ΔT of ~30 K [63].

Based on the above, it is obvious that the developed laminate has the potential for significant thermal energy harvesting. This can be further optimized after the fabrication of in-series or in-parallel interconnected modules of thermoelements to increase the total TE power generation. It is therefore demonstrated that the exploitation of through-thickness TEG GFRP composite laminates could realize effective thermal energy harvesting power by structural components.

The mechanical performance of the multifunctional GFRP laminate was evaluated under flexural loading and compared with the plain reference laminate. Based on representative stress-strain curves of the tested specimens, it can be observed that the functional laminate behaves similarly with the reference GFRP, as shown in Figure 7a. The stress-strain curve of the TEG laminate lies slightly below one of the reference materials. The stress-strain curves of all tested specimens were assessed to calculate the average flexural modulus and strength before and after the GFRP modification. Based on the results presented in Figure 7b, the flexural strength was 371.70 ± 23.43 MPa, and the flexural modulus was 11.41 ± 0.5 GPa for the multifunctional GFRP, while the respective values for the plain GFRP were 420.04 ± 24.47 MPa and 12.89 ± 0.6 GPa. Thus, it can be concluded that the integration of the functional GF ply and the Ag-coated plies resulted in a minor alteration of the response of the functional laminate and a respective reduction of ~11.5% in both the strength and the modulus.

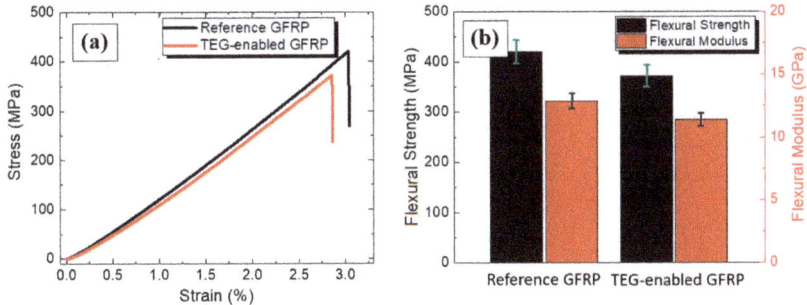

Figure 7. Comparison plots for the reference GFRP and the through-thickness TEG GFRP specimens (a) representative stress-strain curves and (b) average flexural strength and modulus.

The obtained mechanical results disclose relatively equivalent structures according to the flexural strength and the modulus. The requirement of both metallic and TE functional coatings onto specific GF plies contributes to the slightly decreased mechanical properties; however, the laminates maintained to a great extent their advanced properties and can still be used as structural composites.

4. Conclusions

The scope of this research was to introduce through-thickness thermal energy harvesting capabilities to conventional GFRP composite laminates without eliminating their structural integrity. For this reason, an efficient, flexible p-type inorganic-organic TE nanomaterial was synthesized and further processed to produce aqueous TE ink for dip-coating purposes creating hierarchically coated GF UD reinforcement fabrics. The resulting coated functional GF ply was employed for the first time to manufacture a through-thickness TEG-enabled GFRP laminate, which exhibited a power output of 54.22 nW from a single thermoelement upon exposure to a through-thickness ΔT of 100 K. Regarding the mechanical performance, the multifunctional structure displayed slightly decreased values ca. 11.5% of bending strength and flexural modulus with respect to the reference GFRP laminates.

Future developments in flexible, chemically stable, and environmental-friendly TE materials with enhanced ZT values in the range of ~1 at reasonable temperature gradients could dramatically improve the power output characteristics at the material level, oriented to prospect practical applications. Thus, the fabrication at the device level of multiple in-series and/or in-parallel interconnected modules of thermoelements could further stimulate the total generated TE power from composite structures. Thermal energy harvesting and conversion by structural materials with promising power output in the range of several microwatts is sufficient to power up external step-up low power-consuming converters for the energy storage, leading to exploitable energy management and use toward the activation of low-power electronics such as wireless sensor networks, etc. Based on the above, power generation by structural engineering materials designed to be routinely exposed to temperature gradient could emerge as an attractive technology for the realization of large-scale thermal energy harvesting applications in various industrial sectors.

Author Contributions: G.K.: conceptualization, methodology, validation, writing, visualization; C.K.M.: conceptualization, methodology, validation, writing, visualization; L.T.: conceptualization, methodology, validation, A.S.P.: conceptualization, methodology, supervision, N.-M.B.: review and editing, supervision and project administration. All authors have read and agreed to the published version of the manuscript.

Funding: This research is co-financed by Greece and the European Union (European Social Fund-ESF) through the Operational Programme «Human Resources Development, Education and Lifelong Learning 2014–2020» in the context of the project "Advanced Composites for Power Generation in order to Supply Low Energy Requirement Electronic Circuits by Photo-Thermal Energy Harvesting (MIS 5047641).

Institutional Review Board Statement: Not applicable.

Informed Consent Statement: Not applicable.

Data Availability Statement: The data presented in this study are available on request from the corresponding author. The data are not publicly available due to privacy.

Conflicts of Interest: The authors declare no conflict of interest.

References

1. Martín-González, M.; Caballero-Calero, O.; Díaz-Chao, P. Nanoengineering thermoelectrics for 21st century: Energy harvesting and other trends in the field. *Renew. Sustain. Energy Rev.* **2013**, *24*, 288–305. [CrossRef]
2. Ball, A.D.; Gu, F.; Cattley, R.; Wang, X.; Tang, X. Energy harvesting technologies for achieving self-powered wireless sensor networks in machine condition monitoring: A review. *Sensors (Switzerland)* **2018**, *18*, 4113. [CrossRef]
3. Dilhac, J.M.; Monthéard, R.; Bafleur, M.; Boitier, V.; Durand-Estèbe, P.; Tounsi, P. Implementation of thermoelectric generators in airliners for powering battery-free wireless sensor networks. *J. Electron. Mater.* **2014**, *43*, 2444–2451. [CrossRef]
4. Zoui, M.A.; Bentouba, S.; Stocholm, J.G.; Bourouis, M. A review on thermoelectric generators: Progress and applications. *Energies* **2020**, *13*, 3606. [CrossRef]
5. Kraemer, D.; Poudel, B.; Feng, H.P.; Caylor, J.C.; Yu, B.; Yan, X.; Ma, Y.; Wang, X.; Wang, D.; Muto, A.; et al. High-performance flat-panel solar thermoelectric generators with high thermal concentration. *Nat. Mater.* **2011**, *10*, 532–538. [CrossRef] [PubMed]
6. Lee, J.J.; Yoo, D.; Park, C.; Choi, H.H.; Kim, J.H. All organic-based solar cell and thermoelectric generator hybrid device system using highly conductive PEDOT:PSS film as organic thermoelectric generator. *Sol. Energy* **2016**, *134*, 479–483. [CrossRef]
7. Lee, D.; Kim, I.; Kim, D. Hybrid tribo-thermoelectric generator for effectively harvesting thermal energy activated by the shape memory alloy. *Nano Energy* **2021**, *82*, 105696. [CrossRef]
8. Boudouris, B.W. Engineering optoelectronically active macromolecules for polymer-based photovoltaic and thermoelectric devices. *Curr. Opin. Chem. Eng.* **2013**, *2*, 294–301. [CrossRef]
9. Zi, Y.; Wang, Z.L. Nanogenerators: An emerging technology towards nanoenergy. *APL Mater.* **2017**, *5*, 074103. [CrossRef]
10. Liu, L. Feasibility of large-scale power plants based on thermoelectric effects. *New J. Phys.* **2014**, *16*, 123019. [CrossRef]

11. Zappa, W.; Junginger, M.; Van Den Broek, M.; Cover, C.L. Is a 100% renewable European power system feasible by 2050? *Appl. Energy* **2019**, *233–234*, 1027–1050. [CrossRef]
12. Vineis, C.J.; Shakouri, A.; Majumdar, A.; Kanatzidis, M.G. Nanostructured thermoelectrics: Big efficiency gains from small features. *Adv. Mater.* **2010**, *22*, 3970–3980. [CrossRef]
13. Cho, C.; Wallace, K.L.; Tzeng, P.; Hsu, J.H.; Yu, C.; Grunlan, J.C. Outstanding Low Temperature Thermoelectric Power Factor from Completely Organic Thin Films Enabled by Multidimensional Conjugated Nanomaterials. *Adv. Energy Mater.* **2016**, *6*, 1–8. [CrossRef]
14. Snyder, G.J.; Toberer, E.S. Complex thermoelectric materials. *Nat. Mater.* **2008**, *7*, 105–114. [CrossRef]
15. Madan, D.; Wang, Z.; Chen, A.; Juang, R.C.; Keist, J.; Wright, P.K.; Evans, J.W. Enhanced performance of dispenser printed MA n-type Bi2Te 3 composite thermoelectric generators. *ACS Appl. Mater. Interfaces* **2012**, *4*, 6117–6124. [CrossRef] [PubMed]
16. Cao, Z.; Koukharenko, E.; Tudor, M.J.; Torah, R.N.; Beeby, S.P. Flexible screen printed thermoelectric generator with enhanced processes and materials. *Sens. Actuators A Phys.* **2016**, *238*, 196–206. [CrossRef]
17. Shah, K.W.; Wang, S.X.; Soo, D.X.Y.; Xu, J. One-dimensional nanostructure engineering of conducting polymers for thermoelectric applications. *Appl. Sci.* **2019**, *9*, 1422. [CrossRef]
18. Dörling, B.; Ryan, J.D.; Craddock, J.D.; Sorrentino, A.; El Basaty, A.; Gomez, A.; Garriga, M.; Pereiro, E.; Anthony, J.E.; Weisenberger, M.C.; et al. Photoinduced p- to n-type Switching in Thermoelectric Polymer-Carbon Nanotube Composites. *Adv. Mater.* **2016**, *28*, 2782–2789. [CrossRef] [PubMed]
19. Wei, Q.; Mukaida, M.; Kirihara, K.; Naitoh, Y.; Ishida, T. Recent progress on PEDOT-based thermoelectric materials. *Materials (Basel)* **2015**, *8*, 732–750. [CrossRef]
20. Kandemir, A.; Ozden, A.; Cagin, T.; Sevik, C. Thermal conductivity engineering of bulk and one-dimensional Si-Ge nanoarchitectures. *Sci. Technol. Adv. Mater.* **2017**, *18*, 187–196. [CrossRef] [PubMed]
21. Chen, W.Y.; Shi, X.L.; Zou, J.; Chen, Z.G. Wearable fiber-based thermoelectrics from materials to applications. *Nano Energy* **2021**, *81*, 105684. [CrossRef]
22. Allison, L.K.; Andrew, T.L. A Wearable All-Fabric Thermoelectric Generator. *Adv. Mater. Technol.* **2019**, *4*, 1–7. [CrossRef]
23. Jung, K.K.; Jung, Y.; Choi, C.J.; Lee, J.M.; Ko, J.S. Flexible thermoelectric generator with polydimethyl siloxane in thermoelectric material and substrate. *Curr. Appl. Phys.* **2016**, *16*, 1442–1448. [CrossRef]
24. Mytafides, C.K.; Tzounis, L.; Karalis, G.; Formanek, P.; Paipetis, A.S. High-Power All-Carbon Fully Printed and Wearable SWCNT-Based Organic Thermoelectric Generator. *ACS Appl. Mater. Interfaces* **2021**, *13*, 11151–11165. [CrossRef]
25. Kim, S.J.; We, J.H.; Cho, B.J. A wearable thermoelectric generator fabricated on a glass fabric. *Energy Environ. Sci.* **2014**, *7*, 1959–1965. [CrossRef]
26. Wei, J.; Nie, Z.; He, G.; Hao, L.; Zhao, L.; Zhang, Q. Energy harvesting from solar irradiation in cities using the thermoelectric behavior of carbon fiber reinforced cement composites. *RSC Adv.* **2014**, *4*, 48128–48134. [CrossRef]
27. Tzounis, L.; Liebscher, M.; Fuge, R.; Leonhardt, A.; Mechtcherine, V. P- and n-type thermoelectric cement composites with CVD grown p- and n-doped carbon nanotubes: Demonstration of a structural thermoelectric generator. *Energy Build.* **2019**, *191*, 151–163. [CrossRef]
28. Karalis, G.; Tzounis, L.; Lambrou, E.; Gergidis, L.N.; Paipetis, A.S. A carbon fiber thermoelectric generator integrated as a lamina within an 8-ply laminate epoxy composite: Efficient thermal energy harvesting by advanced structural materials. *Appl. Energy* **2019**, *253*, 113512. [CrossRef]
29. Blackburn, J.L.; Ferguson, A.J.; Cho, C.; Grunlan, J.C. Carbon-Nanotube-Based Thermoelectric Materials and Devices. *Adv. Mater.* **2018**, *30*, 1–35. [CrossRef] [PubMed]
30. Kroon, R.; Mengistie, D.A.; Kiefer, D.; Hynynen, J.; Ryan, J.D.; Yu, L.; Müller, C. Thermoelectric plastics: From design to synthesis, processing and structure-property relationships. *Chem. Soc. Rev.* **2016**, *45*, 6147–6164. [CrossRef]
31. Tzounis, L.; Hegde, M.; Liebscher, M.; Dingemans, T.; Pötschke, P.; Paipetis, A.S.; Zafeiropoulos, N.E.; Stamm, M. All-aromatic SWCNT-Polyetherimide nanocomposites for thermal energy harvesting applications. *Compos. Sci. Technol.* **2018**, *156*, 158–165. [CrossRef]
32. Karalis, G.; Mytafides, C.; Polymerou, A.; Tsirka, K.; Tzounis, L.; Gergidis, L.; Paipetis, A.S. Hierarchical Reinforcing Fibers for Energy Harvesting Applications—A Strength Study. *Key Eng. Mater.* **2020**, *827*, 252–257. [CrossRef]
33. Karalis, G.; Tsirka, K.; Tzounis, L.; Mytafides, C.; Koutsotolis, L.; Paipetis, A.S. Epoxy/Glass Fiber Nanostructured p- and n-Type Thermoelectric Enabled Model Composite Interphases. *Appl. Sci.* **2020**, *10*, 5352. [CrossRef]
34. Paluvai, N.R.; Mohanty, S.; Nayak, S.K. Synthesis and Modifications of Epoxy Resins and Their Composites: A Review. *Polym. Plast. Technol. Eng.* **2014**, *53*, 1723–1758. [CrossRef]
35. Gibson, R.F. A review of recent research on mechanics of multifunctional composite materials and structures. *Compos. Struct.* **2010**, *92*, 2793–2810. [CrossRef]
36. De Luca, F.; Clancy, A.J.; Carrero, N.R.; Anthony, D.B.; De Luca, H.G.; Shaffer, M.S.P.; Bismarck, A. Increasing carbon fiber composite strength with a nanostructured "brick-and-mortar" interphase. *Mater. Horiz.* **2018**, *5*, 668–674. [CrossRef]
37. Gkikas, G.; Barkoula, N.M.; Paipetis, A.S. Effect of dispersion conditions on the thermo-mechanical and toughness properties of multi walled carbon nanotubes-reinforced epoxy. *Compos. Part B Eng.* **2012**, *43*, 2697–2705. [CrossRef]
38. Qian, H.; Greenhalgh, E.S.; Shaffer, M.S.P.; Bismarck, A. Carbon nanotube-based hierarchical composites: A review. *J. Mater. Chem.* **2010**, *20*, 4751–4762. [CrossRef]

39. Bekas, D.G.; Paipetis, A.S. Damage monitoring in nanoenhanced composites using impedance spectroscopy. *Compos. Sci. Technol.* **2016**, *134*, 96–105. [CrossRef]
40. Foteinidis, G.; Tsirka, K.; Tzounis, L.; Baltzis, D.; Paipetis, A.S. The role of synergies of MWCNTs and Carbon Black in the enhancement of the electrical and mechanical response of modified epoxy resins. *Appl. Sci.* **2019**, *9*, 3757. [CrossRef]
41. Orfanidis, S.; Papavassiliou, G.; Paipetis, A.S. Microcapsule-based self-healing materials: Healing efficiency and toughness reduction vs. capsule size. *Compos. Part B* **2019**, *171*, 78–86. [CrossRef]
42. Qian, H.; Kucernak, A.R.; Greenhalgh, E.S.; Bismarck, A.; Shaffer, M.S.P. Multifunctional structural supercapacitor composites based on carbon aerogel modified high performance carbon fiber fabric. *ACS Appl. Mater. Interfaces* **2013**, *5*, 6113–6122. [CrossRef]
43. Han, J.H.; Zhang, H.; Chen, M.J.; Wang, D.; Liu, Q.; Wu, Q.L.; Zhang, Z. The combination of carbon nanotube buckypaper and insulating adhesive for lightning strike protection of the carbon fiber/epoxy laminates. *Carbon N. Y.* **2015**, *94*, 101–113. [CrossRef]
44. Sung, D.H.; Kang, G.H.; Kong, K.; Kim, M.; Park, H.W.; Park, Y. Bin Characterization of thermoelectric properties of multifunctional multiscale composites and fiber-reinforced composites for thermal energy harvesting. *Compos. Part B Eng.* **2016**, *92*, 202–209. [CrossRef]
45. Han, S.; Chung, D.D.L. Through-thickness thermoelectric power of a carbon fiber/epoxy composite and decoupled contributions from a lamina and an interlaminar interface. *Carbon N. Y.* **2013**, *52*, 30–39. [CrossRef]
46. Han, S.; Chung, D.D.L. Carbon fiber polymer-matrix structural composites exhibiting greatly enhanced through-thickness thermoelectric figure of merit. *Compos. Part A Appl. Sci. Manuf.* **2013**, *48*, 162–170. [CrossRef]
47. Li, C.; Sun, P.; Liu, C.; Xu, J.; Wang, T.; Wang, W.; Hou, J.; Jiang, F. Fabrication of flexible SWCNTs-Te composite films for improving thermoelectric properties. *J. Alloys Compd.* **2017**, *723*, 642–648. [CrossRef]
48. Choi, J.; Lee, K.; Park, C.R.; Kim, H. Enhanced thermopower in flexible tellurium nanowire films doped using single-walled carbon nanotubes with a rationally designed work function. *Carbon N. Y.* **2015**, *94*, 577–584. [CrossRef]
49. Park, H.; Son, W.; Lee, S.H.; Kim, S.; Lee, J.J.; Cho, W.; Choi, H.H.; Kim, J.H. Aqueous chemical synthesis of tellurium nanowires using a polymeric template for thermoelectric materials. *CrystEngComm* **2015**, *17*, 1092–1097. [CrossRef]
50. Wei, J.; Wen, X.; Zhu, F. Influence of Surfactant on the Morphology and Photocatalytic Activity of Anatase TiO2 by Solvothermal Synthesis. *J. Nanomater.* **2018**, *2018*, 1–7. [CrossRef]
51. Thilagavathi, T.; Geetha, D. Nano ZnO structures synthesized in presence of anionic and cationic surfactant under hydrothermal process. *Appl. Nanosci.* **2014**, *4*, 127–132. [CrossRef]
52. von Lukowicz, M.; Abbe, E.; Schmiel, T.; Tajmar, M. Thermoelectric Generators on Satellites—An Approach for Waste Heat Recovery in Space. *Energies* **2016**, *9*, 541. [CrossRef]
53. ASTM INTERNATIONAL Standard Test Methods for Flexural Properties of Unreinforced and Reinforced Plastics and Electrical Insulating Materials. D790. *Annu. B ASTM Stand.* 2002, pp. 1–12. Available online: https://www.astm.org/Standards/D790 (accessed on 25 April 2021).
54. Roy, A.; Amin, K.R.; Tripathi, S.; Biswas, S.; Singh, A.K.; Bid, A.; Ravishankar, N. Manipulation of Optoelectronic Properties and Band Structure Engineering of Ultrathin Te Nanowires by Chemical Adsorption. *ACS Appl. Mater. Interfaces* **2017**, *9*, 19462–19469. [CrossRef]
55. Tsirka, K.; Karalis, G.; Paipetis, A.S. Raman Strain Sensing and Interfacial Stress Transfer of Hierarchical CNT-Coated Carbon Fibers. *J. Mater. Eng. Perform.* **2018**, *27*, 5095–5101. [CrossRef]
56. Liang, F.; Qian, H. Synthesis of tellurium nanowires and their transport property. *Mater. Chem. Phys.* **2009**, *113*, 523–526. [CrossRef]
57. Choi, J.; Lee, J.Y.; Lee, H.; Park, C.R.; Kim, H. Enhanced thermoelectric properties of the flexible tellurium nanowire film hybridized with single-walled carbon nanotube. *Synth. Met.* **2014**, *198*, 340–344. [CrossRef]
58. Moriarty, G.P.; Wheeler, J.N.; Yu, C.; Grunlan, J.C. Increasing the thermoelectric power factor of polymer composites using a semiconducting stabilizer for carbon nanotubes. *Carbon N. Y.* **2012**, *50*, 885–895. [CrossRef]
59. Zaia, E.W.; Sahu, A.; Zhou, P.; Gordon, M.P.; Forster, J.D.; Aloni, S.; Liu, Y.S.; Guo, J.; Urban, J.J. Carrier Scattering at Alloy Nanointerfaces Enhances Power Factor in PEDOT:PSS Hybrid Thermoelectrics. *Nano Lett.* **2016**, *16*, 3352–3359. [CrossRef]
60. Marshall, G.J.; Mahony, C.P.; Rhodes, M.J.; Daniewicz, S.R.; Tsolas, N.; Thompson, S.M. Thermal Management of Vehicle Cabins, External Surfaces, and Onboard Electronics: An Overview. *Engineering* **2019**, *5*, 954–969. [CrossRef]
61. Khaled, M.; Harambat, F.; Peerhossaini, H. Underhood thermal management: Temperature and heat flux measurements and physical analysis. *Appl. Therm. Eng.* **2010**, *30*, 590–598. [CrossRef]
62. Inayat, S.B.; Rader, K.R.; Hussain, M.M. Thermoelectricity from Window Glasses. *Sci. Rep.* **2012**, *2*, 841. [CrossRef] [PubMed]
63. Lu, Z.; Zhang, H.; Mao, C.; Li, C.M. Silk fabric-based wearable thermoelectric generator for energy harvesting from the human body. *Appl. Energy* **2016**, *164*, 57–63. [CrossRef]

Article

Lumped Element Model for Thermomagnetic Generators Based on Magnetic SMA Films

Joel Joseph [1], Makoto Ohtsuka [2], Hiroyuki Miki [3] and Manfred Kohl [1],*

[1] Institute of Microstructure Technology, Karlsruhe Institute of Technology (KIT), Postfach 3640, D-76021 Karlsruhe, Germany; joel.joseph@kit.edu
[2] Institute of Multidisciplinary Research for Advanced Materials, Tohoku University, Sendai 980-8577, Japan; makoto.ohtsuka.d7@tohoku.ac.jp
[3] Institute of Fluid Science, Tohoku University, Sendai 980-8577, Japan; hiroyuki.miki.c2@tohoku.ac.jp
* Correspondence: manfred.kohl@kit.edu

Abstract: This paper presents a lumped element model (LEM) to describe the coupled dynamic properties of thermomagnetic generators (TMGs) based on magnetic shape memory alloy (MSMA) films. The TMG generators make use of the concept of resonant self-actuation of a freely movable cantilever, caused by a large abrupt temperature-dependent change of magnetization and rapid heat transfer inherent to the MSMA films. The LEM is validated for the case of a Ni-Mn-Ga film with Curie temperature T_C of 375 K. For a heat source temperature of 443 K, the maximum power generated is 3.1 µW corresponding to a power density with respect to the active material's volume of 80 mW/cm^3. Corresponding LEM simulations allow for a detailed study of the time-resolved temperature change of the MSMA film, the change of magnetic field at the position of the film and of the corresponding film magnetization. Resonant self-actuation is observed at 114 Hz, while rapid temperature changes of about 10 K occur within 1 ms during mechanical contact between heat source and Ni-Mn-Ga film. The LEM is used to estimate the effect of decreasing T_C on the lower limit of heat source temperature in order to predict possible routes towards waste heat recovery near room temperature.

Keywords: thermomagnetic energy generators; power generation; waste heat recovery; lumped-element modelling; magnetic shape memory films; Ni-Mn-Ga film; magnetization change; Curie temperature

Citation: Joseph, J.; Ohtsuka, M.; Miki, H.; Kohl, M. Lumped Element Model for Thermomagnetic Generators Based on Magnetic SMA Films. *Materials* **2021**, *14*, 1234. https://doi.org/10.3390/ma14051234

Academic Editor: George Kenanakis

Received: 18 December 2020
Accepted: 2 March 2021
Published: 5 March 2021

Publisher's Note: MDPI stays neutral with regard to jurisdictional claims in published maps and institutional affiliations.

Copyright: © 2021 by the authors. Licensee MDPI, Basel, Switzerland. This article is an open access article distributed under the terms and conditions of the Creative Commons Attribution (CC BY) license (https://creativecommons.org/licenses/by/4.0/).

1. Introduction

With the introduction of the internet of things and wireless sensor networks, the world is moving towards an interconnected network of distributed devices that collect data of their environment and communicate via the internet. In most cases, autonomous operation is mandatory and power supply by batteries or cables is not desired [1,2]. Energy harvesting is considered a solution as it allows producing small amounts of power at site. Recovery of low-grade thermal energy is of special interest, which is mostly rejected as waste heat making up a huge portion of energy lost in the environment. Much of the unrecovered waste heat is in the low temperature regime (10–250 °C) [3–5]. Keeping a high conversion efficiency at small temperature difference is a major challenge, particularly when miniature dimensions are required. Thermoelectric harvesting technology is considered the state of the art in this realm, however, it suffers from scalability due to the need for large heat sinks that may exceed the size of the actual device by far [6–8].

Thermomagnetic energy generators (TMGs) have the potential to overcome the limitations of today's miniaturized thermal energy harvesting systems. The device performance strongly depends on the magnetic material used for energy conversion that should exhibit, among others, a large change of magnetization ΔM at small temperature difference ΔT and a large thermal conductivity. A number of TMG devices make use of ferromagnetic materials, in particular gadolinium (Gd), showing a pronounced second-order ferromagnetic transition near the Curie point close to room temperature [9,10]. An interesting alternative are La-Fe-Si-based materials [11,12]. Recently, spin reorientation has been considered that exploits a temperature-induced change in the magnetic easy axis [13]. Magnetic shape memory alloys (MSMAs) are another promising material candidate that have been tailored to exhibit a large change of magnetization ΔM at small temperature difference ΔT [14–17]. One option is to make use of the steep increase in magnetization at the first-order transformation between non-ferromagnetic martensite and ferromagnetic austenite in metamagnetic alloys, which is highly attractive in the case of small hysteresis [18–20]. Another option is to utilize the large change of magnetization at the second-order transition, e.g., in the Heusler alloy Ni-Mn-Ga, which occurs without hysteresis [21,22].

Another critical aspect of device performance is the engineering of heat intake and dissipation for optimum energy conversion. In recent years several TMG concepts have been introduced that involve energy generation either indirectly via periodic mechanical motion such as rotation and oscillation or directly via thermal-to-magnetic energy conversion. Macroscale TMG demonstrator designs include thermomagnetic oscillators and linear harvesters showing oscillation cycles at a frequency of typically below 1 Hz [9,23]. Millimeter-scale TMG devices have the potential to operate at substantially higher frequencies due to the increased surface-to-volume ratio and, thus, allow for higher power output [10,19]. Recent demonstrators based on Ni-Mn-Ga films show promising results with magnetic power density close to 120 mW/cm^3 with respect to the active layer volume [21]. The electrical power per footprint of the resonant TMG devices reached 50 µW/cm^2 at a temperature change of only 3 K [22]. The large power density relies on rapid heat transfer of the film and the unique concept of resonant self-actuation of a film cantilever oscillating at high frequency in the order of 100 Hz. Keeping the condition of resonant self-actuation is a complex engineering task due to the strong coupling of mechanical, magnetic, thermal and electric performance, which demands for an appropriate model to describe the various interdependencies.

In the following, we present a lumped element modeling (LEM) approach to investigate the interplay of the involved physical properties, in particular, the effects of heat intake and heat dissipation on the local temperature changes of the active material as well as the resulting changes of magnetization and force dynamics on power output. The main novelties of this investigation are describing the experimental performance characteristics of a TMG demonstrator at resonant self-actuation with high accuracy and predicting the effect of decreasing Curie temperature on the lower limit of heat source temperature.

2. Material Properties and Operation Principle

Ni–Mn–Ga films are chosen as a representative MSMA material. Ni–Mn–Ga films of 10 µm thickness are fabricated by RF magnetron sputtering in a high-purity argon atmosphere. The sputtering power is adjusted to control the Ni content and thus the phase transformation temperatures. Heat treatment conditions are tailored for optimal temperature-dependent change of magnetization $\Delta M / \Delta T$ at the ferromagnetic transition. Details on film fabrication and film properties can be found in [24,25] and references therein. Figure 1 shows magnetization versus temperature of a Ni-Mn-Ga film at a low magnetic field of 0.05 T. The material is in martensitic and ferromagnetic state at room temperature. In the temperature range between 330 and 350 K, martensitic phase transformation occurs, whereby the reverse transformation from austenite to martensite at 350 K gives rise to the small step-like feature in Figure 1. At the Curie temperature T_C of about 375 K, magnetization shows a sharp change ΔM due to the ferromagnetic transition.

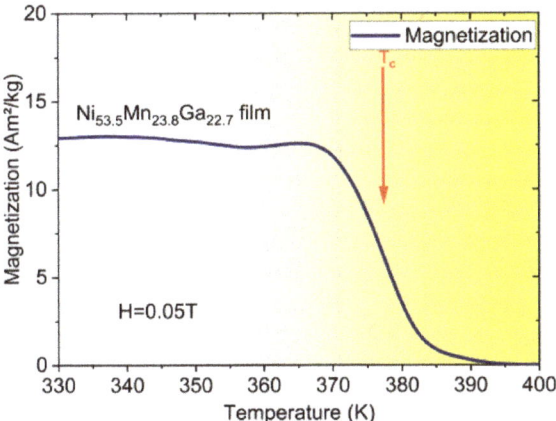

Figure 1. Magnetization versus temperature of a Ni-Mn-Ga film showing a sharp drop at the second-order ferromagnetic transition (Curie temperature T_C) as indicated.

Figure 2 shows a schematic of the TMG device consisting of a cantilever beam attached to a substrate with both an active MSMA film and a pick-up coil attached to its freely movable end. A heated magnet is used to provide the magnetic force that attracts the free end of the cantilever towards the magnetic surface. Heat transfer occurs at mechanical contact causing a ferromagnetic transition in the MSMA film resulting in a reduction in magnetic force. In this case, the elastic force of the cantilever dominates and resets the cantilever. While oscillating back, the MSMA film cools and magnetic attraction force recovers. Consequently, a continuous oscillation occurs, which is sustained by the inertia force of the cantilever front. Thereby, the oscillatory motion supports cooling of the MSMA film. Resonant self-actuation sets in, when cooling and heating cause a sufficiently large change of magnetization and corresponding magnetic actuation force within a cycle. In this case, thermal energy is converted most efficiently into mechanical and magnetic energy, which in turn is converted into electrical energy by the pick-up coil.

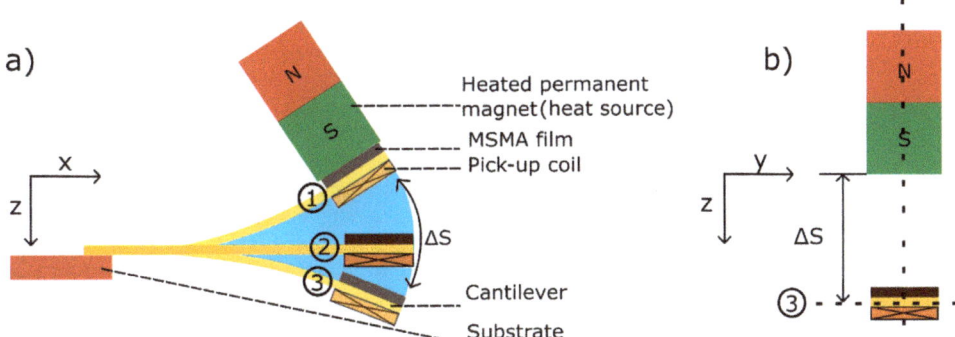

Figure 2. Layout and operation principle of the TMG device. A MSMA film of Ni-Mn-Ga and a pick-up coil are mounted at the cantilever front; (**a**) schematic side view of TMG device for three different deflection states (1)–(3), (**b**) front view for the maximum deflection state 3.

One of the advantages of this TMG concept is that it is self-adapting to operating conditions like the source temperature and ambient temperature. When the source temperature increases, for instance, heat transfer increases causing an increase in oscillation amplitude and resonance frequency [22]. Yet, a major challenge is to maintain the condition of resonant self-actuation for optimum power output. In particular, any change of material parameters and dimensions may affect the resonance condition and, thus, may cause a deviation from optimum operation. Therefore, it is mandatory to understand the complex interplay of mechanical, thermal, and magnetic properties and their effect on the dynamics of heat transfer and cantilever motion. In the following, we discuss how this challenge can be met by lumped element modeling (LEM) of device performance.

3. Lumped Element Model

The LEM has to describe the energy conversion processes in the TMG device by coupling mechanical motion of the cantilever front, magnetization changes of the MSMA film and heat flows during film heating and cooling. Thus, the model should allow for the optimization of the design parameters to obtain maximum power and efficiency by keeping the harvester operating at resonance. The LEM is implemented in SIMSCAPE (R-2018b, 2018, MATLAB, Karlsruhe, Germany) to calculate the mechanical, magnetic and thermal performances, which are described in the following. During simulation, the different physics sections interchange data for each position and time step. The modelling parameters are summarized in the Appendix A.

3.1. Mechanical Section

The mechanical deflection of the cantilever front is approximated as a one-dimensional motion of the center of mass at the movable cantilever front, whereby the lateral displacement due to bending is neglected. When operating the TMG device in vertical direction, the gravitational force acts on the movable mass, but it is negligible compared to magnetic and restoring forces. The net force F_{net} acting on the cantilever front is given by the magnetic force F_{mag}, inertia force, damping force and elastic force of the cantilever:

$$F_{net} = m\frac{d^2x}{dt} + c\frac{dx}{dt} + kx + F_{mag} - F_{em} - F_{air} \quad (1)$$

Thereby, the movable mass m at the cantilever front is taken as [26]:

$$m = \frac{33}{140} \times m_{cant} + m_{coil} + m_{film} \quad (2)$$

When the cantilever front gets in contact with the magnet surface, an additional elastic impact force occurs causing it to suddenly stop its movement at the surface.

The mechanical section of the LEM is depicted schematically in Figure 3. The spring defines the stiffness of the beam cantilever k and the damper represents the structural damping with damping constant c. The impact force at contact is provided by the contact hard stop with contact stiffness k_{cont} and impact damping constant c_{cont}. Numerical values of these parameters used in the LEM simulation are summarized in Table A1. The beam is fixed at one end using a rigid reference. At the opposite freely movable end, the mass acts as a point load. The section takes the magnetic force and electromagnetic damping force as input parameters to compute the effective force at the cantilever front. Thereby, the damping forces caused by structural damping, viscous air damping F_{air} [27] and electromagnetic damping F_{em} [28] are taken into account. Based on the contributing forces, the equation of motion is solved to determine the position and velocity of the center of mass as the output parameters.

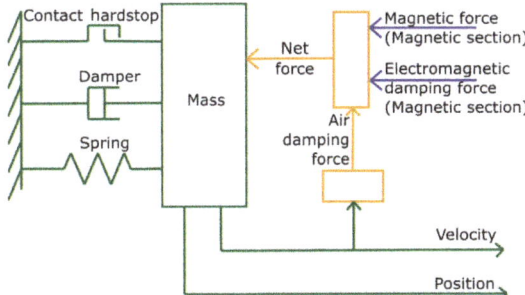

Figure 3. Mechanical section; the magnetic force and electromagnetic damping force from magnetic section are taken as input, while position and velocity are generated as output.

3.2. Magnetic Section

Figure 4 shows a schematic of the magnetic section. Input parameters are the temperature of the MSMA film, the position of the cantilever front and the current of the pick-up coil. Experimental data on the temperature and magnetic field dependence of magnetization of the MSMA film are filed as a lookup table, which allows to select the most appropriate value of magnetization M_T for the given input of magnetic field and temperature. M_T enters into the calculation of the magnetic force F_{mag} on the MSMA film. F_{mag} is computed along the symmetry axis of the magnet in z direction [29]:

$$F_{mag} = V_{mag} M_T \frac{\partial B_z}{\partial z}, \qquad (3)$$

whereby the cantilever deflection is assumed to be a one-dimensional motion. The course of magnetic field B_z and field gradient $\frac{\partial B_z}{\partial z}$ are analytically modeled as a function of position of MSMA film (distance from the magnet surface along z-direction) using equations given in [30]. V_{mag} is the volume of the MSMA film. The electromagnetic damping is computed based on the current in the pick-up coil and the magnetic field. The output parameters are the magnetic force F_{mag}, which is a function of temperature and position of the MSMA film, and the electromagnetic damping force.

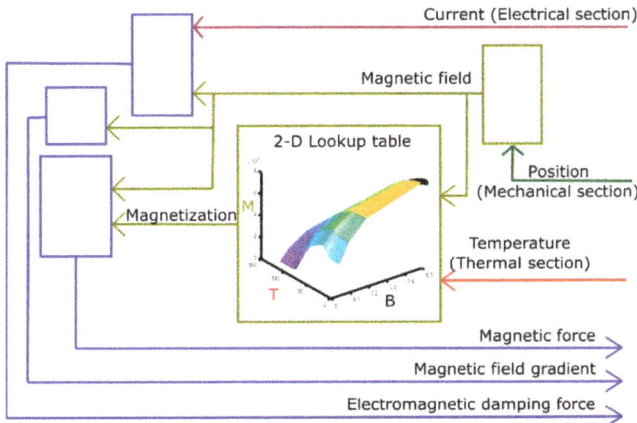

Figure 4. Magnetic section; temperature from thermal section, position from mechanical section and current from electrical section are taken in as inputs. The magnetic section computes magnetic force, magnetic field gradient and electromagnetic damping force as outputs. Magnetization data from material experiments are taken in as a lookup table.

3.3. Thermal Section

The thermal section of the LEM is used to study the heat flows in the TMG device. As shown in Figure 5, heat transfer in the system is modeled by representing the MSMA film, pick-up coil, bonding layers, cantilever front and body as thermal resistances and capacitances [31]. Heat transfer is indicated by bi-directional wire connections, whereby heat flows from higher to lower temperature. The intermittent heat transfer during mechanical contact between MSMA film and heat source is taken into account by control of a variable thermal resistance. For this purpose, a control box is connected to the heat source to model contact and non-contact instances during device operation based on the input of film position computed in the mechanical section. The thermal resistance strongly depends on the contact area and force. An appropriate value of thermal resistance is determined before by adapting simulated temperatures in test structures to corresponding experimental results. We thus obtain an empirical heat transfer coefficient K_{cond} given in Table A1. The contact area is assumed identical to the film area for the case of a polished magnet surface due to the smooth surface finish of the MSMA film. Heat is dissipated from the MSMA film by heat convection to the surrounding air and by heat conduction to the cantilever. It turns out that the temperature of the cantilever front is close to the minimum temperature of the MSMA film during resonant self-actuation showing only a minor temperature change as will be discussed below.

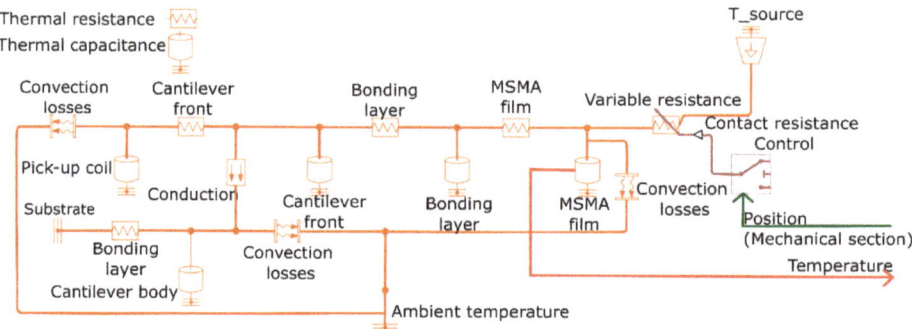

Figure 5. Thermal section; position of the MSMA film determined in the mechanical section is taken in as input and temperature is given out as output. Line connecting each thermal resistance and thermal capacitance represents bi-directional heat flow paths.

The output of the thermal section is the time-dependent temperature of the MSMA film. It is important to note that the TMG device operates above the temperature of ferromagnetic transition of the MSMA film after reaching the condition of resonant self-actuation. In this case, the first order martensite to austenite transformation can be neglected. The thermal LEM section is important to understand the effects of MSMA material, geometry and other parameters on the heat flow characteristics. Optimum heat flow is crucial for resonant self-actuation and operation at optimum power output.

3.4. Electrical Section

Figure 6 shows a schematic of the electrical section. Based on the input of magnetic field gradient and velocity of the pick-up coil, it computes the induced voltage V_{ind} using Faraday's law [32]:

$$V_{ind} = N_{coil} A_{coil} \frac{\partial B_z}{\partial z} \frac{\partial z}{\partial t}, \tag{4}$$

whereby N_{coil} and A_{coil} denote the number of turns and area of the coil. The film velocity is denoted by $\frac{\partial z}{\partial t}$. The output of the electrical section is the induced current, which is determined from V_{ind} and the internal coil resistance as a function of load resistance. The change in flux density due to the change in magnetization is neglected as it is a minor effect in the case of MSMA film.

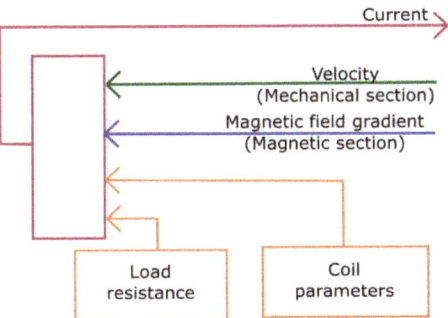

Figure 6. Electrical section; magnetic field gradient and velocity are taken in as inputs and a value for the current is generated as output. Coil parameters and load resistance are constants that are defined before.

4. Performance Analysis

MSMA material and device dimensions are chosen to compare LEM results with experiments on a TMG demonstrator device in order to validate the LEM approach. The MSMA material under study is a Ni-Mn-Ga film having a thickness of 10 µm. The surface area of the film is 2×2 mm^2 and of the cantilever is 2×5.7 mm^2 as illustrated in Figure 2. The thicknesses of the cantilever and of the adhesive layer used to bond the film to the cantilever are 20 µm and about 15 µm, respectively. The cantilever is attached to a supporting ceramic substrate. For bonding, a non-conductive adhesive layer is used to increase heat transfer time and, thus, to retain most of the heat in the MSMA film that is transferred during mechanical contact to the heat source. The operation volume of the harvester is 17.18 mm^3. The temperature of the heatable magnet (heat source temperature) is set to 443 K. The modeling parameters are summarized in Table A1 in the Appendix A.

In the following, the LEM is used to study the time-dependent changes of temperature, magnetic field and corresponding magnetization in the Ni-Mn-Ga film during thermomagnetic cycling. Four stages of cantilever motion may be distinguished as illustrated in Figure 7a. The first stage (1) corresponds to the instant of heat transfer, when the Ni-Mn-Ga film is in mechanical contact to the heat source. (2) Upon deflection from the magnet, the Ni-Mn-Ga film and pick-up coil are accelerated to maximum speed corresponding to a maximum induced voltage. The third stage (3) is reached at the maximum distance to the magnet. (4) A second maximum induced voltage occurs during attraction of the Ni-Mn-Ga film by the magnet, when the pick-up-coil is at maximum speed.

The time-resolved change of temperature in the Ni-Mn-Ga film during resonant self-actuation is depicted in Figure 7b. During one oscillation cycle, the film shows a temperature change of about 10 K within a time duration of about 9 ms. While in mechanical contact (1), the temperature of the film exhibits a steep rise within about 1 ms. The subsequent temperature decrease occurs more slowly while passing through stages (2)–(4). The rather large temperature change is possible due to rapid heat transfer during mechanical contact to the heat source, while heat dissipation through heat conduction and convection is limited. The temperature of the cantilever front below the Ni-Mn-Ga film is close to the minimum temperature of the film and varies by less than 0.5 K. Most of the heat of the Ni-Mn-Ga film is dissipated by heat conduction through the cantilever and, therefore, the temperature of cantilever front represents the heat sink temperature of the TMG device. The temperature change in the pick-up coil is below 0.2 K per cycle.

Figure 7c,d show the corresponding changes of magnetic field and magnetization during thermomagnetic cycling. The time-dependent course of magnetic field reflects the deflection of the Ni-Mn-Ga film. While the film is in mechanical contact to the heat source (stage 1), the magnetic field is at its maximum (about 530 mT), and when the film reaches the maximum distance from the magnet of 1.8 mm, the magnetic field becomes minimal (about 130 mT). Fast-Fourier-Transform analysis of the simulated and experimentally measured deflection yields an oscillation frequency of 114 Hz. The time-dependent course of magnetization is more complex. Figure 7d shows the two cases of the actual cycle of the Ni-Mn-Ga film and the corresponding ideal cycle. In the ideal case of adiabatic conditions, heat exchange only occurs at mechanical contact to the heat source in stage (1) and in stage (3) assuming heat rejection to a heat sink. In this case, sharp changes of magnetization occur in stages (1) and (3) due to rapid temperature change. In stages (2) and (4), the magnetization follows the decrease and increase in magnetic field, respectively. In the actual cycle, a sharp drop in magnetization occurs in stage (1), while in stage (3) only a smooth increase in magnetization occurs. This behavior is due to the non-adiabatic operating conditions, as the film starts cooling as soon as it leaves the magnet surface and continues cooling during stages (2)–(4).

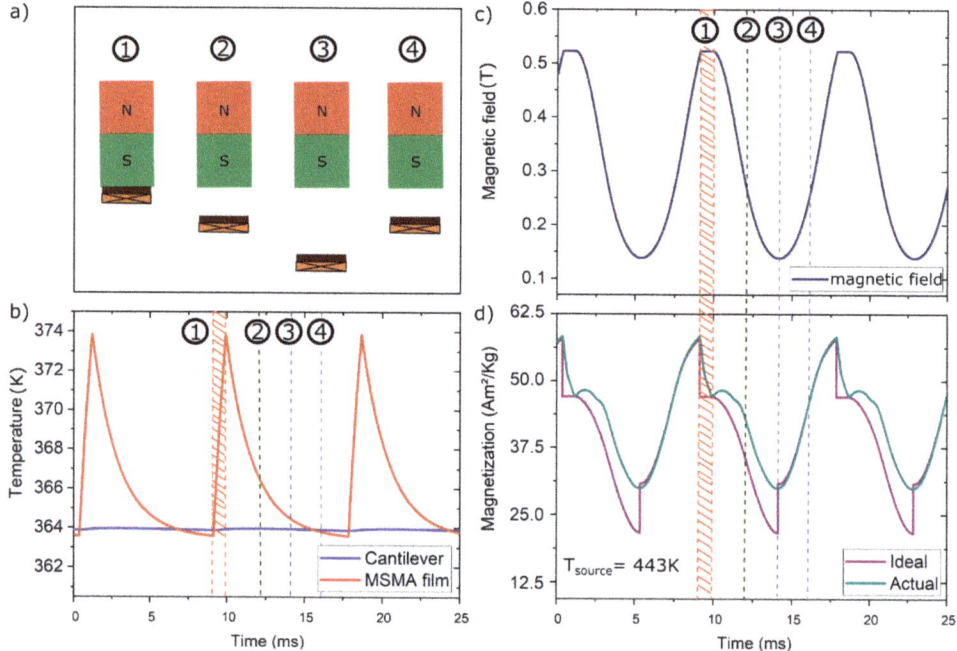

Figure 7. (a) Schematic illustration of cantilever motion to distinguish four stages (1)–(4) and time-resolved changes of (b) temperature in the Ni-Mn-Ga film and cantilever front, (c) magnetic field at the position of the Ni-Mn-Ga film and (d) magnetization of the Ni-Mn-Ga film.

Figure 8a,b show the simulated and experimental electrical power of the TMG device versus time, respectively. Presented results are determined under stationary operation conditions, which is reached after a few minutes of continuous resonant oscillation and then persists for hours without noticeable change of performance. Two power maxima occur during one thermomagnetic cycle at the stages (2) and (4), where the Ni-Mn-Ga film deflects at maximum speed causing a direction-dependent positive and negative peak of alternating current, respectively. The first peak at stage (2) is slightly smaller compared to the peak at stage (4) due to the position-dependent difference in magnetic field. The maximum power generated is 3.1 µW corresponding to a power density with respect to the active material's volume of 80 mW/cm^3. The simulated course of electrical power agrees with the experiment with an accuracy better than 10 %, which validates the LEM approach.

Additionally, time-dependent deflection, stroke and frequency (not shown here) show very good agreement.

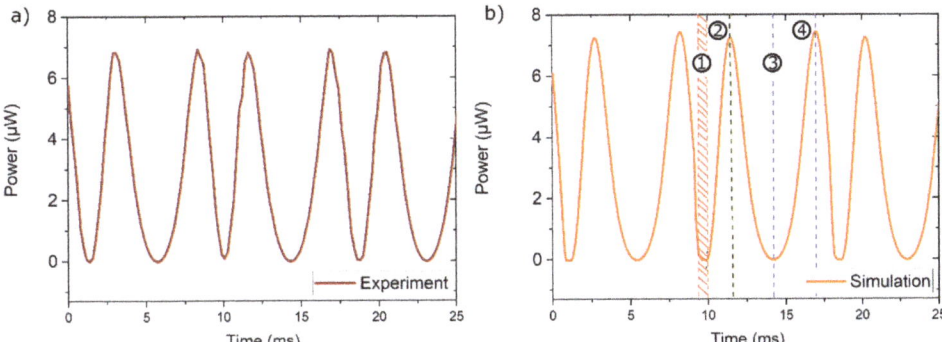

Figure 8. Experiment (**a**) and LEM simulation (**b**) of electrical power versus time of a TMG device based on a Ni-Mn-Ga film of 10 µm thickness. The four stages of cantilever motion (1)–(4) are highlighted (compare Figure 7a).

5. Discussion

The LEM approach allows to describe the coupled thermo-magneto-mechanical performance of the TMG device and to analyze each physical section. Here, we focus on the temperature change of the MSMA film, the time-dependence of magnetic field at the position of the film and on the corresponding film magnetization, which causes the change of magnetic actuation force required to maintain resonant self-actuation. The investigated TMG device is actuated by a Ni-Mn-Ga film of 10 µm thickness. In this case, time-dependent temperature characteristics reveal a rapid temperature increase of 10 K within about 1 ms, followed by a temperature decrease within about 8 ms. This performance results from the rapid heat intake during mechanical contact between magnet and Ni-Mn-Ga film, and the slower heat dissipation by convection and conduction via the bonding layer and cantilever. The much slower temperature decrease results from the non-conductive adhesive between cantilever front and Ni-Mn-Ga film, which acts as a thermal boundary layer and helps to retain most of the input heat. Optimum heat dissipation occurs when input and output heat flows are matching over time. One limitation of the model is that we have to assume the thermal contact resistance between magnet and film and also the convective heat transfer coefficient, as these parameters are hard to measure and difficult to deduce theoretically due to the number of unknown variables. Therefore, these parameters are adapted to the experiment, see Table A1 in the Appendix A. By this approach, realistic values are obtained that match with corresponding data in literature [33].

The heat intake depends on the impact force at contact to the magnet and, thus, on the actuation force, which is determined by the magnetic field gradient and magnetization. The maximum magnetization at contact with the magnet is 4.7×10^5 A/m, which gives rise to a maximum attraction force of 5.48 mN. During contact, the magnetization rapidly drops by about 10^5 A/m while the temperature increases adiabatically from about 364 K by about 10 K. The corresponding ratio $\Delta M / \Delta T$ is about 10^4 $Am^{-1} K^{-1}$, which is consistent with the magnetization characteristic shown in Figure 1. It turns out that optimum operation occurs at the onset of ferromagnetic transition, where the magnetization is still high. Thus, sufficiently large magnetic forces are retained to complete the thermomagnetic cycle. The spring constant of the cantilever is adjusted to counterbalance the actuation and inertia forces, which gives rise to the rather high operation frequency of 114 Hz.

The LEM simulation reveals an average electrical power output of 3.1 µW for a Ni-Mn-Ga film with Curie temperature T_C of 375 K and a heat source temperature of 443 K in agreement with the experiment. The material's T_C sets the lower limit of heat source temperature T_{source} required to achieve resonant self-actuation. Previous investigations show that resonant self-actuation can be uphold for decreasing T_{source} down to about 400 K [17]. Power generation from waste heat at lower temperatures requires MSMA materials with lower T_C. This gives rise to the interesting question, how the lower limit of T_{source} changes for decreasing T_C. In order to address this question, the material's T_C is decreased stepwise and the limit of T_{source} is evaluated in each case by LEM simulations without changing $\Delta M/\Delta T$ and extrapolating heat transfer coefficients based on the dependence on T_{source} between 443 and 400 K. Under these assumptions, we obtain the characteristics of T_{source} limit and corresponding power output shown in Figure 9. When decreasing T_C down to 315 K, the LEM simulation predicts that resonant self-actuation can be uphold down to about 338 K, while the power output decreases to about 0.5 µW. This performance is mainly limited by the temperature-dependent magnetization change of the material $\Delta M/\Delta T$ and heat transfer.

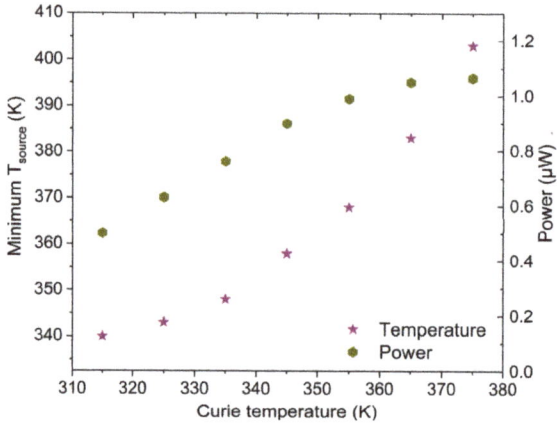

Figure 9. Minimum source temperature required for operation at various Curie temperatures.

6. Conclusions

We present a LEM approach to describe the coupled thermo-magneto-mechanical performance of a thermomagnetic generator (TMG) device that makes use of the large temperature-dependent change of magnetization at the ferromagnetic transition T_C of a MSMA film. The LEM is divided in sections to describe the various physical performances. The key requirement for maximum energy conversion and, thus, optimum power output is to meet the condition of resonant self-actuation, which is accurately described by the LEM. The LEM approach is validated by comparing LEM simulation results with experiments on a TMG demonstrator device using a Ni-Mn-Ga film of 10 µm thickness having a T_C of 375 K. For a heat source temperature of 443 K, the maximum power generated is 3.1 µW corresponding to a power density with respect to the active material's volume of 80 mW/cm³. A time-dependent investigation of temperature of the MSMA film reveals rapid temperature changes of about 10 K caused by the heat flow during mechanical contact between magnet and Ni-Mn-Ga film and by matching of heat conduction via the bonding layer and cantilever. From the magnetization change ΔM and adiabatic temperature increase ΔT during contact the ratio $\Delta M/\Delta T$ is determined to be about 10^4 Am^{-1}K^{-1}. The simulated time dependence of magnetization indicates that optimum operation occurs at the onset of ferromagnetic transition, where the magnetization is high enough to generate sufficiently large magnetic forces to uphold resonant self-actuation. As the LEM is useful to analyze and optimize the device performance, it could be a useful tool to make performance predictions. Here, we estimate the effect of decreasing T_C on the lower limit of heat source temperature T_{source} and corresponding power output, which indicates that resonant self-actuation is possible down to source temperatures of about 338 K. Further optimization would require MSMA materials with enhanced thermomagnetic properties $\Delta M/\Delta T$ and thermal interfaces with improved thermal contact to enable waste heat recovery near room temperature.

Author Contributions: Conceptualization, M.K.; design, M.K. and J.J.; fabrication, J.J., H.M. and M.O.; software and model validation, J.J.; investigation, J.J., H.M. and M.O.; writing—original draft preparation, J.J.; writing—review and editing, M.K.; visualization, J.J.; supervision, project administration and funding acquisition, M.K. All authors have read and agreed to the published version of the manuscript.

Funding: This work was funded by the German Science Foundation (DFG) by the project "Thervest II" and partly supported by the Core-to-Core Program A "Advanced Research Networks" of the Japanese Science Foundation (JSPS).

Institutional Review Board Statement: Not applicable.

Informed Consent Statement: Not applicable.

Data Availability Statement: Data sharing is not applicable to this article.

Conflicts of Interest: The authors declare no conflict of interest.

Appendix A. Model Parameters

Table A1 summarizes all parameters used for LEM simulation. The parameters depend on the MSMA material and the detailed operation conditions of the TMG device. Table A1 also shows specific values used for LEM simulation of a TMG demonstrator device using a Ni-Mn-Ga film of 10 µm thickness with Curie temperature T_C of 375 K.

Table A1. Parameters used for LEM simulation.

Parameter	TMG Device Based on a Ni-Mn-Ga Film	Reference
Length of MSMA film	2 mm	This work
Width of MSMA film	2 mm	This work
Thickness of MSMA film	10 µm	This work
Density of MSMA material	8020 kg/m^3	[34]
Length of the cantilever beam	5.7 mm	This work
Width of the cantilever beam	2 mm	This work
Thickness of the cantilever beam	20 µm	This work
Density of cantilever material	8500 kg/m^3	[35]
Length of bonding layer	2 mm	This work
Width of bonding layer	2 mm	This work
Thickness of bonding layer	10 µm	This work
Density of bonding layer	1250 kg/m^3	[36]
Young's modulus of cantilever material (E)	1×10^{11} N/m^2	[37]
Contact stiffness beam-magnet (k_{cont})	1×10^4 N/m	This work
Structural damping (c)	1.12×10^{-5} Ns/m	This work
Impact damping (c_{cont})	0.1 Ns/m	This work
Thermal conductivity of MSMA material	23.2 W/mK	[38]
Specific heat capacity of MSMA material	490 J/kgK	[38]
Thermal conductivity of cantilever material	109 W/mK	[39]
Specific heat capacity of cantilever material	400 J/kgK	[35]
Thermal conductivity of bonding layer	0.33 W/mK	[36]
Specific heat capacity of bonding layer	2100 J/kgK	[36]
Area of thermal contact	4 mm^2	This work
Max. Conductive heat transfer coefficient (K_{cond})	8400 W/m^{2}K	This work
Convective heat transfer coefficient (K_{conv})	150 W/m^{2}K	This work
Remanent magnetic field	1.07 T	This work
Number of turns of coil	400	This work

Table A1. *Cont.*

Parameter	TMG Device Based on a Ni-Mn-Ga Film	Reference
Area of coil	1.96×10^{-6} m^2	This work
Electrical resistance of coil (internal resistance)	250 Ω	This work
Electrical load resistance	400 Ω	This work
Operation temperature	443 K	This work

References

1. Matiko, J.W.; Grabham, N.J.; Beeby, S.P.; Tudor, M.J. Review of the application of energy harvesting in buildings. *Meas. Sci. Technol.* **2014**, *25*, 012002. [CrossRef]
2. Vullers, R.; Schaijk, R.; Visser, H.; Penders, J.; Hoof, C. Energy harvesting for autonomous wireless sensor networks. *IEEE Solid-State Circuits Mag.* **2010**, *2*, 29–38. [CrossRef]
3. Kishore, R.A.; Priya, S. A review on design and performance of thermomagnetic devices. *Renew. Sustain. Energy Rev.* **2018**, *81*, 33–44. [CrossRef]
4. Bucsek, A.N.; Nunn, W.; Jalan, B.; James, R.D. Energy Conversion by Phase Transformation in the Small-Temperature-Difference Regime. *Annu. Rev. Mater. Res.* **2020**, *50*, 1–36. [CrossRef]
5. Kitanovski, A. Energy Applications of Magnetocaloric Materials. *Adv. Energy Mater.* **2020**, *10*, 1903741. [CrossRef]
6. Bierschenk, J.L. Optimized Thermoelectrics For Energy Harvesting Applications. In *Energy Harvesting Technologies*; Priya, S., Inman, D.J., Eds.; Springer: Boston, MA, USA, 2009; pp. 337–350. ISBN 9780387764634.
7. Min, G. Thermoelectric Energy Harvesting. In *Energy Harvesting for Autonomous Systems*; Beeby, S.P., White, N., Eds.; Artech House: Norwood, MA, USA, 2010; pp. 135–155. ISBN 978-1-59693-718-5.
8. Snyder, G.J.; Toberer, E.S. Complex thermoelectric materials. *Nat. Mater.* **2008**, *7*, 105–114. [CrossRef]
9. Kishore, R.A.; Singh, D.; Sriramdas, R.; Garcia, A.J.; Sanghadasa, M.; Priya, S. Linear thermomagnetic energy harvester for low-grade thermal energy harvesting. *J. Appl. Phys.* **2020**, *127*, 044501. [CrossRef]
10. Ujihara, M.; Carman, G.P.; Lee, D.G. Thermal energy harvesting device using ferromagnetic materials. *Appl. Phys. Lett.* **2007**, *91*, 1–4. [CrossRef]
11. Ma, Z.; Chen, H.; Liu, X.; Xing, C.; Wu, M.; Wang, Y.; Liu, P.; Ou, Z.; Shen, J.; Taskaev, S.V.; et al. Thermomagnetic Generation Performance of Gd and La(Fe, Si)13Hy/In Material for Low-Grade Waste Heat Recovery. *Adv. Sustain. Syst.* **2021**, 2000234. [CrossRef]
12. Waske, A.; Dzekan, D.; Sellschopp, K.; Berger, D.; Stork, A.; Nielsch, K.; Fähler, S. Energy harvesting near room temperature using a thermomagnetic generator with a pretzel-like magnetic flux topology. *Nat. Energy* **2019**, *4*, 68–74. [CrossRef]
13. Wetzlar, K.P.; Keller, S.M.; Phillips, M.R.; Carman, G.P. A unifying metric for comparing thermomagnetic transduction utilizing magnetic entropy. *J. Appl. Phys.* **2016**, *120*, 244101. [CrossRef]
14. Webster, P.J.; Ziebeck, K.R.A. 1.5.5.1 Crystallographic structure. In *Alloys and Compounds of d-Elements with Main Group Elements Part 2*; Springer: Berlin/Heidelberg, Germany, 1988; pp. 75–79.
15. Ullakko, K.; Huang, J.K.; Kantner, C.; Handley, R.C.O. Large magnetic-field-induced strains in Ni2MnGa single crystals. *Appl. Phys. Lett.* **1996**, *69*, 1966–1968. [CrossRef]
16. Kainuma, R.; Imano, Y.; Ito, W.; Sutou, Y.; Morito, H.; Okamoto, S.; Kitakami, O.; Oikawa, K.; Fujita, A.; Kanomata, T.; et al. Magnetic-field-induced shape recovery by reverse phase transformation. *Nature* **2006**, *439*, 957–960. [CrossRef] [PubMed]
17. Chernenko, V.A.; Ohtsuka, M.; Kohl, M.; Khovailo, V.V.; Takagi, T. Transformation behavior of Ni-Mn-Ga thin films. *Smart Mater. Struct.* **2005**, *14*, S245–S252. [CrossRef]
18. Srivastava, V.; Song, Y.; Bhatti, K.; James, R.D. The direct conversion of heat to electricity using multiferroic alloys. *Adv. Energy Mater.* **2011**, *1*, 97–104. [CrossRef]
19. Gueltig, M.; Ossmer, H.; Ohtsuka, M.; Miki, H.; Tsuchiya, K.; Takagi, T.; Kohl, M. High frequency thermal energy harvesting using magnetic shape memory films. *Adv. Energy Mater.* **2014**, *4*, 1400751. [CrossRef]
20. Post, A.; Knight, C.; Kisi, E. Thermomagnetic energy harvesting with first order phase change materials. *J. Appl. Phys.* **2013**, *114*, 033915. [CrossRef]
21. Gueltig, M.; Wendler, F.; Ossmer, H.; Ohtsuka, M.; Miki, H.; Takagi, T.; Kohl, M. High-Performance Thermomagnetic Generators Based on Heusler Alloy Films. *Adv. Energy Mater.* **2017**, *7*, 1601879. [CrossRef]
22. Joseph, J.; Ohtsuka, M.; Miki, H.; Kohl, M. Upscaling of Thermomagnetic Generators Based on Heusler Alloy Films. *Joule* **2020**, *4*, 2718–2732. [CrossRef]
23. Deepak, K.; Varma, V.B.; Prasanna, G.; Ramanujan, R.V. Hybrid thermomagnetic oscillator for cooling and direct waste heat conversion to electricity. *Appl. Energy* **2019**, *233–234*, 312–320. [CrossRef]
24. Ohtsuka, M.; Sanada, M.; Matsumoto, M.; Takagi, T.; Itagaki, K. Shape memory behavior of Ni-Mn-Ga sputtered films under a magnetic field. *Mater. Trans.* **2003**, *44*, 2513–2519. [CrossRef]

25. Suzuki, M.; Ohtsuka, M.; Suzuki, T.; Matsumoto, M.; Miki, H. Fabrication and Characterization of Sputtered Ni2MnGa Thin Films. *Mater. Trans.* **1999**, *40*, 1174–1177. [CrossRef]
26. Rubin, S. Concepts in Shock and Data Analyses. In *Shock and Vibration Handbook*; Harris, C.M., Piersol, A.G., Eds.; McGraw-Hill Education-Europe: London, UK, 1962; Volume 15, pp. 65–66. ISBN 0071370811.
27. Sumali, H.; Carne, T.G. Air-drag Damping on Micro-Cantilever Beams. In Proceedings of the International Modal Analysis Conference, Orlando, FL, USA, 5 February 2007; pp. 1–7.
28. Onorato, P.; De Ambrosis, A. Magnetic damping: Integrating experimental and theoretical analysis. *Am. J. Phys.* **2012**, *80*, 27–35. [CrossRef]
29. Engel, A.; Friedrichs, R. On the electromagnetic force on a polarizable body. *Am. J. Phys.* **2002**, *70*, 428–432. [CrossRef]
30. Camacho, J.M.; Sosa, V. Alternative method to calculate the magnetic field of permanent magnets with azimuthal symmetry. *Rev. Mex. Fis. E* **2013**, *59*, 8–17.
31. Sidebotham, G. Thermal Circuits. In *Heat Transfer Modeling: An Inductive Approach*; Springer International Publishing: Cham, Switzerland, 2015; pp. 3–29. ISBN 978-3-319-14514-3.
32. Greiner, W. Faraday's Law of Induction. In *Classical Electrodynamics*; Springer: New York, NY, USA, 1998; pp. 237–249. ISBN 978-1-4612-0587-6.
33. Karwa, R. Thermal contact resistance. In *Heat and Mass Transfer*; Springer Nature: Singapore, 2017; pp. 101–104. ISBN 9781498715294.
34. Tickle, R.; James, R.D. Magnetic and magnetomechanical properties of Ni2MnGa. *J. Magn. Magn. Mater.* **1999**, *195*, 627–638. [CrossRef]
35. Beiss, P. Non ferrous materials. In *Powder Metallurgy Data*; Landolt-Börnstein-Group VIII Advanced Materials and Technologies; Springer: Berlin/Heidelberg, Germany, 2003; Volume 91, pp. 460–470.
36. Assmus, W.; Brühne, S.; Charra, F.; Chiarotti, G.; Fischer, C.; Fuchs, G.; Goodwin, F.; Gota-Goldman, S.; Guruswamy, S.; Gurzadyan, G.; et al. *Springer Handbook of Condensed Matter and Materials Data*; Martienssen, W., Warlimont, H., Eds.; Springer: Berlin/Heidelberg, Germany, 2005; ISBN 978-3-540-30437-1.
37. Gere, J.M.; Timoshenko, S. *Mechanics of Materials*, 4th ed.; PWS: Boston, MA, USA, 1997; ISBN 9780534934293.
38. Söderberg, O.; Aaltio, I.; Ge, Y.; Heczko, O.; Hannula, S.P. Ni-Mn-Ga multifunctional compounds. *Mater. Sci. Eng. A* **2008**, *481–482*, 80–85. [CrossRef]
39. Young, H.; Freedman, R.A. *Sears and Zemansky's University Physics with Modern Physics*, 14th ed.; Pearson Education Limited: Harlow, UK, 2016.

Article

Preparation of Photoactive Transition-Metal Layered Double Hydroxides (LDH) to Replace Dye-Sensitized Materials in Solar Cells

Sajid Naseem [1,2,*], Bianca R. Gevers [3], Frederick J. W. J. Labuschagné [3] and Andreas Leuteritz [1,*]

1. Processing Department, Leibniz Institute of Polymer Research Dresden, 01069 Dresden, Germany
2. Institute of Materials Science, Technical University (TU) Dresden, 01069 Dresden, Germany
3. Department of Chemical Engineering, Institute of Applied Materials, University of Pretoria, Pretoria 0002, South Africa; bianca.gevers@tuks.co.za (B.R.G.); johan.labuschagne@up.ac.za (F.J.W.J.L.)
* Correspondence: naseem@ipfdd.de (S.N.); leuteritz@ipfdd.de (A.L.); Tel.: +49-351-4658-689 (S.N.)

Received: 14 August 2020; Accepted: 28 September 2020; Published: 1 October 2020

Abstract: This work highlights the use of Fe-modified MgAl-layered double hydroxides (LDHs) to replace dye and semiconductor complexes in dye-sensitized solar cells (DSSCs), forming a layered double hydroxide solar cell (LDHSC). For this purpose, a MgAl-LDH and a Fe-modified MgAl LDH were prepared. X-ray diffraction spectroscopy (XRD), scanning electron microscopy (SEM), and energy-dispersive X-ray (EDX) spectroscopy were used to analyze the structural properties, morphology, and success of the Fe-modification of the synthesized LDHs. Ultraviolet-visible (UV-Vis) absorption spectroscopy was used to analyze the photoactive behavior of these LDHs and compare it to that of TiO_2 and dye-sensitized TiO_2. Current-voltage (I–V) solar simulation was used to determine the fill factor (FF), open circuit voltage (V_{OC}), short circuit current (I_{SC}), and efficiency of the LDHSCs. It was shown that the MgFeAl-LDH can act as a simultaneous photoabsorber and charge separator, effectively replacing the dye and semiconductor complex in DSSCs and yielding an efficiency of 1.56%.

Keywords: layered double hydroxide solar cell (LDHSC); photoactive material; UV-Vis absorption; dye sensitized solar cell (DSSC); photoactive layered double hydroxide (LDH); transition metal modification; optical bandgap analysis; renewable energy; photovoltaic device design; iron (Fe) modified MgFeAl LDH

1. Introduction

With increasing environmental changes because of the wide use of fossil fuels, research into alternative sustainable energy resources is becoming ever more important. Sunlight is an abundantly available energy source that can be harnessed for renewable energy generation. Over the past couple of decades, research into materials suitable for the harvesting of solar light has been on the rise to overcome the pollution problems created by our society [1]. A multitude of approaches has been considered to try to overcome these problems, and the use of solar cells to harvest widely available sunlight and turn it into electricity is one that has shown very good results [2,3]. A large number of different types of solar cells have been developed over the years and research is continuing to find better functioning cells and cells that make use of new materials developed specifically with the aim to be easily manufacturable, sustainable, and cost-effective [4]. Efficiency and final cost are two key factors to consider in the design of new types of solar cells or materials for them. Silicon solar cells are the oldest and most widely used solar cells and have been forerunners in the field, achieving the highest efficiencies but coming with the drawback of high cost and difficult preparation [5,6]. Some low cost alternatives for silicon solar cells exist, dye-sensitized solar cells (DSSCs) being one of them. However, DSSCs have only achieved low efficiencies compared to silicon solar cells [7–9].

Since their inception in 1991 by O'Regan and Grätzel [8], DSSCs have received a lot of research attention to try to improve their performance. The improvements attempted have encompassed all parts of the solar cell: the dyes used, the anode material (originally TiO_2), the counter electrode, and the electrolyte. In DSSCs, light absorption leads to photoexcitation of the dye, subsequent electron injection into the conduction band of the (typically) TiO_2 anode, and regeneration of the oxidized dye by accepting an electron from the redox system which itself is reduced on the Pt counter electrode [8,10–12]. Problems frequently encountered in DSSCs are the cost, long-term stability of the dyes used, and the evaporation of solvents from and leakage of the electrolyte. Some of the problems associated with the electrolyte have been attempted to be solved by using gel, quasi-solid-state, or ionic liquids. A large range of dyes have also been tested for the sake of improvements in oxidation stability or for matching the dye appropriately to the electrolyte [13].

In DSSCs wide bandgap semiconductors are typically used to facilitate appropriate injection of separated charges into the conduction band of these materials [13]. The semiconductor applied to the working electrode is thus typically only active in the UV spectrum and while it can absorb light in this range and separate charges, its main function is frequently to act as a support structure for the dye with high surface area and to effectively transport the charges. Some research has been conducted on the suitability of layered double hydroxides (LDHs) to replace commonly used ZnO or TiO_2 as wide bandgap semiconductors.

LDHs are anionic clays with a formula of $[M_{1-x}^{2+}M_x^{3+}(OH)_2]^{x+} \cdot [(A^{n-})_{x/n}\, yH_2O]^x$, where M^{2+}, M^{3+}, and A^{n-} are divalent metal cations, trivalent metal cations, and interlayer anions respectively [14]. The study of LDHs has increased widely in different applications because of their wide range of applications and tunable properties [15]. LDHs could also be potential materials for photovoltaic applications because of their promising tailorability with respect to light absorption [16]. Especially Zn- and Ti-based LDHs are good UV-absorbers [17]. For application in DSSCs, LDHs have been used as mixed metal oxides (MMOs) bonded or sintered to ITO on their own or deposited on other wideband semiconductors such as TiO_2 [18]. They have also found application in quasi-solid-state electrolytes [18–22]. However, currently, very few studies exist that examine the use of LDHs and their MMO or layered double oxide (LDO) derivatives in DSSCs.

Results obtained using liquid electrolytes and containing LDH-MMOs/LDOs as electron conducting materials have shown efficiencies of up to 5.68% (TiO_2@NiAl-LDO) [23]. Most frequently however, reported results achieve up to 1% efficiency or less, even with changing frequently studied parameters, such as the influence of calcination temperature [24,25] and different dyes [26]. The higher efficiencies reported include the use of a ZnSn-MMO sensitized with D205 (1.28% efficiency) [25], ZnTi-MMO sensitized with C101 (1.57% efficiency) [26] and $Zn_{73}Al_{27}O$ sensitized with N719 (1.02% efficiency) [27]. The possibility to change the photoanode material, the dye, and the electrolyte, makes it possible to alter and improve the design and efficiency of DSSCs. For this purpose some researchers have exchanged the dye with CdS quantum dots to increase light absorption (3.92% efficiency reached) [28], some have exfoliated and sensitized LDHs with anthraquinone sulfonate (0.2% efficiency reached) [29], and others have come up with different designs which only loosely resemble the DSSC concept [30], and some used LDHs as part of solid electrolytes (8.4% efficiency reached) [21] or combined the LDH as a p-type/n-type heterojunction with polythiophene (0.0032% efficiency reached) [31].

One of the tailoring options we have found to remain unexplored in solar cells using liquid electrolytes is the use of photoactive LDHs as a combined photoabsorber and separated charge generator and conductor. In previous work [16], we have shown that it is possible to modify and tailor the absorbance of simple MgAl-LDHs based on the transition metals (Fe, Co, Ni, Cu, and Zn) used. Different synthesis methods for these transition-metal modified MgAl-LDHs were considered (co-precipitation and urea hydrolysis). It was shown that the effect of the synthesis method significantly alters the morphology and some of the properties of the LDHs, yielding nanostructured materials for the co-precipitated LDHs [32] and large, thin platelets for the urea hydrolysis synthesized materials [33].

Fe-modified MgAl-LDH synthesized using urea hydrolysis, hereby showed the best and strongest absorbance in the MgFeAl-LDH.

In this work, it was thus explored whether such a UV-Vis photoactive MgFeAl-LDH could replace the need for dye-sensitization in liquid electrolyte solar cells. For this purpose, four cells were prepared:

1. A first reference cell containing only MgAl-LDH to determine whether this material on its own could function as a photoabsorber (SCN1).
2. A second reference cell containing plain MgAl-LDH sensitized with dye (Coumarin 153) (SCN2).
3. A third reference cell containing dye-sensitized (Coumarin 153) TiO_2 (SCN3).
4. A cell containing 5 mol% Fe modified MgFeAl-LDH which has been shown to absorb MgFeAl-LDH (SCN4).

The cells prepared will be referred to as LDHSCs (layered double hydroxide solar cells) in the text.

2. Materials and Methods

2.1. Materials

$Al(NO_3)_3 \cdot 9H_2O$, $Fe(NO_3)_3 \cdot 9H_2O$, and $Mg(NO_3)_2 \cdot 6H_2O$ were purchased from ABCR GmbH (Karlsruhe, Germany). Urea, absolute ethanol, EL-HPE high-performance electrolyte, Coumarin 153, and indium tin oxide (ITO) coated square glass slide with surface resistivity of 8–12 Ω/sq were purchased from Sigma Aldrich. TiO_2 (SSA: 30 m^2/g, APS: 30–40 nm, purity: 99%) was purchased from Nanostructured & Amorphous Materials Inc. (Houston, TX, USA). Chemically pure (CP) or analytical grade (AR) reactants were used for all experiments without treatment. Distilled water was used in the LDH synthesis.

2.2. Synthesis of MgAl- and MgFeAl-LDHs

MgAl- and MgFeAl–LDHs were synthesized using urea hydrolysis as described in previous work [33]. A MII^+: $MIII^+$ molar ratio of 2:1 was used in the synthesis. Fe was substituted as Fe/(Al+Fe) = 0.05 (all on a molar basis). Solutions of $Mg(NO_3)_2 \cdot 6H_2O$, $Al(NO_3)_3 \cdot 9H_2O$ and $Fe(NO_3)_3 \cdot 9H_2O$ salts were prepared in distilled water and mixed well in a round bottom flask. The solution was heated to 100 °C, and stirred at this temperature for 48 h. After reaction, the mixture was cooled to room temperature, filtered, and thoroughly washed with distilled water. The filtered material was dried in an oven at 70 °C for 24 h.

2.3. Preparation of the Solar Cells

The simplest preparation possible was chosen for the cells. While calcination is a frequent tool in the preparation of solar cells in order to reduce contact resistance between the semiconductor and the conductive material it is deposited on, calcination of the material onto the substrate was excluded in this work in order to retain the photo-response of the MgFeAl–LDH in the UV-Vis region as previously identified [16] and to exclude any possible changes in the material characteristics resulting from calcination. Two types of cells were prepared: dye-sensitized cells and cells making use only of the photoabsorptive capacity of the LDHs. For all four cells, 5 mg of either TiO_2 or LDH was suspended in 5 mL of absolute ethanol and ground to a paste in a mortar and pestle. The paste was then drop-cast onto ITO coated glass. This paste was allowed to dry fully prior to any additional steps. For the dye-sensitized cells, 0.0096 mL of Coumarin 153 was subsequently dropped onto the material after drying. The prepared photo-anodes were then combined with a Pt-coated ITO glass counter electrode. This electrode was obtained by sputter coating glass with a thin layer of Pt with a SCD 500 sputter coater from Baltec. Finally, 0.0192 mL high-performance electrolyte was injected between the two electrodes to complete the cell. The actual amounts of MgAl–LDH deposited onto the surface of ITO glass were 2.10 mg for MgAl–LDH, 2.17 mg for MgAl–LDH sensitized with dye, 2.15 mg for TiO_2, and 2.21 mg for MgFeAl–LDH, thus achieving close comparability in material deposition. The cells will be referred to as SCN1, SCN2, SCN3, and SCN4, respectively. The structure and detailed composition of the cells are shown in Table 1, while Figure 1 shows a schematic of the LDHSC preparation procedure utilized.

Table 1. Layered double hydroxides (LDH)-based and TiO$_2$-based solar cell compositions and structures.

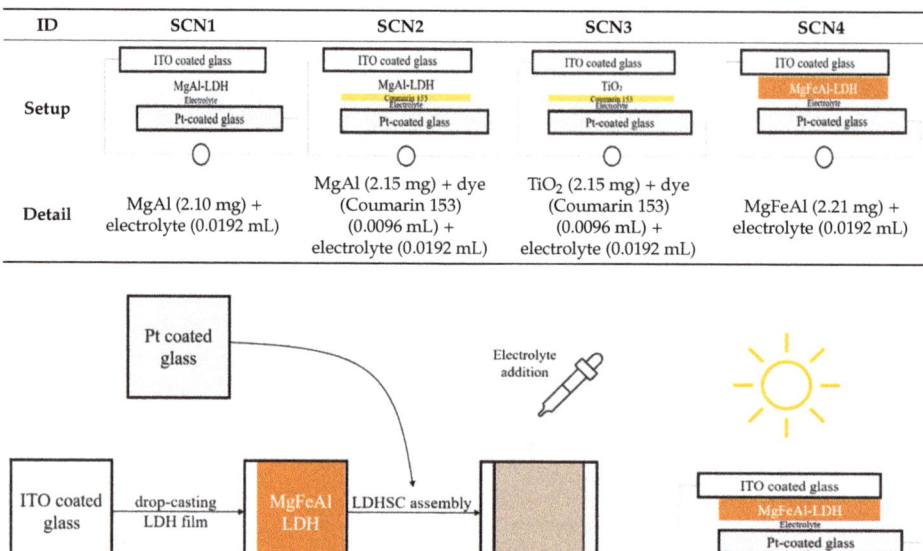

ID	SCN1	SCN2	SCN3	SCN4
Setup	ITO coated glass / MgAl-LDH / Electrolyte / Pt-coated glass	ITO coated glass / MgAl-LDH / Coumarin 153 Electrolyte / Pt-coated glass	ITO coated glass / TiO$_2$ / Coumarin 153 Electrolyte / Pt-coated glass	ITO coated glass / MgFeAl-LDH / Electrolyte / Pt-coated glass
Detail	MgAl (2.10 mg) + electrolyte (0.0192 mL)	MgAl (2.15 mg) + dye (Coumarin 153) (0.0096 mL) + electrolyte (0.0192 mL)	TiO$_2$ (2.15 mg) + dye (Coumarin 153) (0.0096 mL) + electrolyte (0.0192 mL)	MgFeAl (2.21 mg) + electrolyte (0.0192 mL)

Figure 1. Schematic representation of the preparation of a layered double hydroxide solar cell (LDHSC).

2.4. Characterization Methods

X-ray diffraction spectroscopy (XRD) was performed on a Panalytical X'Pert PRO X-ray diffractometer (Malvern Panalytical, Malvern, UK) in θ–θ configuration, equipped with a Fe-filtered Co-Kα radiation (1.789 Å) and with an X'Celerator detector and variable divergence- and fixed receiving slits. Samples were prepared according to the standardized Panalytical backloading system, which provides nearly random distribution of the particles. The data were collected in the angular range 5° ≤ 2θ ≤ 80° with a step size 0.008° 2θ and a 13 s scan step time. The phases were identified using X'Pert Highscore plus software. Scanning electron microscopy (SEM) was done using a Zeiss Ultra Plus (Carl Zeiss Microscopy GmbH, Jena, Germany) at 3.00 keV and 6.00 keV for LDH and TiO$_2$ respectively. Energy dispersive X-ray spectroscopy (EDX) was done using a QUANTAX FlatQUAD from Bruker Nano GmbH (Berlin, Germany) at 6.00 keV. SEM and EDX samples were prepared in a suspension of ethanol and dropped onto a silicon wafer. The samples were carbon-coated before analysis. UV-Vis absorption spectra of all the samples were obtained using a Lambda 800 from Perkin Elmer (Hamburg, Germany). Bandgap values of all the samples were determined using the absorption spectrum fitting (ASF) method [16,34]. The pellets were prepared by grinding a mixture of KBr and the LDHs or TiO$_2$ (70% KBr and 30% LDH, dye sensitized TiO$_2$ and TiO$_2$) for 60 s. The powder was then pressed into pellets using a pellet press with a pressure of 8 t that was applied for 2 min. The final pellet weighed 300 mg and was 1-mm thick. Pellets of the dye sensitized (TiO$_2$ and MgAl) were prepared in the same way. The dye-sensitized powder was obtained in the same way as used in the cell reparation procedure but before injecting the electrolyte. The current-voltage (I–V) curves were obtained using an I–V tester (Jmida SCT-110, JmidaTechnology, Richmond, TX, USA) under simulated AM 1.5 sunlight with an output power of 100 mW/cm^2 using an I-V solar simulator (Trisol 300 mm solar Simulator, OAI, San Jose, CA, USA) as the light source. The active area of solar cells was about 0.125 cm^2.

3. Results and Discussion

Figure 2 shows the XRD patterns of MgAl- and MgFeAl-LDH. Narrow, high-intensity reflections could be observed corresponding to the (003), (006), and (009) reflections of carbonate intercalated MgAl-LDH in the R$\bar{3}$m space group as described in [32] with high crystallinity and order. The crystallinity of the MgFeAl-LDH was observed to be lower than that of MgAl-LDH, which is expected to be a result of the substitution of Fe causing the formation of small amounts of amorphous material that hinders the crystal growth of the LDH and in turn reduce the crystallinity of the material, as previously observed [34,35], and cause the formation of smaller platelets. Overall, good crystallinity and no appreciable amounts of impurity phases being visible indicated the achievement of Fe-substitution as desired [33]. The XRD observations could be corroborated to the SEM micrographs of the materials showing smaller platelets, more particulate matter and less well-defined platelets for MgFeAl-LDH (Figure 3).

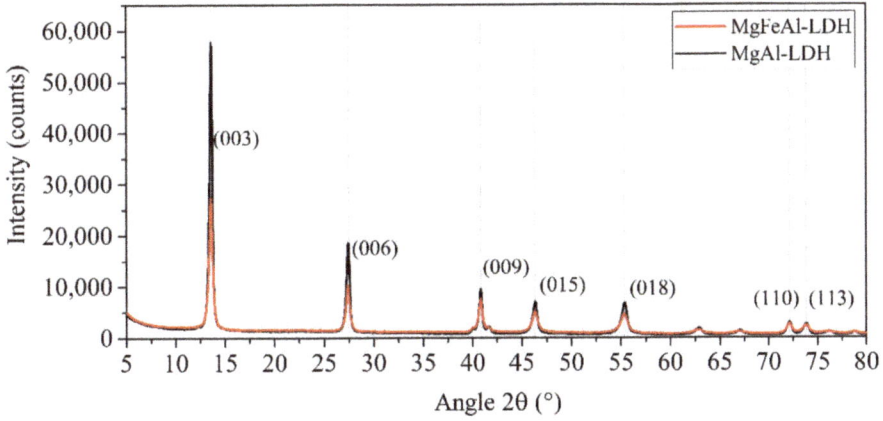

Figure 2. XRD patterns of the MgAl- and MgFeAl–LDH.

Since these LDHs were prepared using urea-hydrolysis, the platelets were expected to be thin, large, and well-crystallized [36] in comparison to materials formed through co-precipitation. MgAl-LDH better portrayed this hexagonal platelet structure. The well-crystallized nature of these materials is a function of their synthesis procedure, where urea is slowly dissolved in the reaction mixture and leads to a constant low degree of supersaturation [31,35]. The small amounts of impurity-type phases visible, sticking on the Fe-modified LDH platelets were not picked up by XRD as impurities. These phases would thus either need to be crystalline phases present at less than 2% (due to the XRD crystalline phase detection limit) or be of amorphous type. Likely such impurity phases would be oxides or hydroxides of the metals used. Because of no large amorphous bulges being visible on the XRD patterns, only small amounts of an amorphous impurity phase could have been present. Figure 3 also shows the SEM micrograph obtained for TiO_2, which was confirmed to be a nano-sized material, as specified by the manufacturer.

EDX analysis (Figure 4) showed that Fe was well-dispersed in the MgFeAl-LDH, although some accumulation of the transition metal was visible, as well as some accumulation of Al and Mg. It is expected that brightly colored spots (indicating a greater accumulation of the metal) are the result of some amorphous material or very small crystalline phase formed, as previously discussed. It could be shown that a good distribution of the Fe in the LDHs layers was achieved with only very small amounts of residue material on the LDH platelets visible.

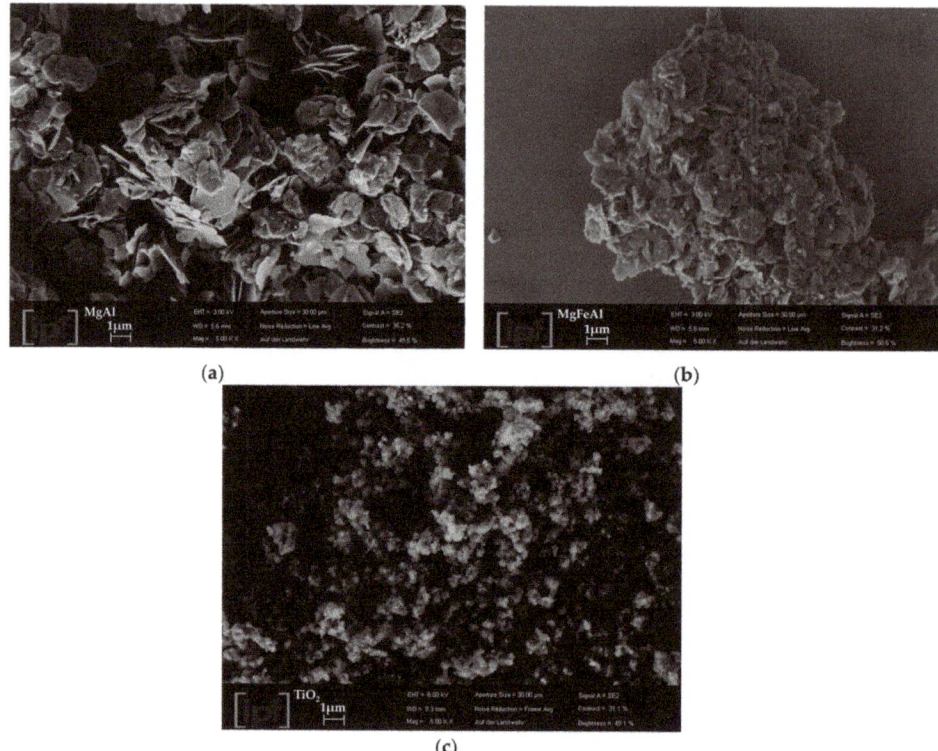

Figure 3. SEM micrographs of the synthesized LDHs (**a**) MgAl-LDH; (**b**) MgFeAl-LDH and (**c**) TiO$_2$.

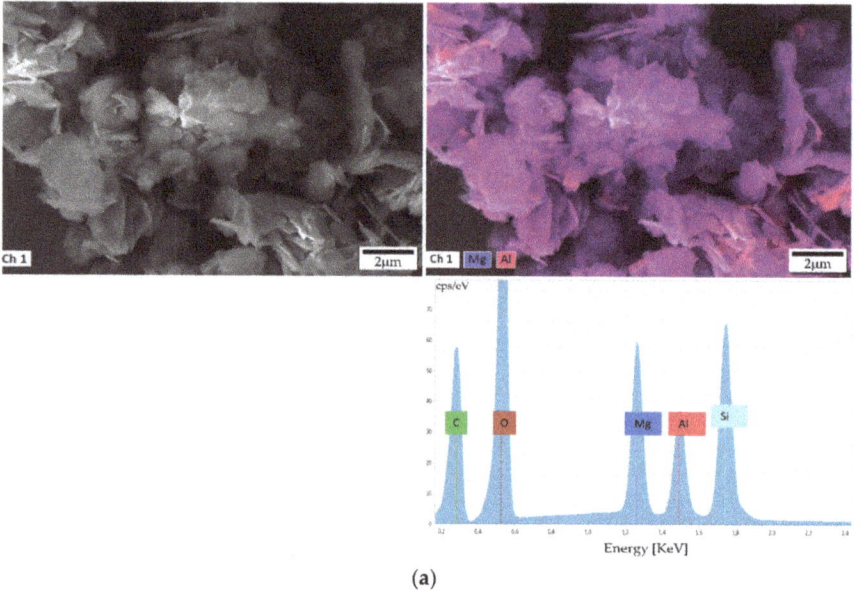

(**a**)

Figure 4. Cont.

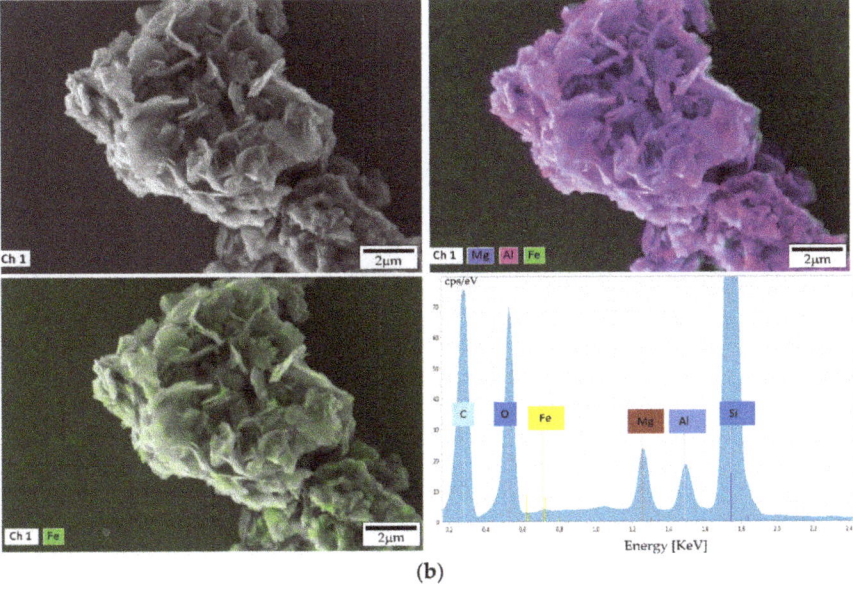

Figure 4. EDX analysis of synthesized LDHs (**a**) MgAl-LDH and (**b**) MgFeAl-LDH.

Figure 5 shows the UV-Vis absorption spectra of dye (Coumarin 153), TiO_2, TiO_2 sensitized with Coumarin 153, MgAl-LDH, MgAl-LDH sensitized with Coumarin 153 and MgFeAl-LDH. MgAl-LDH only showed a small absorption band in the UV-region. MgFeAl-LDH exhibited a strong absorption band spanning the UV-Vis region. UV-Vis absorbance was greatly enhanced by the Fe-substitution in comparison to MgAl-LDH. Comparison of the UV-Vis results of the LDHs and the TiO_2 (photosensitized and plain) showed that the Fe-substituted LDH absorbed considerably more light in the visible region. Coumarin 153 sensitization led to an increase in absorption between approximately 350 nm and 500 nm, as expected because of the absorption range of the dye itself. All LDHs showed a "background" absorbance, as has been previously noted [16].

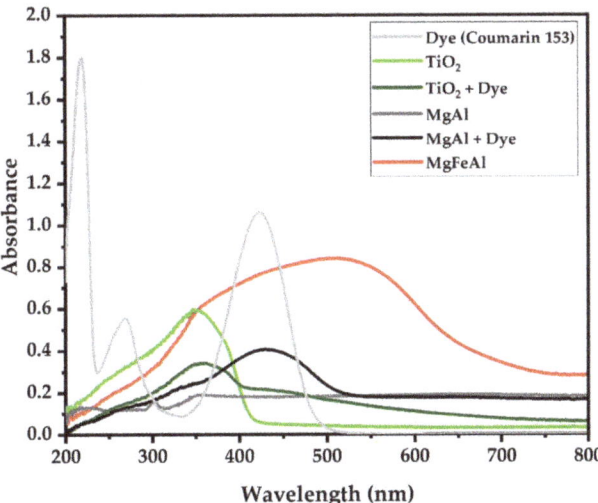

Figure 5. UV-Vis absorption spectra of dye, TiO_2, TiO_2+dye, MgAl, MgAl+dye, and MgFeAl.

The bandgap of all materials was determined using the ASF method as described by Ghobadi et al. [37], the results of which are shown in Figure 6.

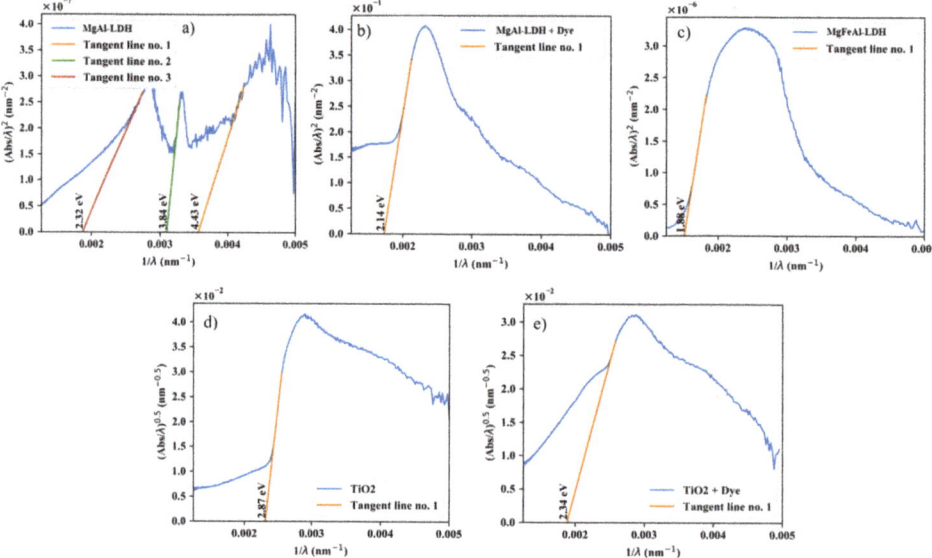

Figure 6. Bandgap determination of (**a**) MgAl-LDH (2.32 eV, 3.84 eV and 4.43 eV); (**b**) MgAl-LDH+dye (2.14 eV); (**c**) MgFeAl-LDH (1.88 eV); (**d**) TiO_2 (2.87 eV); and (**e**) TiO_2+dye (2.34 eV) using the ASF method.

For the use of the ASF method, it was assumed that the LDHs act as direct bandgap semiconductors. LDHs can act as direct and indirect bandgap semiconductors [38], especially with transition-metal modification. However, using indirect transition types, the bandgaps obtained could not be consolidated with the UV-Vis spectra. The results obtained with the assumption of direct transitions matched those that would have been obtained using, for example, the cut-off wavelength method very well. More conservative results were obtained using the ASF method, which are given here. LDHs have been described to have a complex band structure and consist of multiple absorption bands [16]. Such a structure was visible for MgAl-LDH, albeit the total absorbance not being very high for this material. The absorption bands were very small and almost invisible in comparison to the other materials as shown in Figure 7. MgAl-LDH displayed three small absorption bands with bandgaps of 2.32 eV, 3.84 eV, and 4.43 eV. MgFeAl-LDH had a bandgap of 1.88 eV. Upon modification with Coumarin 153, the UV-Vis absorption of the MgAl-LDH+dye complex was enhanced, leading to a modified material with a bandgap of 2.14 eV. The TiO_2 used for comparative purposes consisted of 89 vol% anatase and 11 vol% rutile. Anatase is an indirect bandgap semiconductor and rutile a direct bandgap semiconductor [24,39]. The combination of these two leads to a lowering in the bandgap of the overall material [40]. For the bandgap determination, because the material consisted of 89 vol% anatase, indirect transitions were used. Through the ASF method, the bandgap of TiO_2 and the TiO_2+dye complex were determined to be 2.87 eV and 2.34 eV respectively, about 0.5 eV to 1 eV higher than that of the MgFeAl-LDH.

Finally, the photovoltaic potential of the LDHSCs was studied using a current-voltage (I-V) tester and a solar simulator. The values obtained from the I-V tester under solar simulation are shown in Table 2. The IV graphs obtained of the prepared cells are shown in Figure 7.

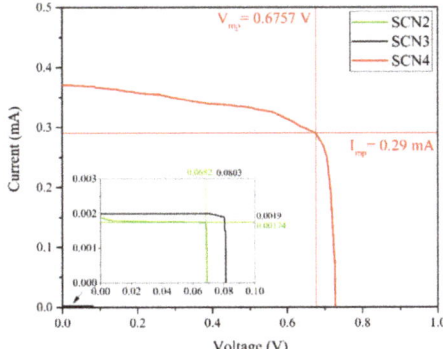

Figure 7. Current (I)-voltage (V) characteristics curves of LDH solar cells (LDHSC).

Under the simulated AM 1.5 illumination, SCN4 showed an open circuit voltage (V_{OC}) of 726 mV. In comparison, SCN2 and SCN3 only achieved an open circuit voltage of 69 mV and 81 mV, respectively. SCN1 showed no functionality, the results were thus excluded from the table. Without the use of Coumarin 153, MgFeAl (SCN4) surpassed the open circuit voltages of the other cells by multifold and achieved a power conversion efficiency of 1.56%. SCN1, SCN2, and SCN3 showed no or very low efficiencies of 0%, 0.0009%, and 0.0012%, respectively.

Table 2. Photovoltaic performance of the layered double hydroxide solar cells (LDHSCs) tested. SCN2 (MgAl-LDH+dye), SCN3 (TiO$_2$+dye), and SCN4 (MgFeAl-LDH). The fill factor (FF), open circuit voltage (V_{OC}), short circuit current (I_{SC}), and efficiency (η) are shown.

	FF	V_{OC} (mV)	I_{SC} (mA)	η (%)
SCN2	0.945	69	0.00182	0.0009
SCN3	0.941	81	0.002	0.0012
SCN4	0.727	726	0.371	1.56

The MgFeAl-LDH-based cell without dye-sensitization thus significantly outperformed the other cells tested, which can be correlated to its superior UV-Vis absorption behavior (as shown in Figure 5) both in intensity of the absorption band and increase in the absorption range, extending far into the visible. Typically, semiconductors are calcined onto the electrode to reduce contact resistance between the two materials. It is believed that the omission of this step and lower overall UV-Vis absorption led to the low efficiency of the TiO$_2$-based cells. The MgAl-LDH-based cell without dye-sensitization showed no functionality, which is believed to have resulted from the minor absorption capacity of the material in comparison to MgFeAl-LDH. The material had three small absorption bands as identified in Figure 6a. Only one of them was well defined, albeit a small band, at 300 nm and two of them were broader and less well-defined. The dye-sensitized MgAl-LDH material showed a better efficiency, although not better than the dye-sensitized TiO$_2$, which had a lower overall UV-Vis absorption range and intensity. This is believed to have resulted from a mismatch between the dye and MgAl-LDH as can be observed in the UV-Vis spectra in Figure 5, where the absorption band of the dye and small absorption band of the MgAl-LDH do not overlap significantly, only the ill-defined bands show some overlap with the dye. Efficient electron injection into the conduction band of the LDH would thus likely have been hindered. For TiO$_2$, an appropriate overlap of the absorption band of the dye and the semiconductor can be observed in Figure 5, which is mirrored in the increased performance of this material. This material was also of smaller particle size, thus facilitating better contact with the electrode.

The results from the I-V testing of the cells show that the replacement of the TiO$_2$+dye complex in DSSCs with a UV-Vis photoabsorbing LDH can achieve conversion efficiencies of up to 1.56%. Some of the main problems in DSSCs are the cost and environmental concerns associated with the use of dyes as photosensitizers and electrolyte leakage [13]. The dyes are typically added to increase absorbance efficiency of the semiconductor in the Vis-light region. Use of an UV-Vis absorbing LDH showed that it is possible to circumvent the necessity of dye-sensitization by the LDH acting as an effective simultaneous photoabsorber and charge separator—with far greater absorbance than the TiO_2+dye complex—that is able to induce an electron flow. It is believed that the Fe incorporated into the LDH facilitated this flow of electrons by acting as a redox couple.

4. Conclusions

With the use of UV-Vis photoabsorbing LDHs it was possible to replace the dye+semiconductor complex used in DSSCs, giving an alternative to dye-sensitization. The resulting dye-free LDHSC comprised of a simple design that achieved a 1.56% efficiency. Even with its simple set-up, it surpassed the efficiency of many other, much more complex, DSSCs containing MMOs/LDOs derived from LDHs and made the use of expensive dye redundant. Using LDHs instead of dye+semiconductor complexes could open up research into low-cost photovoltaic devices.

As with typical DSSCs, there remain many different options to increase the efficiency of LDHSCs. It remains to be determined how stable these separated charges are or through which mechanism they are transferred to the respective electrodes and what the materials' stability through interaction of the electrolyte would be. One of the key elements to this performance is the Fe-substitution in LDH. This phenomenon could be further explored by using other Fe concentrations. Better efficiency is also expected through decreased contact resistance in the cell. Further work on the application of photoactive LDH in this way and also in other cell design concepts is in progress and improvements in the design and efficiency are expected.

Author Contributions: Conceptualization, S.N.and A.L.; methodology, S.N., B.R.G., and A.L.; validation, S.N., A.L., B.R.G., and F.J.W.J.L.; formal analysis, S.N. and B.R.G.; investigation S.N. and B.R.G.; resources, A.L. and F.J.W.J.L.; writing—original draft preparation, S.N.; writing—review and editing, S.N., B.R.G., A.L., and F.J.W.J.L.; visualization, S.N. and B.R.G.; supervision, S.N. and A.L.; project administration, S.N. and A.L. All authors have read and agreed to the published version of the manuscript.

Funding: This research received no external funding.

Acknowledgments: The authors acknowledge the Leibniz Institute of Polymer Research (IPF) Dresden, Germany for providing the research facilities. Sajid Naseem also acknowledges the Higher Education Commission (HEC) Pakistan and German Academic Exchange Service (DAAD) Germany for supporting his scholarship. The authors also thank David Viljoen for providing the software to draw the bandgap figures.

Conflicts of Interest: The authors declare no conflict of interest.

References

1. Lior, N. Energy resources and use: The present situation and possible paths to the future. *Energy* **2008**, *33*, 842–857. [CrossRef]
2. Barnham, K.W.; Mazzer, M.; Clive, B. Resolving the energy crisis: Nuclear or photovoltaics? *Nat. Mater.* **2006**, *5*, 161–164. [CrossRef]
3. Kamat, P.V. Meeting the Clean Energy Demand: Nanostructure Architectures for Solar Energy Conversion. *J. Phys. Chem. C* **2007**, *111*, 2834–2860. [CrossRef]
4. Fraunhofer ISE - Annual Report 2019/20. Available online: https://www.ise.fraunhofer.de/ (accessed on 25 July 2020).
5. Green, M.A.; Emery, K.; Hishikawa, Y.; Warta, W.; Dunlop, E.D. Solar cell efficiency tables (version 44). *Prog. Photovoltaics: Res. Appl.* **2014**, *22*, 701–710. [CrossRef]
6. Saga, T. Advances in crystalline silicon solar cell technology for industrial mass production. *NPG Asia Mater.* **2010**, *2*, 96–102. [CrossRef]
7. Grätzel, M. Solar Energy Conversion by Dye-Sensitized Photovoltaic Cells. *Inorg. Chem.* **2005**, *44*, 6841–6851. [CrossRef]

8. O'Regan, B.; Grätzel, M.; Gr, M. A low-cost, high-efficiency solar cell based on dye-sensitized colloidal TiO2 films. *Nature* **1991**, *353*, 737–740. [CrossRef]
9. Hagfeldt, A.; Grätzel, M. Molecular Photovoltaics. *Accounts Chem. Res.* **2000**, *33*, 269–277. [CrossRef]
10. Hagfeldt, A.; Boschloo, G.; Sun, L.; Kloo, L.; Pettersson, H. Dye-Sensitized Solar Cells. *Chem. Rev.* **2010**, *110*, 6595–6663. [CrossRef]
11. Grätzel, M. Photoelectrochemical cells. *Nature.* **2001**, *414*, 338–344. [CrossRef]
12. Venkatesan, S.; Lee, Y.-L. Nanofillers in the electrolytes of dye-sensitized solar cells – A short review. *Co-ord. Chem. Rev.* **2017**, *353*, 58–112. [CrossRef]
13. Sharma, K.; Sharma, V.; Sharma, S.S. Dye-Sensitized Solar Cells: Fundamentals and Current Status. *Nanoscale Res. Lett.* **2018**, *13*, 381. [CrossRef] [PubMed]
14. Forano, C.; Costantino, U.; Prévot, V.; Gueho, C.T. Layered double hydroxides (LDH). In *Developments in clay science*; Bergaya, F., Lagaly, G., Eds.; Elsevier: Amsterdam, The Netherlands, 2013; Volume 5, pp. 745–782.
15. Yu, J.; Wang, Q.; O'Hare, D.; Sun, L. Preparation of two dimensional layered double hydroxide nanosheets and their applications. *Chem. Soc. Rev.* **2017**, *46*, 5950–5974. [CrossRef] [PubMed]
16. Gevers, B.R.; Naseem, S.; Sheppard, C.J.; Leuteritz, A.; Labuschagné, F.J.W.J. Modification of layered double hydroxides using first-row transition metals for superior UV-Vis-NIR absorption and the influence of the synthesis method used. 2020. Available online: https://10.26434/chemrxiv.11815443.v1. (accessed on 10 August 2020). Preprint.
17. Naseem, S.; Lonkar, S.P.; Leuteritz, A.; Labuschagné, F.J.W.J. Different transition metal combinations of LDH systems and their organic modifications as UV protecting materials for polypropylene (PP). *RSC Adv.* **2018**, *8*, 29789–29796. [CrossRef]
18. George, G.; Saravanakumar, M. Synthesising methods of layered double hydroxides and its use in the fabrication of dye Sensitised solar cell (DSSC): A short review. *IOP Conf. Series: Mater. Sci. Eng.* **2017**, *263*, 32020. [CrossRef]
19. Bastianini, M.; Vivani, R.; Nocchetti, M.; Costenaro, D.; Bisio, C.; Oswald, F.; Meyer, T.B.; Marchese, L. Effect of iodine intercalation in nanosized layered double hydroxides for the preparation of quasi-solid electrolyte in DSSC devices. *Sol. Energy* **2014**, *107*, 692–699. [CrossRef]
20. Ho, H.-W.; Cheng, W.-Y.; Lo, Y.-C.; Wei, T.-C.; Lu, S.-Y. Layered Double Hydroxides as an Effective Additive in Polymer Gelled Electrolyte based Dye-Sensitized Solar Cells. *ACS Appl. Mater. Interfaces* **2014**, *6*, 17518–17525. [CrossRef]
21. Wang, X.; Deng, R.; Kulkarni, S.A.; Wang, X.; Pramana, S.S.; Wong, C.C.; Grätzel, M.; Uchida, S.; Mhaisalkar, S. Investigation of the role of anions in hydrotalcite for quasi-solid state dye-sensitized solar cells application. *J. Mater. Chem. A* **2013**, *1*, 4345. [CrossRef]
22. He, H.; Zhu, J.; Wang, N.; Luo, F.; Yang, K. Composite Gel Polymer Electrolytes Containing Layered Mg-Al Hydrotalcite for Quasi-Solid Dye-Sensitized Solar Cells. *J. Electrochem. Soc.* **2013**, *161*, H17–H20. [CrossRef]
23. Foruzin, L.J.; Rezvani, Z.; Nejati, K. TiO2@NiAl-Layered double oxide nanocomposite: An excellent photoanode for a dye sensitized solar cell. *Sol. Energy* **2019**, *186*, 106–112. [CrossRef]
24. Zhang, L.; Liu, J.; Xiao, H.; Liu, D.; Qin, Y.; Wu, H.; Li, H.; Du, N.; Hou, W. Preparation and properties of mixed metal oxides based layered double hydroxide as anode materials for dye-sensitized solar cell. *Chem. Eng. J.* **2014**, *250*, 1–5. [CrossRef]
25. Liu, S.; Liu, N.; Liu, J.; Wang, D.; Zhu, Y.; Li, H.; Du, N.; Xiao, H.; Hao, X.; Liu, J. The prospective photo anode composed of zinc tin mixed metal oxides for the dye-sensitized solar cells. *Colloids Surfaces A: Physicochem. Eng. Asp.* **2018**, *547*, 111–116. [CrossRef]
26. Liu, S.; Liu, J.; Wang, T.; Wang, C.; Ge, Z.; Liu, J.; Hao, X.; Du, N.; Xiao, H. Preparation and photovoltaic properties of dye-sensitized solar cells based on zinc titanium mixed metal oxides. *Colloids Surfaces A: Physicochem. Eng. Asp.* **2019**, *568*, 59–65. [CrossRef]
27. Xu, Z.; Shi, J.; Haroone, M.S.; Chen, W.; Zheng, S.; Lu, J. Zinc-aluminum oxide solid solution nanosheets obtained by pyrolysis of layered double hydroxide as the photoanodes for dye-sensitized solar cells. *J. Colloid Interface Sci.* **2018**, *515*, 240–247. [CrossRef]
28. Khodam, F.; Amani-Ghadim, A.; Aber, S. Preparation of CdS quantum dot sensitized solar cell based on ZnTi-layered double hydroxide photoanode to enhance photovoltaic properties. *Sol. Energy* **2019**, *181*, 325–332. [CrossRef]

29. Lee, J.H.; Chang, J.; Cha, J.-H.; Jung, D.-Y.; Kim, S.S.; Kim, J.M. Anthraquinone Sulfonate Modified, Layered Double Hydroxide Nanosheets for Dye-Sensitized Solar Cells. *Chem. Eur. J* **2010**, *16*, 8296–8299. [CrossRef]
30. Liu, X.; He, Y.; Zhang, G.; Wang, R.; Zhou, J.; Zhang, L.; Gu, J.; Jiao, T. Preparation and High Photocurrent Generation Enhancement of Self-Assembled Layered Double Hydroxide-Based Composite Dye Films. *Langmuir* **2020**, *36*, 7483–7493. [CrossRef]
31. Schwenzer, B.; Neilson, J.R.; Sivula, K.; Woo, C.; Fréchet, J.M.J.; Morse, D.E. Nanostructured p-type cobalt layered double hydroxide/n-type polymer bulk heterojunction yields an inexpensive photovoltaic cell. *Thin Solid Films* **2009**, *517*, 5722–5727. [CrossRef]
32. Gevers, B.R.; Naseem, S.; Leuteritz, A.; Labuschagné, F.J.W.J. Comparison of nano-structured transition metal modified tri-metal MgMAl–LDHs (M = Fe, Zn, Cu, Ni, Co) prepared using co-precipitation. *RSC Adv.* **2019**, *9*, 28262–28275. [CrossRef]
33. Naseem, S.; Gevers, B.R.; Boldt, R.; Labuschagné, F.J.W.J.; Leuteritz, A. Comparison of transition metal (Fe, Co, Ni, Cu, and Zn) containing tri-metal layered double hydroxides (LDHs) prepared by urea hydrolysis. *RSC Adv.* **2019**, *9*, 3030–3040. [CrossRef]
34. Parida, K.; Satpathy, M.; Mohapatra, L. Incorporation of Fe3+ into Mg/Al layered double hydroxide framework: Effects on textural properties and photocatalytic activity for H2 generation. *J. Mater. Chem.* **2012**, *22*, 7350–7357. [CrossRef]
35. Chmielarz, L.; Kuśtrowski, P.; Rafalska-Łasocha, A.; Dziembaj, R. Influence of Cu, Co and Ni cations incorporated in brucite-type layers on thermal behaviour of hydrotalcites and reducibility of the derived mixed oxide systems. *Thermochim. Acta* **2002**, *395*, 225–236. [CrossRef]
36. He, J.; Wei, M.; Li, B.; Kang, Y.; Evans, D.G.; Duan, X. Preparation of Layered Double Hydroxides. Springer Science and Business Media LLC, 2005; Volume 119, pp. 89–119.
37. Ghobadi, N.; Moradian, R. Strong localization of the charge carriers in CdSe nanostructural films. *Int. Nano Lett.* **2013**, *3*, 47. [CrossRef]
38. Xu, S.-M.; Yan, H.; Wei, M. Band Structure Engineering of Transition-Metal-Based Layered Double Hydroxides toward Photocatalytic Oxygen Evolution from Water: A Theoretical–Experimental Combination Study. *J. Phys. Chem. C* **2017**, *121*, 2683–2695. [CrossRef]
39. Zhang, J.; Zhou, P.; Liu, J.; Yu, J. New understanding of the difference of photocatalytic activity among anatase, rutile and brookite TiO2. *Phys. Chem. Chem. Phys.* **2014**, *16*, 20382–20386. [CrossRef] [PubMed]
40. Scanlon, D.O.; Dunnill, C.W.; Buckeridge, J.; Shevlin, S.A.; Logsdail, A.J.; Woodley, S.M.; Catlow, C.R.A.; Powell, M.J.; Palgrave, R.G.; Parkin, I.P.; et al. Band alignment of rutile and anatase TiO2. *Nat. Mater.* **2013**, *12*, 798–801. [CrossRef] [PubMed]

© 2020 by the authors. Licensee MDPI, Basel, Switzerland. This article is an open access article distributed under the terms and conditions of the Creative Commons Attribution (CC BY) license (http://creativecommons.org/licenses/by/4.0/).

MDPI
St. Alban-Anlage 66
4052 Basel
Switzerland
Tel. +41 61 683 77 34
Fax +41 61 302 89 18
www.mdpi.com

Materials Editorial Office
E-mail: materials@mdpi.com
www.mdpi.com/journal/materials

www.ingramcontent.com/pod-product-compliance
Lightning Source LLC
LaVergne TN
LVHW070401100526
838202LV00014B/1363